Advances in

BOTANICAL RESEARCH

VOLUME 58

Advances in

BOTANICAL RESEARCH

Series Editors

JEAN-CLAUDE KADER — *Laboratoire Physiologie Cellulaire et Moléculaire des Plantes, CNRS, Université de Paris, Paris, France*

MICHEL DELSENY — *Laboratoire Génome et Développement des Plantes, CNRS IRD UP, Université de Perpignan, Perpignan, France*

BIOSYNTHESIS OF VITAMINS IN PLANTS

Vitamins A, B1, B2, B3, B5

PART A

Volume Editors

FABRICE RÉBEILLÉ and ROLAND DOUCE

Laboratoire de Physiologie Cellulaire Végétale
CEA - CNRS UMR5168 - INRA USC1200 - Université Joseph Fourier Grenoble I
Institut de Recherches en Technologies et Sciences pour le Vivant
CEA-Grenoble 17, rue des Martyrs
38054 Grenoble Cedex 9, France

VOLUME 58

ELSEVIER

AMSTERDAM • BOSTON • HEIDELBERG • LONDON
NEW YORK • OXFORD • PARIS • SAN DIEGO
SAN FRANCISCO • SINGAPORE • SYDNEY • TOKYO
Academic Press is an imprint of Elsevier

Academic Press is an imprint of Elsevier
32 Jamestown Road, London NW17BY, UK
225 Wyman Street, Waltham, MA 02451, USA
525 B Street, Suite 1900, San Diego, CA 92101-4495, USA
Radarweg 29, PO Box 211, 1000 AE Amsterdam, The Netherlands

First edition 2011

ISBN: 978-0-12-386479-6
ISSN: 0065-2296

For information on all Academic Press publications
visit our Web site at www.elsevierdirect.com

Printed and bound by CPI Group (UK) Ltd, Croydon, CR0 4YY

Transferred to Digital Print 2012

CONTENTS

CONTRIBUTORS TO VOLUME 58

ADELBERT BACHER *Institute of Food Chemistry, University of Hamburg, Hamburg, Germany; Ikosatec GmbH, Garching, Germany*

CHRISTOPHER I. CAZZONELLI *ARC Centre of Excellence in Plant Energy Biology, Research School of Biology, Australian National University, Canberra, ACT 0200, Australia*

ABBY J. CUTTRISS *Molecular Biosciences and Bioengineering, University of Hawai'i at Mānoa, Honolulu, HI, USA; Department of Biological Sciences, Lehman College, The City University of New York, Bronx, New York, USA*

MARKUS FISCHER *Institute of Food Chemistry, University of Hamburg, Hamburg, Germany; Ikosatec GmbH, Garching, Germany*

JUTTA HAGER *Institut of Biologie des Plantes, UMR8618 CNRS/ Université de Paris sud 11, Bâtiment 630, Université de Paris sud 11, 91405 Orsay CEDEX, France*

SHENGCHUN LI *Institut of Biologie des Plantes, UMR8618 CNRS/ Université de Paris sud 11, Bâtiment 630, Université de Paris sud 11, 91405 Orsay CEDEX, France*

GRAHAM NOCTOR *Institut of Biologie des Plantes, UMR8618 CNRS/ Université de Paris sud 11, Bâtiment 630, Université de Paris sud 11, 91405 Orsay CEDEX, France*

BARRY J. POGSON *ARC Centre of Excellence in Plant Energy Biology, Research School of Biology, Australian National University, Canberra, ACT 0200, Australia*

MARIA RAPALA-KOZIK *Department of Analytical Biochemistry, Faculty of Biochemistry, Biophysics and Biotechnology, Jagiellonian University, Krakow, Poland*

ALISON G. SMITH *Department of Plant Sciences, University of Cambridge, Cambridge, United Kingdom*

MICHAEL E. WEBB *School of Chemistry and Astbury Centre for Structural Molecular Biology, University of Leeds, Leeds, United Kingdom*

ELEANORE T. WURTZEL *Department of Biological Sciences, Lehman College, The City University of New York, Bronx, New York, USA*

PREFACE

VITAMINS: A PLANT AFFAIR

All organisms need to synthesize, transform and interconvert a myriad of molecules to enable them to grow and reproduce. All these reactions are catalysed by enzymes (the living tools) which facilitate chemical modifications of substrates owing to their specific binding properties. In many cases, suitable coenzymes (nicotinamide adenine dinucleotide [NAD^+], nicotinamide adenine dinucleotide phosphate [$NADP^+$], flavin adenine dinucleotide [FAD], flavin mononucleotide [FMN], pyridoxal 5′-phosphate, biotin, coenzyme A, etc.) may assist in biochemical transformations. Some of these coenzymes may be more or less tightly bound to enzymes as part of prosthetic groups (biotin, FMN, etc.). Coenzymes may also be loosely bound to enzymes as detachable molecules. In that case, they are acting as substrates, being often recycled through other set of reactions ($NAD(P)^+$, folates, ascorbate, etc.).

Vitamin (a combination word from *vita* and *amine*) are by definition dietary substances required for good health and normal development that are only synthesized by microorganisms and plants. During the course of animal evolution, the ability to biosynthesize these compounds has been lost and, instead, elaborate uptake mechanisms have been developed. There are 13 recognized vitamins, involved in various catalytic functions. The largest number of vitamins serve as precursors to coenzymes (vitamins B_1 [thiamine], B_2 [riboflavin], B_3 [niacin], B_5 [pantothenic acid], B_6 [pyridoxine], B_9 [folic acid]) or as coenzymes themselves (vitamins B_8 [biotin], B_{12} [cobalamin], C [ascorbic acid], K [phylloquinone, menaquinone]). Some of these vitamins, especially the hydrophobic (vitamins A [retinol, pro-vitamin A carotenoids], E [tocopherols, tocotrienols] and D [ergocalciferol, cholecalciferol]), cannot be truly considered as coenzymes: vitamins A and D display hormonal effects in the human body, and vitamin E has a protective role in membranes by scavenging free radicals. Vitamins are involved in almost all important cellular functions, displaying protective (antioxidant) functions or participating to numerous metabolisms, including the energetic metabolism (respiration, photosynthesis) and the metabolisms of sugars, amino acids, fatty acids and nucleic acids. The daily amount of vitamins required for a good health depends on the considered vitamin and fluctuates widely, from a few micrograms (B_{12}, D, K) to several milligrams (B_3, B_5, C). Vitamin deficiencies are quite common in low-resource countries but also occur in developed countries due to bad food habits. Well-known vitamin-related diseases include, among others, blindness (vitamin A), beriberi (vitamin B_1), pellagra (vitamin B_3), anaemia (vitamins B_6 and B_9), scurvy (vitamin C), rickets (vitamin D) or neural tube defects (vitamin B_9). In addition, antioxidant vitamins (such as A, C, E and B_6) have protective roles as efficient quenchers of reactive oxygen species.

Plants synthesize an impressive diverse array of natural products including vitamins, and plants are considered as a major nutritional source for these essential molecules. Plants are able to synthesize 12 out of the 13 vitamins. Indeed, plants have no cobalamin-dependent proteins and use for methionine synthesis an alternate catalytic mechanism that does not need vitamin B_{12}. Vitamin B_{12} is only synthesized in prokaryotes, and humans primarily obtained it from animal food, thanks to the intestine flora of herbivores. Two of the vitamins (vitamins A and D) have 'hormonal' functions in animals, which functions do not exist in plants. Plants do not synthesize vitamin A, but carotenoids. Some of these carotenoids are pro-vitamin A, which are transformed in retinol once assimilated by animals. Vitamin D (D_2 and D_3) is formed from the precursors ergosterol (mainly present in fungal cells) and cholesterol (mainly present in mammalian cells) following sun exposure (UV radiation). Although vitamins D_2 and D_3 can be found in low amounts in the membranes of some *Solanaceous* plants, higher plants are not considered as a source of vitamin D and plant food cannot compensate insufficient synthesis in the human body. Thus, the plant kingdom is a recognized dietary source for 11 out of the 13 vitamins.

As many vitamins are only required in trace quantities, their biosynthesis is normally strictly controlled and the involved enzymes are generally produced in very small amounts. This is why it has been extremely difficult to elucidate their complete biosynthetic pathways, and it still remains the case that several steps within the biosynthesis of vitamins are poorly understood (e.g. thiazole ring scaffolding). However, the advent of modern recombinant DNA techniques, coupled with the completion of many genome projects, made possible to decipher pathways in plants, thus allowing now a more complete understanding of how these molecules are made. The general picture emerging from these recent data indicates that the metabolic web represented by these molecules is of a rare complexity. Indeed, not only may the synthesis of vitamins require some 10 enzymatic steps but also several of these metabolic routes are split between various compartments of the plant cell, adding a further level of complexity when compared to prokaryotes. Since all cell compartments need their vitamins, this situation implies transport and trafficking of intermediates and end products of the pathways. Today, there is no explanation for such compartmentalization.

The actual understanding of how these biosynthetic pathways operate can be exploited for health and wealth creation. Vitamin synthesis is largely restricted to plants and microorganisms, a biochemical feature that can be harnessed for the development of specific pesticides (bactericides, herbicides, fungicides, etc.). Taking into account the health problems related to vitamin deficiencies, together with an increase in the use of vitamin supplements for human and animal nutrition, there is also a requirement, from a nutritional and commercial standpoint, to enhance the production of many of these vitamins. Overproduction of the vitamins can be achieved in a number of ways, by removing transcriptional controls, overproduction of key enzymes that represent bottlenecks in the pathways of biosynthesis, suppression of metabolic feedbacks, limitation of the catabolism and increase of the storage. It is clear that the 'optimization' of these systems requires a complete understanding of (i) their endogenous regulation and (ii) their integration within the metabolism as a whole.

This book includes comprehensive and authoritative reviews from leading experts on vitamins in plants, and we are thankful for their time and effort. The aim of this book is to collect and interpret the rapid growing experimental information on vitamins in plants, especially in the challenging area of their biosynthesis. We also hope that this book may be useful as a starting point for those graduates and undergraduate students and researchers wishing to pursue special studies in this field.

FABRICE RÉBEILLÉ AND ROLAND DOUCE

CONTENTS OF VOLUMES 35–57

Series Editor (Volumes 35–44)

J.A. CALLOW
School of Biosciences, University of Birmingham,
Birmingham, United Kingdom

Contents of Volume 35

Contents of Volume 36

Contents of Volume 37

ANTHOCYANINS IN LEAVES

Edited by **K. S. Gould and D. W. Lee**

Contents of Volume 38

Contents of Volume 39

Contents of Volume 40

Contents of Volume 41

Contents of Volume 42

Contents of Volume 43

Contents of Volume 44

Series Editors (Volume 45–)

JEAN-CLAUDE KADER
Laboratoire Physiologie Cellulaire et Moléculaire des Plantes, CNRS, Université de Paris, Paris, France
MICHEL DELSENY
Laboratoire Génome et Développement des Plantes, CNRS IRD UP, Université de Perpignan, Perpignan, France

Contents of Volume 45

RAPESEED BREEDING

Contents of Volume 46

Contents of Volume 47

INCORPORATING ADVANCES IN PLANT PATHOLOGY

Contents of Volume 48

Contents of Volume 49

Contents of Volume 50

Contents of Volume 51

Contents of Volume 52

Contents of Volume 53

Contents of Volume 54

Contents of Volume 55

Contents of Volume 56

Contents of Volume 57

Carotenoids

ABBY J. CUTTRISS,*,† CHRISTOPHER I. CAZZONELLI,‡
ELEANORE T. WURTZEL† AND BARRY J. POGSON‡,[1]

*Molecular Biosciences and Bioengineering,
University of Hawai'i at Mānoa, Honolulu, HI, USA
†Department of Biological Sciences, Lehman College,
The City University of New York, Bronx, New York, USA
‡ARC Centre of Excellence in Plant Energy Biology,
Research School of Biology, Australian National University,
Canberra, ACT 0200, Australia

[1]Corresponding author: E-mail: barry.pogson@anu.edu.au

Advances in Botanical Research, Vol. 58 0065-2296/11 $35.00

DOI: 10.1016/B978-0-12-386479-6.00005-6

ABSTRACT

Carotenoid pigments provide fruits and flowers with distinctive red, orange and yellow colours as well as a number of aromas, which make them commercially important in agriculture, food, health and the cosmetic industries. Carotenoids comprise a large family of C_{40} polyenes that are critical for the survival of plants and animals alike. β-carotene and its derivatives contain unmodified β-ionone groups, which serve as precursors for vitamin A and are therefore essential dietary components for mammals. Significant progress has been made towards producing staple food crops with elevated provitamin A carotenoids, an important first step in alleviating worldwide vitamin A deficiency. Recent insights into the regulatory processes that control carotenoid composition and content may further advance biofortification projects.

ABBREVIATIONS

βLCY	lycopene β-cyclase
εLCY	lycopene ε-cyclase
βOH	β-hydroxylase
εOH	ε-hydroxylase
ABA	abscisic acid
CCD	carotenoid cleavage dioxygenases
CRTISO	carotenoid isomerase
CsZCD	crocus zeaxanthin 7,8(7′,8′)-cleavage dioxygenase
DMAPP	dimethylallyl diphosphate
DXP	deoxy-D-xylulose 5-phosphate
DXS	deoxy-D-xylulose 5-phosphate synthase
GGPP	geranylgeranyl diphosphate
IPP	isopentenyl diphosphate
MEP	methylerythritol 4-phosphate
MVA	mevalonic acid
NCED	9-cis-epoxycarotenoid dioxygenase
NPQ	non-photochemical quenching
NXS	neoxanthin synthase

PDS	phytoene desaturase
PSY	phytoene synthase
VDE	violaxanthin de-epoxidase
ZDS	ζ-carotene desaturase
ZEP	zeaxanthin epoxidase
Z-ISO	15-cis-ζ-carotene isomerase

I. BIOLOGICAL FUNCTION

A. DIETARY CAROTENOIDS

Carotenoids are a vital component of mammalian diets, providing precursors for vitamin A biosynthesis. Antioxidants and their dietary uptake can pigment the tissues of animals such as fish, crustaceans and birds. Vitamin A (all-*trans*-retinol) is generated from unmodified β-ring containing provitamin A carotenoids in the diet (von Lintig, 2010), of which β-carotene (two nonhydroxylated β-ionone rings), is the most efficient, because it can generate up to two retinol molecules. α-carotene and β-cryptoxanthin also contain provitamin A potential, but only have a single nonhydroxylated β-ring (Davis *et al.*, 2008).

Vitamin A deficiency is responsible for a number of disorders that range from impaired iron mobilization, growth retardation and blindness to a depressed immune response, as well as increased susceptibility to infectious disease (Sommer and Davidson, 2002). Between 140 and 250 million children are at risk of vitamin A deficiency (Underwood, 2004); 250,000–500,000 become blind every year and half will die within 12 months after losing their sight (http://www.who.int/nut/vad.htm). Simply improving the vitamin A status of children, by increasing the uptake of provitamin A (e.g. β- and α-carotene), can reduce overall child mortality by 25% (http://www.unicef.org/immunization/facts_vitamina.html).

Low serum levels of vitamin A (less than 0.7 μmol L^{-1}) can be used as a population-based indicator of health risks (Underwood, 2004). Recommended daily allowances for vitamin A range from 300–600 μg for children to 900–1300 μg for adults of retinol activity equivalents (retinol and provitamin A carotenoids; Fig. 1). There is no recommended daily allowance for provitamin A carotenoids, as the conversion efficiency remains imprecise; however, between 3 and 6 mg of β-carotene daily is sufficient to maintain healthy serum carotenoid levels, as would five or more servings of fruits and vegetables per day (Panel on Micronutrients, 2001).

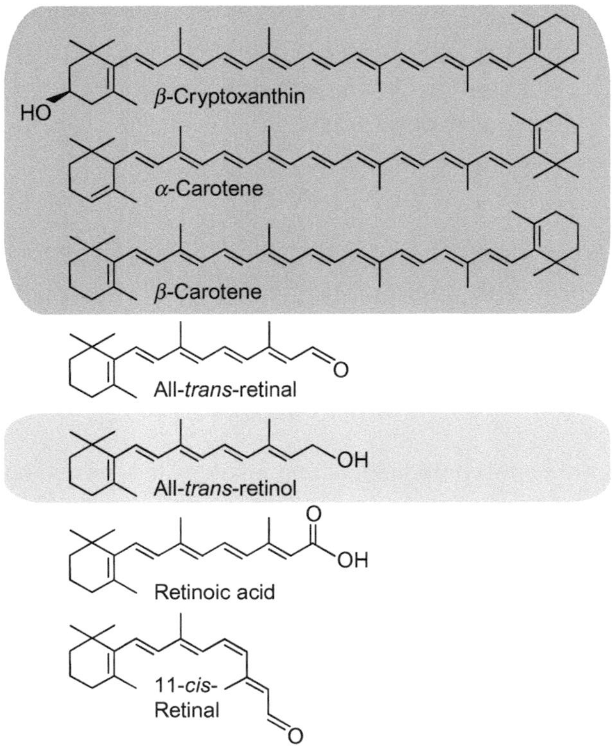

Fig. 1. Vitamin A and carotenoid precursor structures. Common dietary provitamin A carotenoids with unmodified β-ionone rings (highlighted in orange/dark grey) are processed to form C_{20} retinoids, including all-*trans*-retinol (vitamin A, highlighted in yellow/light grey), all-*trans*-retinal, retinoic acid and 11-*cis*-retinal, a photoreceptor chromophore. (For interpretation of the references to colour in this figure legend, the reader is referred to the Web version of this chapter.)

B. CAROTENOIDS IN PHOTOSYNTHETIC ORGANISMS

Carotenoids play a variety of crucial roles in photosynthetic organisms. Carotenoids are involved in photosystem assembly where they contribute to harvesting light in a broader range of wavelengths in the blue region of the visible light spectrum and subsequently transfer the energy to chlorophyll (Fig. 2). The distinctive yellow colours of light-harvesting carotenoids become visible during autumn when chlorophyll degrades. The colour of carotenoids, typically ranging from pale yellow to red is defined by the number of conjugated double bonds along the C_{40} backbone as well as other structural and oxygenic modifications that impart different spectral properties. Carotenoids also provide protection from excessive light via

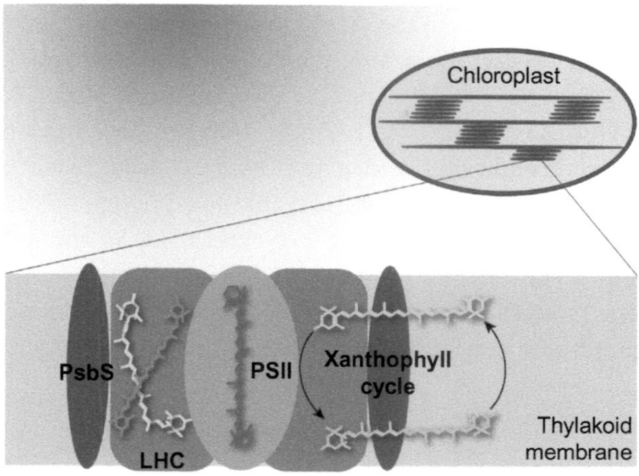

Fig. 2. Photoprotective carotenoids in chloroplast membranes and proteins. Carotenoids accumulate in chloroplast thylakoid membranes, as indicated by this simplified schematic. Xanthophylls, such as lutein, zeaxanthin, violaxanthin and neoxanthin, accumulate in light-harvesting complex proteins (LHC) where they have a structural role and contribute to light harvesting. β-carotene molecules in the photosystem II reaction centre (PSII) could quench singlet oxygen or possibly have a role in electron transfer. In high light, zeaxanthin is formed from violaxanthin via the xanthophyll cycle. Zeaxanthin, lutein, PsbS and specific antenna proteins all contribute to non-photochemical quenching of chlorophyll fluorescence; note, the exact locations of each are not depicted in this cartoon.

energy dissipation and free radical detoxification, which limits damage to membranes and proteins (DellaPenna and Pogson, 2006).

Plants need to maintain a balance between absorbing sufficient light for photosynthetic processes and avoiding oxidative damage caused by high light. Complementary photoprotective mechanisms are employed to minimize photodamage induced by exposure to high light and these include (1) the harmless dissipation of excess energy via non-photochemical quenching (NPQ) that is mediated by certain xanthophylls (zeaxanthin, antheraxanthin and lutein), (2) quenching of triplet chlorophylls by carotenoids, (3) accumulation of antioxidants (ascorbate, tocopherols and carotenoids) and (4) activation of antioxidant enzymes such as ascorbate peroxidase that de-toxify free radicals, as well as repair damaged proteins (Bailey and Grossman, 2008; Niyogi, 1999).

The physiological relevance of xanthophylls is exemplified by the bleaching, delayed greening, viviparous and semi-lethal phenotypes observed in several carotenoid- and NPQ-deficient mutants (Neill *et al.*, 1986; Niyogi *et al.*, 1997; Pogson *et al.*, 1998; Robertson *et al.*, 1966; Treharne *et al.*, 1966;

Wurtzel, 2004). Alterations in the carotenoid pool size make the xanthophyll cycle affect plant fitness. Increasing the xanthophyll cycle pool by overexpressing the bacterial *βOH* gene (*chyB*) enhances stress tolerance in Arabidopsis (Johnson *et al.*, 2008). Zeaxanthin prevents oxidative damage of the thylakoid membranes and plants with reduced zeaxanthin exhibit increased sensitivity to light stress (Havaux and Niyogi, 1999; Verhoeven *et al.*, 2001). Conversely, a lycopene β-cyclase (βLCY) mutant that lacks zeaxanthin but accumulates additional lutein and α-carotene (suppressor of zeaxanthin-less1, *szl1*) exhibits a partially restored quenching efficiency, suggesting that lutein may substitute for zeaxanthin (Li *et al.*, 2009).

II. DISTRIBUTION

Carotenoids are synthesized by all photosynthetic organisms, some bacteria and fungi. Other organisms, such as humans, must acquire carotenoids through dietary intake. For instance, the commercially significant pigment astaxanthin is primarily synthesized by microorganisms, such as the green alga *Haematococcus pluvialis* and is accumulated by fish such as salmon, thus colouring their flesh red. In the case of lobster and other crustaceans, astaxanthin's spectral properties are modified by the protein, crustacyanin, which results in blue pigmentation that shifts to red upon cooking, which causes protein-pigment denaturation (Britton *et al.*, 1997). Flamingos can also make use of carotenoids cosmetically and when the birds applied canthaxanthin-rich secretions onto their feathers, their courting behaviour became more frequent during mating seasons due to a visually more attractive breeding partner (Amat *et al.*, 2010). Humans have been using carotenoids and their derivatives, such as bixin, as food additives, as well as for cosmetic purposes (Bouvier *et al.*, 2003a).

Curious exceptions to the lack of synthesis of carotenoids by animals include the synthesis of carotenoids in the human protist parasites, *Plasmodium* and *Toxoplasma* (Tonhosolo *et al.*, 2009), which is explained by the existence of a remnant plastid, known as an apicoplast. An aphid genome was found to encode enzymes for carotenoid biosynthesis, which was the result of lateral gene transfer from a fungus, thus making aphids the only known animal to date capable of synthesizing their own carotenoids (Moran and Jarvik, 2010).

Carotenoid accumulation relies on the presence of structures capable of storing and retaining carotenoids. During the transformation of a chloroplast into a chromoplast, carotenoids become localized in plastoglobuli before incorporation into the chromoplast (Tevini and Steinmuller, 1985). Carotenoids within plastoglobuli exhibit much higher light stability than carotenoids within chloroplast membranes, suggesting that pigments are better protected from light destruction in these structures (Merzlyak and Solovchenko, 2002). Cyanobacterial mutants with inactivated plastoglobulin-like genes are especially sensitive to altered light regimes, and the plastoglobulin-like peptides accumulate to a greater extent in wild-type cultures that are exposed to high light (Cunningham *et al.*, 2010). Chromoplasts also accumulate carotenoids in lipoprotein structures (Bartley and Scolnik, 1995; Vishnevetsky *et al.*, 1999) that are sequestered as crystals. For example, in a novel cauliflower mutant with orange curd, *Or*, β-carotene accumulates in the plastids of the pith and curd as sheets, ribbons and crystals (Li *et al.*, 2001; Lu *et al.*, 2006).

There are other plastid organelles capable of storing carotenoids. These include the 'colourless' amyloplasts, which store starch granules (Kirk and Tiliney-Bassett, 1978). Lutein is the predominant carotenoid present in many seed amyloplasts such as wheat (Hentschel *et al.*, 2002; Howitt *et al.*, 2009), whereas maize exhibits great diversity in terms of pigment composition (Harjes *et al.*, 2008). Leucoplasts are characteristic of mature root cells and accumulate trace levels of neoxanthin and violaxanthin, which amount to only 0.03–0.07% of the levels in light-grown leaves (Parry and Horgan, 1992). Elaioplasts are specialized lipid-storing plastids and provide an ideal hydrophobic sink for accumulation of carotenoids. The dark-grown etioplast is distinguished by the prolamellar body, a uniformly curved lattice of tubular membranes, which contains several of the biochemical building blocks required for the chloroplast (Gunning and Jagoe, 1967) including the xanthophylls, lutein and violaxanthin (Joyard *et al.*, 1998). The Arabidopsis *crtiso* (*ccr2*) mutant accumulates tetra-*cis*-lycopene and lacks a prolamellar body. Thus, a mutation in carotenoid biosynthesis apparently disrupts membrane curvature and stabilization of the prolamellar body (Park *et al.*, 2002). The absence of this structure in *CRTISO* mutants suggests that different carotenoids either directly or indirectly impede formation of the membrane lattices, which results in a delay in plastid development and greening upon exposure to light. These data demonstrate an important role for carotenoids in plastid differentiation (Park *et al.*, 2002).

III. CAROTENOID BIOSYNTHESIS

A. ISOPRENOID PRECURSORS

Isoprenoids (or terpenoids) are a large and diverse class of naturally occurring organic chemicals derived from five-carbon isoprene units. Carotenoids are derived from two isoprene isomers, isopentenyl diphosphate (IPP) and dimethylallyl diphosphate (DMAPP). The same precursors are used to make a diverse range of compounds that include tocopherols, chlorophylls, phylloquinone, gibberellins, abscisic acid (ABA), monoterpenes and plastoquinone. The biosynthesis of isoprenoid precursors has been covered in detail elsewhere (Rodriguez-Concepcion, 2010).

Two distinct pathways exist for IPP production: the mevalonic acid (MVA) pathway and the mevalonate-independent, methylerythritol 4-phosphate (MEP) pathway (Lange *et al.*, 2000). The plastid-localized MEP pathway combines glyceraldehyde-3-phosphate and pyruvate to form deoxy-D-xylulose 5-phosphate (DXP), a reaction catalysed by DXP synthase (DXS). A number of steps are then required to form geranylgeranyl diphosphate (GGPP), the precursor to carotenoid biosynthesis. The Arabidopsis *Cla1* mutant, in which the *DXS* gene of the MEP pathway is disrupted, is photobleached because of the absence of protective carotenoids (Araki *et al.*, 2000; Estevez *et al.*, 2000). Conversely, overexpression of *PSY* (phytoene synthase) resulted in increased carotenoid accumulation and a concomitant accumulation of the DXS enzyme (Rodriguez-Villalon *et al.*, 2009).

B. CAROTENE SYNTHESIS

1. Phytoene synthase

The first committed step is the condensation of two molecules of GGPP to produce phytoene (Fig. 3). This reaction is catalysed by PSY in higher plants and bacteria (CrtB; Armstrong, 1994). *PSY* is a single-copy gene in Arabidopsis but present in multiple copies in other plants such as rice, maize and cassava, all of which have three copies that are expressed in different tissues and show differential responses to environmental stimuli (Arango *et al.*, 2010; Li *et al.*, 2008a,b; Welsch *et al.*, 2008). PSY is a rate-limiting step and a dosage effect of the maize Y1 allele was noted as early as 1940 (Randolph and Hand, 1940). Overexpression of an exogenous daffodil PSY in rice endosperm leads to phytoene accumulation, the first instance of carotenoid engineering in rice (Burkhardt *et al.*, 1997).

2. Desaturases (PDS and ZDS)

Phytoene is produced as a 15-*cis* isomer, which is subsequently converted to all-*trans* isomer derivatives (Beyer *et al.*, 1989; Chen *et al.*, 2010). Two desaturases, phytoene desaturase (PDS) and ζ-carotene desaturase (ZDS), catalyse a series of dehydrogenation reactions by introducing four double bonds to form lycopene. Desaturation is linked to a plastidic respiratory

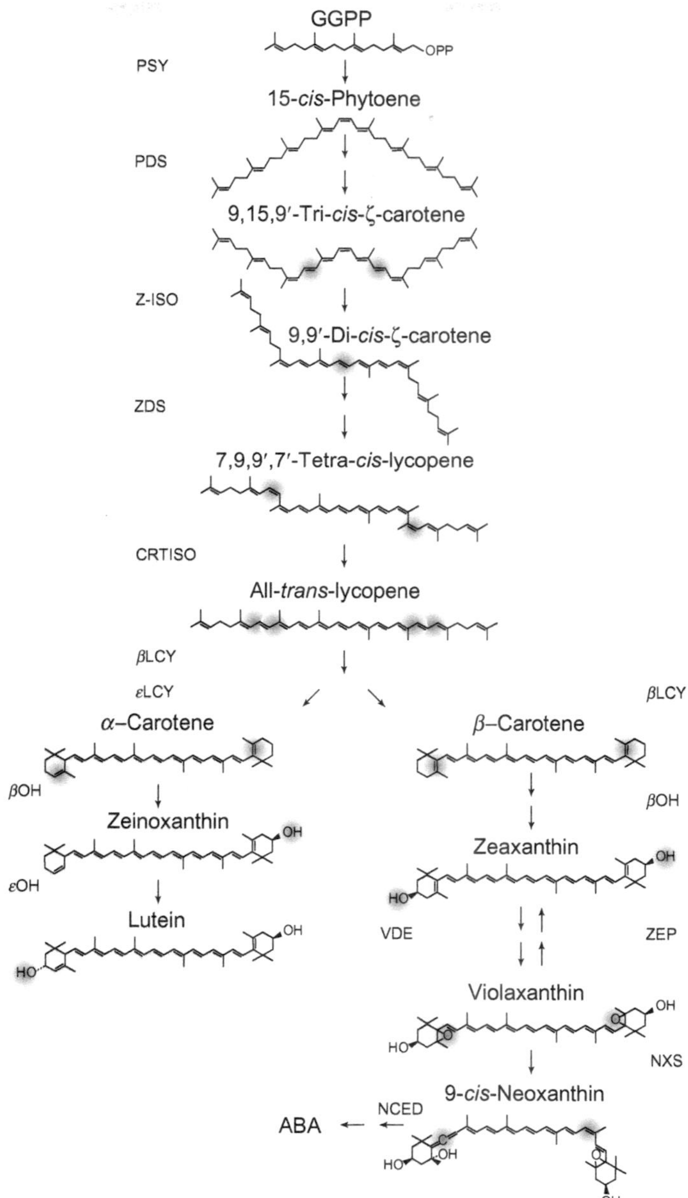

Fig. 3. Carotenoid biosynthetic pathway in higher plants. The pathway shows the primary reactions found in nearly all plant species. Grey shaded areas on carotenoid structures indicate site of activity for each biosynthetic enzyme. ABA, abscisic acid; βLCY, lycopene β-cyclase; βOH, β-hydroxylase; CRTISO, carotenoid isomerase; εLCY, lycopene ε-cyclase; εOH, ε-hydroxylase; NCED, 9-*cis*-epoxycarotenoid dioxygenase; NXS, neoxanthin synthase; PDS, phytoene desaturase; PSY, phytoene synthase; VDE, violaxanthin de-epoxidase; ZDS, ζ-carotene desaturase; ZEP, zeaxanthin epoxidase; Z-ISO, 15-cis-ζ-carotene isomerase.

redox chain (Nievelstein *et al.*, 1995) and evidence for a quinone requirement was demonstrated in daffodil and Arabidopsis (Beyer, 1989; Norris *et al.*, 1995).

3. Isomerases (Z-ISO and CRTISO)

Recent biochemical evidence confirms that the desaturation reactions in plants proceed via various *cis* intermediates, including 9,15,9′-tri-*cis*-ζ-carotene, 9,9′-di-*cis*-ζ-carotene and 7,9,9′-tri-*cis*-neurosporene (Chen *et al.*, 2010; Isaacson *et al.*, 2004). Thus, all-*trans*-lycopene, the preferred substrate for the cyclases, is produced by the desaturases in concert with two isomerases. The first isomerase was identified in Arabidopsis and tomato (Isaacson *et al.*, 2002; Park *et al.*, 2002). Lesions in *CRTISO* result in accumulation of *cis*-carotene isomers in dark-grown plants (Park *et al.*, 2002). Characterization of the maize recessive *y9* mutant demonstrated that, like *crtiso* mutants, the phenotype could be rescued by light exposure, to form 9,9′-di-*cis*-zeta-carotene, the substrate for ZDS (Li *et al.*, 2007). The *Z-ISO* gene was identified in both maize and Arabidopsis and found to be similar to *NnrU* (for nitrite and nitric oxide reductase U), which is required for bacterial denitrification, a pathway that produces nitrogen oxides as alternate electron acceptors for anaerobic growth. An *Escherichia coli* assay proved that Z-ISO was capable of 15-*cis* bond isomerization in 9,15,9′-tri-*cis*-ζ-carotene (Chen *et al.*, 2010).

In the Arabidopsis *CRTISO* (*ccr2*) and *Z-ISO* mutants, *cis* intermediates are photoisomerized in the light, which raises questions about the necessity of carotenoid isomerases in plants and why there are four genes required for the synthesis of lycopene in plants but only one in bacteria. In chromoplasts, CRTISO activity is required for all-*trans*-lycopene accumulation, regardless of the light regime, because the *tangerine* mutant accumulates tetra-*cis*-lycopene in the light (Isaacson *et al.*, 2002). Carotenoids are deposited in a crystalline form in tomato chromoplasts and these may be more resistant to photoisomerization. Further, although the biosynthetic pathway proceeds in chloroplasts, a delayed greening and substantial reduction in lutein occurs in mutants defective in CRTISO in Arabidopsis and some chlorosis occurs in rice and tomato leaves (Fang *et al.*, 2008; Isaacson *et al.*, 2002; Park *et al.*, 2002). Thus, carotenoid synthesis in dark-grown tissues absolutely requires isomerase activity. Such tissues include the endosperm, a target for provitamin A carotenoid biofortification.

4. Cyclases

After lycopene, the carotenoid biosynthetic pathway divides into two branches, distinguished by different cyclic end groups, namely beta or epsilon. Two β-rings form the β,β branch (β-carotene and its derivatives) with one β- and one ε- forming the β,ε branch (α-carotene and its derivatives). βLCY introduces a β-ionone ring to either end of all-*trans*-lycopene to produce β-carotene, whereas both the β-cyclase and ε-cyclase enzymes are required to form α-carotene (Cunningham and Gantt, 2001). Curiously, mutated maize endosperm tissue lacking βLCY activity was also found to accumulate lactucaxanthin (ε,ε-ring) and other unusual carotenes, including δ-carotene, and ε-carotene. The ratio of *βLCY:εLCY* transcripts correlated with the accumulation of different cyclization products in embryo and endosperm tissues (Bai *et al.*, 2009). *εLCY* expression is important in controlling pathway flux to carotenes with higher provitamin A value and the breeding alleles that have been developed for breeding high-provitamin A maize (Harjes *et al.*, 2008).

Other cyclase activities include the capsanthin–capsorubin synthase (CCS) (Lefebvre *et al.*, 1998) in capsicum that cyclizes lycopene to produce the κ-cyclic carotenoids, capsanthin and capsorubin. CCS was found to contain a noncovalently bound flavin adenine dinucleotide (FAD), though it was only required for activity in the presence of NADPH, which functions as the FAD reductant. The CCS flavoproteins catalyse reactions with no net redox change as the reaction did not transfer hydrogen from the dinucleotide cofactors to β-carotene or capsanthin. Thus, FAD in its reduced form could be implicated in the stabilization of the carbocation intermediate (Mialoundama *et al.*, 2010).

C. XANTHOPHYLL SYNTHESIS

Xanthophylls are oxygenated derivatives of carotenes and play important roles in photoprotection and light-harvesting antennae formation (Niyogi, 1999).

1. Hydroxylases

Nearly all xanthophylls in higher plants have hydroxyl moieties on the 3-carbon in the β- or α-carotene rings to form zeaxanthin and lutein, respectively. There are two distinct hydroxylation reactions of the ε- and β-rings, confirmed by the identification of the ε-hydroxylase (εOH) locus, *lut1* (Pogson *et al.*, 1996), and the β-hydroxylase (βOH) genes in higher plants (Sun *et al.*, 1996). βOH enzymes are ferredoxin dependent and contain an iron-coordinating histidine cluster that is required for activity (Bouvier *et al.*, 1998). In contrast, εOH is a plastid-targeted cytochrome P450-type monooxygenase with a distinctly different enzymatic mechanism from the βOHs (Tian *et al.*, 2004).

βOH activity is an important provitamin A biofortification target, as hydroxylation or any other modification of β-ionone rings depletes vitamin A potential. Thus, reduced hydroxylase activity will result in fewer β-rings modifications, thereby maintaining β-carotene pool and maximum vitamin A potential. Of the six loci encoding this enzyme, one locus, *HYD3*, was found to be critical for maize endosperm β-carotene levels and alleles were identified in a population of 51 maize lines (Vallabhaneni *et al.*, 2009) and further association and linkage population studies in maize found that this gene was indeed responsible for a QTL associated with β-carotene accumulation (Yan *et al.*, 2010), and in combination with ϵ*LCY* alleles (Harjes *et al.*, 2008), it is now possible to use molecular markers to select for high-provitamin A carotenoid maize seeds.

2. Zeaxanthin epoxidase and violaxanthin de-epoxidase

An epoxide group is introduced into both rings of zeaxanthin by zeaxanthin epoxidase (ZEP) to form violaxanthin. Under high light stress, the reverse reaction is rapidly undertaken by the violaxanthin de-epoxidase (VDE; Yamamoto, 1979). Light is critical in modulating the interconversion of zeaxanthin and violaxanthin. Under normal light conditions, when the incident light can be safely utilized for photosynthetic electron transport, ZEP converts zeaxanthin to violaxanthin by introducing 5,6-epoxy groups to the 3-hydroxy-β-rings. However, when incident light is in excess, VDE converts a substantial pool of violaxanthin to zeaxanthin (Pfundel *et al.*, 1994).

VDE is soluble and inactive at neutral pH, but following acidification (below pH 6.5) it attaches to the thylakoid membrane and its violaxanthin substrate (Hager and Holocher, 1994). The thylakoid membrane lipid monogalactosyldiacylglycerol is needed for optimal VDE activity when assayed *in vitro* and it requires ascorbate as a reductant (Schaller *et al.*, 2010). Structural analyses revealed that at neutral pH, VDE is monomeric and its active site occluded within a lipocalin barrel, but acidification causes the barrel to open and the enzyme dimerizes. The carotenoid substrate could fit in a channel linking the two active sites of the dimer enabling de-epoxidation of both violaxanthin β-rings, thus forming zeaxanthin (Arnoux *et al.*, 2009). Site-directed mutagenesis of amino acid residues lying in close contact with the two substrates supported the proposed substrate-binding sites and identified two residues, Asp-177 and Tyr-198, that are required for catalytic activity (Saga *et al.*, 2010).

ZEP mutants, *aba1,* are deficient in ABA and display a partially de-etiolated phenotype, including reduced hypocotyl growth, cotyledon expansion and the development of true leaves during late skotomorphogenic growth. However, other ABA-deficient mutants lack this phenotype and

ABA application did not rescue the skotomorphogenesis, though it could be phenocopied by the addition of fluridone, a carotenoid inhibitor that blocks PDS activity. Thus, ZEP appears to have a role in skotomorphogenic growth (Barrero *et al.*, 2008).

3. Neoxanthin synthase

Conversion of violaxanthin to neoxanthin is performed by the enzyme neoxanthin synthase (NXS), which was unequivocally identified in a novel ABA-deficient Arabidopsis mutant, *aba4*. The predicted gene product is a novel chloroplast membrane protein, and constitutive expression of *ABA4* in Arabidopsis led to increased accumulation of *trans*-neoxanthin. Significantly reduced levels of ABA were synthesized in dehydrated *aba4* mutants, demonstrating that ABA biosynthesis in response to stress must occur mainly via neoxanthin isomer precursors (North *et al.*, 2007). Detached *aba4.1* leaves were more sensitive to oxidative stress than the wild type and *aba4.1 npq1* double mutants, lacking both zeaxanthin and neoxanthin, underwent stronger PSII photoinhibition (Dall'Osto *et al.*, 2007).

D. CLEAVAGE PRODUCTS

Characterization of the carotenoid-cleavage gene family has yielded some interesting results in recent years. The enzyme products are varyingly referred to as carotenoid-cleavage dioxygenases (CCD) or 9-*cis*-epoxycarotenoid dioxygenases (NCED), reflecting the first characterized member of this gene family (Schwartz *et al.*, 1997; Tan, 1997). The nine members of the gene family in Arabidopsis show different substrate specificity and tissue distribution (Schwartz *et al.*, 2001, 2003; Tan *et al.*, 2003). The CCD gene family is responsible for the formation of vitamin A, phytohormones (e.g. ABA and strigolactones), coloured spices (e.g. saffron and bixin) and novel signalling molecules as well as plant volatiles used in the perfume industry (Fig. 4).

1. Vitamin A

Vitamin A is a C_{20} cleavage product of carotenoids, which, in addition to its retinoid derivatives, is essential for animal survival and vitamin A biosynthesis has recently been reviewed in detail (von Lintig, 2010). Cleavage of β-carotene was postulated as an important step in the formation on vitamin A, but it was not until 2000 that a β-carotene 15,15$'$-dioxygenase was cloned from *Drosophila melanogaster* (von Lintig and Vogt, 2000) and chicken (Wyss *et al.*, 2000). The deduced amino acid sequence showed homology to the maize carotenoid dioxygenase, VP14, involved in the synthesis of ABA.

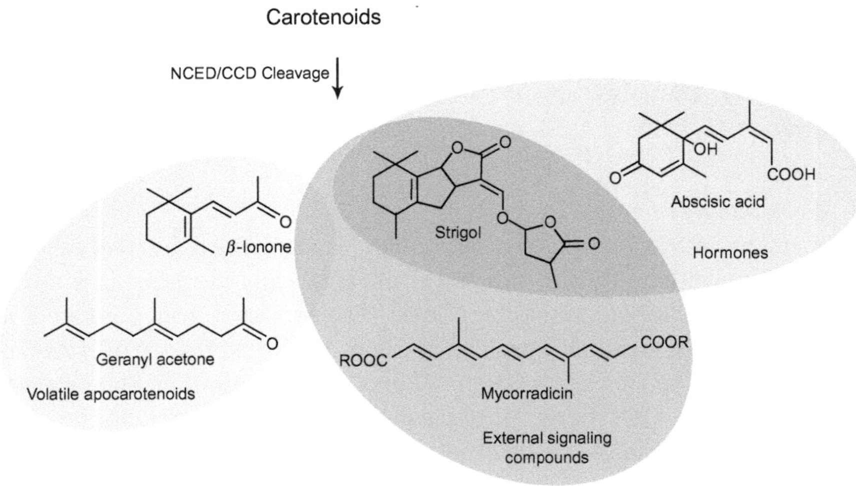

Fig. 4. Carotenoid cleavage products have diverse roles. Carotenoids are cleaved by 9-*cis*-epoxycarotenoid dioxygenase (NCED) or carotenoid cleavage dioxygenase (CCD) enzymes and further modified to form apocarotenoids with diverse functions. Geranyl acetone and β-ionone are volatile apocarotenoids that are commonly used in fragrance manufacture. Mycorradicin is involved in recruiting beneficial fungi. Strigolactones such as strigol enhance the germination of harmful parasitic plant seeds and modulate shoot branching as well as stimulate beneficial mycorrhizal fungi symbiosis. Abscisic acid mediates plant stress responses, playing an important role in controlling stomatal aperture and transpiration as well as promoting seed development and dormancy.

Any carotenoid containing an unmodified β-ionone ring has provitamin A activity; thus, β-carotene is one of the most active because a single β-carotene molecule is cleaved to form two all-*trans*-retinal molecules, which are reduced to form all-*trans*-retinol (vitamin A). All retinoids are derived from this compound and maintain the characteristic β-ionone ring. Different end groups or β-ionone ring modifications characterize the various retinoids. For example, retinoic acid (or 11-*cis*-retinal), which is required for reproduction, embryonic development, cell differentiation, immunity and other biological processes, binds to opsin to provide a chromophore for the visual pigments that mediate phototransduction (von Lintig, 2010).

2. *Phytohormones*

The plant hormone ABA is primarily involved in plant stress responses, seed development and dormancy (Seo and Koshiba, 2002). ABA is a cleavage product of 9-*cis*-violaxanthin and/or 9′-*cis*-neoxanthin, an idea that was first proposed by Taylor and Smith (1967). Cleavage of 9′-*cis*-neoxanthin by

NCED produces xanthoxin and was first identified in the maize *viviparous14* (*vp14*) mutant (Schwartz *et al.*, 1997; Tan, 1997). Xanthoxin is followed in the pathway by a number of further modified products that are required to produce ABA (Seo and Koshiba, 2002). For the ABA signal to be transmitted, it must first bind a receptor molecule. The putative identification of such receptors has been the topic of recent controversy, though the recent crystal structure of a PYR/PYL (pyrabactin resistance/pyrabactin resistance-like) or RCAR (regulatory component of ABA receptor) protein appears to resolve this question (Park *et al.*, 2009). ABA-bound PYR/PYL/RCAR protein inhibits a phosphatase 2C that is known to participate in ABA signalling (Ma *et al.*, 2009).

Strigolactones are carotenoid-derived terpenoid lactones that inhibit shoot branching and can be exuded from plant roots to recruit beneficial mycorrhizal fungi. This apocarotenoid signal has been hijacked by parasitic plant seeds to encourage germination (Dun *et al.*, 2009; Matusova *et al.*, 2005). Such a signal was initially proposed after novel *CCD* mutants were found to exhibit increased shoot branching in Arabidopsis *max4* and pea *rms1* mutants (Sorefan *et al.*, 2003). MAX3 (CCD7) (Booker *et al.*, 2004) and MAX4 (CCD8) can sequentially cleave β-carotene to form the C_{18} compound 13-apo-carotenone (Schwartz *et al.*, 2004). The recent discovery that both rice and pea branching mutants were deficient in strigolactones resolved years of speculation about the nature of the branching signal. It has been shown that strigolactone application restores the wild-type branching phenotype in pea *CCD8* mutants, confirming that strigolactones are necessary and sufficient to inhibit shoot branching in plants. Further, the *CCD8* mutants exhibited additional typical strigolactone-deficient phenotypes including alterations to mycorrhizal symbiosis and parasitic weed interaction (Gomez-Roldan *et al.*, 2008). Concurrent studies confirmed that synthetic strigolactone application inhibits tillering in rice *D10* (*CCD8*) and *D17* (*CCD7*) mutants as well as rescuing the equivalent Arabidopsis mutants. An elegant indirect assay confirmed that these mutants were deficient in strigolactone synthesis, as root exudates did not stimulate germination of parasitic *Striga* seeds to the same extent as wild-type exudates (Umehara *et al.*, 2008). The *CCD7* knockdown in tomato exhibited increased branching, but a metabolic screen did not identify any significant changes in root carotenoid substrate. However, C_{13} cyclohexenone and C_{14} mycorradicin apocarotenoids were reduced in response to mycorrhizal colonization, indicating that CCD7 is required for arbuscular mycorrhiza-induced apocarotenoid synthesis (Vogel *et al.*, 2010).

Other components of the strigolactone biosynthetic pathway have been identified, including *MAX1*, which encodes a cytochrome p450 that modifies an apocarotenoid product of the CCD7 and CCD8 cleavage reactions to produce another mobile intermediate (Booker *et al.*, 2005). *MAX2/RMS4/D3* encode F-box proteins and the mutants are not rescued by exogenous strigolactones and are thus predicted to have a role in signalling via ubiquitin-mediated protein degradation (Beveridge *et al.*, 1996; Stirnberg *et al.*, 2002). Additional steps have been identified in rice, including another high-tillering rice mutant, *d27*, which does not exude strigolactones. *D27* is chloroplast localized, though its enzymatic activity has not been described. Crosses with *d10* (*CCD8*) are not additive and the *d27* mutant can be rescued by strigolactone application, thus is thought to be required for the biosynthesis of strigolactones (Lin *et al.*, 2009). The *D14* gene encodes a α/β-fold hydrolase, and the *d14* mutant is strigolactone insensitive, but exhibits increased tillering and does not show an additive phenotype when crossed with *d10* (Arite *et al.*, 2009). Characterization of this curious mutant could provide insights into strigolactone signalling or have a role in producing a bioactive strigolactone-derived hormone.

Strigolactone and ABA composition were analysed in plants treated with inhibitors of specific carotenoid-cleavage enzymes. Strigolactone content was reduced in plants treated with the CCD inhibitor, D2, but root ABA levels were maintained. Conversely, plants with genetically or chemically inhibited ABA biosynthesis also had reduced strigolactones and a concomitant reduction in *CCD7* and *CCD8* transcript abundance, implying a potential cross-talk role for ABA in the regulation of strigolactone biosynthesis (Lopez-Raez *et al.*, 2010). Finally, strigolactone biosynthesis and the concomitant branching phenotype are responsive to phosphate deficiency in Arabidopsis (Kohlen *et al.*, 2010). The role of strigolactones in controlling plant morphology and response to the environment has become an exciting area of active research.

3. *Bixin, saffron and plant volatiles*

Carotenoid cleavage metabolites are vital for plants and animals. They are also highly prized in the food and cosmetic industries. Bixin (annatto) is a red-coloured, di-carboxylic monomethyl ester apocarotenoid, traditionally derived from the plant *Bixa orellana.* Bouvier and colleagues identified a lycopene cleavage dioxygenase, bixin aldehyde dehydrogenase and norbixin carboxyl methyltransferase that are required to produce bixin from lycopene. Co-transforming the appropriate constructs into *E. coli,* engineered to produce lycopene, resulted in bixin production at a level of 5 mg g^{-1} dry weight (Bouvier *et al.*, 2003a).

Saffron, another commercially important coloured compound, can attribute the majority of its characteristic colour, flavour and aroma to the accumulation of carotenoid derivatives. A crocus (*Crocus sativus*) zeaxanthin 7,8(7′,8′)-cleavage dioxygenase (CsZCD) was cloned and found to be targeted to the chromoplast and initiated the production of the cleavage products. Another enzyme, 9,10(9′,10′)-cleavage dioxygenase was also cloned and found to be a less specific cleavage enzyme (Bouvier *et al.*, 2003b).

Beta-ionone is the predominant norisoprenoid volatile in the mature stigma tissue. Four *CCD* genes were isolated from crocus that were capable of cleaving β-carotene at the 9,10(9′,10′) positions to yield β-ionone, though with different expression patterns indicative of sub-functionalization (Rubio *et al.*, 2008). Differential expression was also observed for *βLCY* genes, *CstLcyB1* and *CstLcyB2a*. The CstLcyB2a is chromoplast specific and conspicuously absent in crocus species with low apocarotenoid content, suggesting that it encodes an important step in determining the accumulation of β-carotene substrate that is required to produce the distinctive saffron apocarotenoids (Ahrazem *et al.*, 2010).

4. *Novel-signalling molecules*

A putative novel signal has been observed in Arabidopsis *bps1* mutants, which are developmentally defective but the shoot can be rescued if the roots are removed or carotenoid biosynthesis is chemically blocked with norflurazon. It is hypothesized that an unknown substance moves constitutively from the root to the shoot to arrest growth, and this is supported by experiments demonstrating that mutant roots are sufficient to arrest wild-type shoot development (Van Norman *et al.*, 2004). *BYPASS1* encodes a novel protein of unknown function that is widespread in plant genomes (Sieburth and Lee, 2010), though the tobacco homologue contains a transmembrane domain and GFP fusion proteins were endoplasmic reticulum associated (Kang *et al.*, 2008). It is likely that more novel carotenoid-derived signalling molecules remain to be identified.

IV. REGULATION OF CAROTENOID BIOSYNTHESIS

A. TRANSCRIPTIONAL REGULATION

Carotenoid composition is responsive to environmental stimuli, oxidative stress, redox poise and metabolite feedback regulation. In general, increases in carotenoid accumulation, be it during fruit ripening, flower development or production of stress-induced carotenoids in algae, coincide with increased transcript abundance of some key (but not all) steps in the pathway.

Phytoene biosynthesis is a rate-limiting step in carotenogenesis and transcript abundance can dramatically alter carotenoid pool size, thus making PSY a logical target in the study of carotenoid regulation. Changes in transcript abundance are particularly evident during morphogenic changes from etioplast to chloroplast or chloroplast to chromoplast. PSY transcript abundance is upregulated during photomorphogenesis via a phytochrome-mediated (red-light) pathway, a response that is abolished in the *phyA* mutant (Welsch *et al.*, 2000, 2008). Phytochrome-mediated light signals regulate carotenoid biosynthesis in plants by way of phytochrome-interacting factor 1 (PIF1), which directly binds to the *PSY* promoter, thus repressing *PSY* expression. Light-triggered degradation of PIFs by photoactivated phytochromes during deetiolation permits *PSY* expression, which enables rapid production of carotenoids (Toledo-Ortiz *et al.*, 2010).

Further evidence that *PSY* controls metabolic flux was obtained by paclobutrazol treatment, which inhibits gibberellin synthesis and enables deetiolation despite the absence of light. PSY activity and carotenoid levels increased in the dark following treatment with paclobutrazol, and this increase was supported by feedback regulation of DXS protein abundance. Overexpression of *DXS* alone in etiolated tissue did not increase carotenoid accumulation; however, *PSY* overexpression resulted in increased carotenoid accumulation and a concomitant post-transcriptional accumulation of DXS (Rodriguez-Villalon *et al.*, 2009).

PSY is present as a single copy in Arabidopsis, but additional homologues have been identified in tomato, poplar and cereal crops such as rice, wheat and maize (Chaudhary *et al.*, 2010; Howitt *et al.*, 2009; Li *et al.*, 2008a,b; Welsch *et al.*, 2008). *PSY* homologues respond differently to abiotic stimuli and have unique tissue specificities though their function remains redundant. For example, salt and drought induce *PSY3* transcript abundance in maize roots, which correlated with increased carotenoid flux and ABA in maize roots (Li *et al.*, 2008a). Rapid disappearance of *PSY2* and *PSY3* mRNA after rewatering suggests mRNA instability or strict control of transcription (Li *et al.*, 2008a). Similar responses were observed in rice *PSY* homologues (Welsch *et al.*, 2008). Cassava also has three sub-functionalized *PSY* genes; however, it was not *PSY3*, but a *PSY1* paralogue that responded to abiotic stress (Arango *et al.*, 2010). Perhaps the most dramatic enhancement of carotenoid accumulation has been achieved in the oil seeds of canola (*Brassica napus*) and Arabidopsis, where overexpression of *PSY* in seeds resulted in a 43- to 50-fold increase in total carotenoid content (Lindgren *et al.*, 2003; Shewmaker *et al.*, 1999). *PSY* overexpression in Arabidopsis seedlings did not alter carotenoid content. However, non-photosynthetic calli and roots overexpressing *PSY* accumulated 10- to 100-fold more carotenoids than corresponding wild-type tissues, predominantly β-carotene and its derivatives, which were deposited as crystals. Similarly, overexpression of the bacterial *PSY*, *crtB*, in white carrot roots also initiated carotenoid crystal formation (Maass *et al.*, 2009).

The complexity of carotenoid regulation is further demonstrated by the analysis of the *PSY* promoter where a *cis*-acting motif (ATCTA) was identified to be involved in mediating the transcriptional regulation of photosynthetic genes, including *PSY* (Welsch *et al.*, 2003). Manipulation of *RAP2.2*, APETALA2 transcription factors that bind to the *PSY* promoter, resulted in only minor carotenoid alterations in root calli (Welsch *et al.*, 2007).

The relative activities of the εLCY and βLCY at the branch point of the pathway have a major regulatory role in modulating the ratio of lutein to that of the β-branch carotenoids (Cuttriss *et al.*, 2007). CRTISO is a major regulatory node at the branch point of the biosynthetic pathway (Cazzonelli *et al.*, 2009; Isaacson *et al.*, 2004). A chromatin-modifying histone methyltransferase enzyme (SET DOMAIN GROUP 8, SDG8) has been shown to be necessary for maintaining *CRTISO* gene expression (Cazzonelli *et al.*, 2009). The *CRTISO* and *SDG8* promoters show overlapping patterns of expression specifically in the shoot apical meristem and pollen, which are active sites of cell division and epigenetic programming (Cazzonelli and Pogson, 2010). The absence of SDG8 reduces *CRTISO* transcript abundance, thereby altering carotenoid flux through the pathway, which might potentially impair strigolactone biosynthesis. This was the first report implicating epigenetic regulatory mechanisms in the control of carotenoid composition (Cazzonelli *et al.*, 2009).

Allelic variation is another important source of carotenoid regulation. For example, alternative splicing of the *PSY-A1* allele altered the relative abundance of functional *PSY* transcript and appeared to be a major QTL determinant of flour colour in bread wheat (Howitt *et al.*, 2009). This was reiterated by a detailed analysis of natural genetic variation in maize. Association analysis, linkage mapping, expression analysis and mutagenesis confirmed that variation at the εLCY locus altered flux partitioning. Four polymorphisms controlled 58% of the variation between α- and β-branch accumulation, thus enabling the selection of alleles that confer high-provitamin A status for improved maize varieties (Harjes *et al.*, 2008). Natural variation in βOH activity also has a significant impact on carotenoid composition (Vallabhaneni *et al.*, 2009; Yan *et al.*, 2010). Multiple control points both within the carotenoid pathway and MEP precursor pathway were identified in maize, and the timing of gene expression was found to be critical in determining carotenoid composition (Vallabhaneni and Wurtzel, 2009).

B. METABOLITE FEEDBACK

Feedback regulation by ABA increases *PSY3* gene expression in rice and plays a critical role in the formation of a positive feedback loop that mediates abiotic stress-induced ABA formation (Welsch *et al.*, 2008). The *βLCY* gene from the eubacterium *Erwinia herbicola* and daffodil (*Narcissus pseudonarcissus*) flowers were introduced into the tomato plastid genome resulting in increased accumulation of xanthophyll cycle pigments in leaves and *β*-carotene in fruits. Surprisingly, transplastomic tomatoes showed $> 50\%$ increase in total carotenoid accumulation (Apel and Bock, 2009), which may be due to a carotenoid product or intermediate feedback.

Lutein levels are altered when the higher plant desaturases and isomerases are bypassed, and thus *cis*-carotene intermediates are not produced (Misawa *et al.*, 1994). Similarly, the absence of CRTISO or specific carotene isomers results in less lutein (Isaacson *et al.*, 2002; Park *et al.*, 2002). The mechanism of this flux partitioning is unclear, though flux through the two branches can be determined by *εLCY* mRNA levels (Cuttriss *et al.*, 2007; Harjes *et al.*, 2008; Pogson *et al.*, 1996; Pogson and Rissler, 2000) and recent experiments indicate that both CRTISO (*ccr2*) and SDG8 (*ccr1*) mutants have aberrant *εLCY* transcript levels. It is thus possible that feedback may account for at least part of the reduction in lutein (Cazzonelli *et al.*, 2009; Cuttriss *et al.*, 2007).

C. CATABOLISM

Accumulation of carotenoids in photosynthetic tissue requires a balance between their rate of synthesis and catabolism. Recent $^{14}CO_2$ uptake data demonstrates that synthesis, and by inference, turnover, is much more rapid than expected (Beisel *et al.*, 2010). The incorporation of ^{14}C into different carotenoids was not uniform and varied between mutants and under high light (Beisel *et al.*, 2010), implying active degradation both enzymatically and by oxidative damage.

Studies in Arabidopsis, strawberry (*Fragaria ananassa*) and chrysanthemum (*Chrysanthemum morifolium*) petals have all demonstrated that the pool of carotenoids is determined in part by CCD catalysed degradation (Auldridge *et al.*, 2006; Garcia-Limones *et al.*, 2008; Ohmiya *et al.*, 2006). In Arabidopsis seeds, loss of CCD function leads to significantly higher carotenoid levels (Auldridge *et al.*, 2006).

CCD1 expression levels in strawberry correlate with ripening and a decrease in lutein content, which suggests that lutein could constitute the main natural substrate of FaCCD1 activity (Garcia-Limones *et al.*, 2008). High expression of CCD1 associated with certain maize alleles was correlated with low carotenoid levels in maize endosperm (Vallabhaneni *et al.*, 2010). Petal colour in chrysanthemums is also regulated by CCD activity; white petals contain elevated transcript levels of *CmCCD4a*, which catabolizes the yellow carotenoid pigments (Ohmiya *et al.*, 2006). Curiously, when *CCD1* was overexpressed in high carotenoid golden rice lines (GR2), there appeared to be little impact on carotenoid levels in the endosperm. In fact, a similar carotenoid content was observed in both GR2 and antisense lines. Surprisingly, *in vitro* analyses suggested that apocarotenoids were the primary substrates of OsCCD1 (Ilg *et al.*, 2010).

D. STORAGE CAPACITY

Carotenoid biosynthesis appears to take place largely at the chloroplast envelope and, in some cases, the thylakoid membrane (Joyard *et al.*, 2009). Storage capacity is a major determinant of carotenoid pool size; the *high pigment2* (*hp2*) tomato mutant (DEETIOLATED1, a negative regulator of light signalling) has a larger plastid and thus increased pigmentation (Kolotilin *et al.*, 2007). Similarly, the *hp3* tomato mutant (ZE) revealed an ABA deficiency, an enlarged plastid compartment and 30% more carotenoids in mature fruit (Galpaz *et al.*, 2008). Plastid differentiation is an important mechanism in determining storage capacity, as demonstrated by the cauliflower (*Brassica oleracea*) Orange (*Or*) gene that creates a metabolic sink to accumulate β-carotene in the chromoplast (Li *et al.*, 2001; Li and Van Eck, 2007; Lu *et al.*, 2006). During the chloroplast to chromoplast transformation process, carotenoids become localized in plastoglobuli (Steinmuller and Tevini, 1985). Carotenoids within plastoglobuli exhibit much higher light stability than carotenoids within chloroplast membranes (Merzlyak and Solovchenko, 2002).

V. NUTRITION

A. RICE

Golden rice (*Oryza sativa*) was developed to alleviate vitamin A deficiency as this important staple crop does not typically accumulate any carotenoids in edible endosperm tissue. Daffodil *PSY* and bacterial desaturases (*crtI, Erwinia uredovora*) were targeted to endosperm tissue, where they produced up to 1.6 $\mu g\ g^{-1}$ carotenoids, predominantly β-carotene due to endogenous cyclase activity (Ye *et al.*, 2000). A second generation line ‘Golden Rice 2’ overcame a metabolic bottleneck by incorporating a more active *PSY* gene from maize, which substantially improved carotenoid biosynthesis, with some lines accumulating up to 37 μg g^{-1} (Paine *et al.*, 2005). More recent work has focused on transgene stability and the transformation of high-yielding cultivars (Datta *et al.*, 2006, 2007). A dietary study of Golden Rice confirmed that deuterium-labelled [^{2}H]*â*-carotene produced by these plants could be converted to retinol and is thus an effective biofortification strategy (Tang *et al.*, 2009).

B. MAIZE

Zea mays is an essential staple cereal crop that naturally accumulates provitamin A carotenoids in the endosperm of the seed. There are vast diverse collections from which to source favourable alleles for plant breeding programmes. Such collections have been extensively utilized to identify important regulatory points in determining provitamin A potential. A significant QTL analysis determined that *PSY1* was responsible for 6.6–27.2% of phenotypic variation in carotenoid content (Chander *et al.*, 2008). Genetic variation in *εLCY* was responsible for 58% of the variation in flux between the two branches of the pathway and is critical for driving provitamin A levels (Harjes *et al.*, 2008). Two recent studies identified different *β*OH alleles of one locus that were important in determining the extent of *β*-ionone ring hydroxylation, and thus loss of provitamin A activity (Vallabhaneni *et al.*, 2009; Yan *et al.*, 2010). The most favourable alleles were found in temperate varieties and will be bred into tropical maize germplasm to help alleviate vitamin A deficiency in third world countries (Yan *et al.*, 2010). Recent studies also identified additional control points that offer future possibilities for further enhancing carotenoid levels in maize (Vallabhaneni *et al.*, 2010; Vallabhaneni and Wurtzel, 2009). Transgenic approaches to maize biofortification have also played a significant role in modifying *β*-carotene content (Aluru *et al.*, 2008; Zhu *et al.*, 2008) and laid the foundation for targeting alternative approaches. Analyses of tropical varieties (Menkir *et al.*, 2008) and sweet corn (Fanning *et al.*, 2010) have identified further diversity for carotenoid enhancement projects.

C. WHEAT

Triticum spp. endosperm colour is an important agronomic trait and thus has been the focus of several QTL studies. Lutein is the predominant carotenoid in wheat endosperm tissue and is frequently heavily esterified (Atienza *et al.*, 2007; Howitt *et al.*, 2009). A targeted molecular marker was developed for the *PSY1* gene on wheat chromosome 7A, and found to co-segregate with yellow pigmentation in a collection of Chinese wheat cultivars (He *et al.*, 2008). Further, the total carotenoid pool size was found to be modulated by *εLCY* alleles and/or *PSY-A1* spice variants (Howitt *et al.*, 2009). Transgenic wheat expressing endosperm-specific PSY1 from maize and bacterial CRTI (desaturases) produced a 10.8-fold increase (up to 4.96 μg g^{-1} dry weight) in total seed carotenoid content (Cong *et al.*, 2009). Thus, both targeted breeding and transgenic approaches are likely to improve wheat lutein content, which is correlated with protection against age-related macular degeneration (AMD) of the eye—the leading cause of blindness in the developed world. Whether such strategies can increase provitamin A levels in wheat has not been reported thus far.

D. CASSAVA

Manihot esculenta is an important staple crop, especially in arid regions such as sub-Saharan Africa, though it is nutrient poor and typically accumulates very little provitamin A. Analysis of diversity collections identified landraces that accumulate lycopene (5 mg kg^{-1}) or β-carotene (4 mg kg^{-1}) (Nassar *et al.*, 2007) and such variation was harnessed to identify natural *PSY* alleles that altered metabolic flux (Welsch *et al.*, 2010). Cassava has three *PSY* genes, one of which (*PSY1*) responded strongly to abiotic stress (Arango *et al.*, 2010). A single nucleotide polymorphism in *PSY2* was found to co-segregate with yellow-rooted cultivars in a breeding population that accumulated between 6.0 and 11.5 μg g^{-1} carotenoids in fresh tissue. This genetic variant was used to successfully produce transgenic cassava with increased carotenoid accumulation in the roots (Welsch *et al.*, 2010). Bioavailability of β-carotene in cassava was analysed and found to be as efficacious as β-carotene supplementation; thus, biofortification of cassava is a valid approach to alleviating vitamin A deficiencies (Howe *et al.*, 2009)

E. SORGHUM

Sorghum bicolor is a major staple crop grown in semiarid regions due to its drought tolerance, which makes it a good candidate for biofortification. Yellow endosperm varieties contain provitamin A carotenoids and diverse collections of sorghum landraces have been analysed to quantify pigment diversity, including a collection of 164 landraces from Niger and Nigeria (Fernandez *et al.*, 2009). Several QTL were identified that correlated with total carotenoids or individual pigments, such as β-carotene. A strong QTL that accounted for between 11% and 15% of phenotypic variation was associated with *PSY3*, thus pinpointing a focal point for breeding high-provitamin A sorghums (Fernandez *et al.*, 2008).

F. BANANA AND PLANTAIN

Banana and plantain (*Musa* spp.) are tropical crops and some of the most highly consumed fruits in the world. They have a high genetic diversity, as exemplified by the Embrapa international germplasm collection of more than 400 accessions, including wild diploids, triploids and tetraploids; however, they are not readily bred. Analysis of pigment composition identified 42 high pigment lines that accumulate between 1.06 and 19.24 $\mu g\ g^{-1}$ of total carotenoids. Genetic variability was estimated using Diversity Arrays Technology molecular markers to establish a biofortification programme (Amorim *et al.*, 2009). A similar study identified broad pigment diversity but limited accumulation of mineral micronutrients in a 171 genotype collection (Davey *et al.*, 2009).

G. SWEET POTATO

Proof of the biofortification principle was established in Kenya where consumption of the orange-fleshed sweet potato (*Ipomoea batatas*) increased the vitamin A status of women and children (Hagenimana *et al.*, 1999). A similar study in South Africa demonstrated a reduction in vitamin A deficiency of children (van Jaarsveld *et al.*, 2005). However, analysis of carotenoid degradation in stored sweet potato, which is typically dried and stored for months, indicated losses of around 70% of the total carotenoid pool after 4 months' storage in Uganda. This demonstrates the necessity for establishing diversity in carotenoid-rich agricultural products and underlines the difficulty in maintaining provitamin A intake outside of the growing season (Bechoff *et al.*, 2010).

H. POTATO

Another staple food crop with limited micronutrient content is potato (*Solanum tuberosum*). Potato has been successfully fortified to produce provitamin A carotenoids. Overexpression of three bacterial genes for β-carotene synthesis (*CrtB, CrtI* and *CrtY*, encoding PSY, PDS and βLCY, respectively) from Erwinia were targeted to the tuber. The transgenic lines accumulated up to 47 $\mu g\ g^{-1}$ β-carotene (Diretto *et al.*, 2007). Detailed transcript analyses of lines carrying various combinations of transgenes found that β-cyclase had the greatest impact on regulating the amount of carotenoid accumulation (Diretto *et al.*, 2010).

VI. CONCLUSIONS

The essential roles that carotenoids play in human health, as well as plant photosynthesis, photoprotection and reproduction, make them obvious candidates for enhancement and manipulation. To this end, molecular genetics, in concert with classical biochemistry, has facilitated an advanced understanding of the biosynthetic pathway. Breakthroughs in understanding the regulation of carotenoid accumulation are paving the way for improving the provitamin A content of staple food crops that would otherwise be of low nutritional value. This is of utmost importance for developing countries, where food storage is a problem and effective agriculture practices are still being developed. Further characterisation of regulatory processes that determine carotenoid accumulation, composition and storage capacity, as well as developing new transgenic technologies and breeding varieties, will all continue to strengthen biofortification projects in diverse crop species.

ACKNOWLEDGEMENTS

The authors acknowledge the support of Professor David Christopher, the New Zealand Foundation for Research Science and Technology to A. J. C.; funding from the United States National Institutes of Health (GM081160) and New York State to E. T. W. and A. J. C; and the Australian Research Council Centre of Excellence in Plant Energy Biology (CE0561495) to C. I. C. and B. J. P.

REFERENCES

Ahrazem, O., Rubio-Moraga, A., Lopez, R. C. and Gomez-Gomez, L. (2010). The expression of a chromoplast-specific lycopene beta cyclase gene is involved in the high production of saffron's apocarotenoid precursors. *Journal of Experimental Botany* **61**, 105–119.

Aluru, M., Xu, Y., Guo, R., Wang, Z. G., Li, S. S., White, W., Wang, K. and Rodermel, S. (2008). Generation of transgenic maize with enhanced provitamin A content. *Journal of Experimental Botany* **59**, 3551–3562.

Amat, J. A., Rendón, M. A., Garrido-Fernández, J., Garrido, A., Rendón-Martos, M. and Pérez-Gálvez, A. (2010). Greater flamingos Phoenicopterus roseus use uropygial secretions as make-up. *Behaviour Ecology and Sociobiology* **50**, 55–60.

Amorim, E. P., Vilarinhos, A. D., Cohen, K. O., Amorim, V. B. O., dos Santos-Serejo, J. A., Oliveira e Silva, S., Pestana, K. N., dos Santos, V. J., Paes, N. S., Monte, D. C. and dos Reis, R. V. (2009). Genetic diversity of carotenoid-rich bananas evaluated by Diversity Arrays Technology (DArT). *Genetics and Molecular Biology* **32**, 96–103.

Apel, W. and Bock, R. (2009). Enhancement of carotenoid biosynthesis in transplastomic tomatoes by induced lycopene-to-provitamin A conversion. *Plant Physiology* **151**, 59–66.

Araki, N., Kusumi, K., Masamoto, K., Niwa, Y. and Iba, K. (2000). Temperature-sensitive Arabidopsis mutant defective in 1-deoxy-D-xylulose-5-phosphate synthase within the plastid non-mevalonate pathway of isoprenoid biosynthesis. *Physiologia Plantarum* **108**, 19–24.

Arango, J., Wust, F., Beyer, P. and Welsch, R. (2010). Characterization of phytoene synthases from cassava and their involvement in abiotic stress-mediated responses. *Planta* **232**, 1251–1256.

Arite, T., Umehara, M., Ishikawa, S., Hanada, A., Maekawa, M., Yamaguchi, S. and Kyozuka, J. (2009). d14, a strigolactone-insensitive mutant of rice, shows an accelerated outgrowth of tillers. *Plant & Cell Physiology* **50**, 1416–1424.

Armstrong, G. A. (1994). Eubacteria show their true colors: Genetics of carotenoid pigment biosynthesis from microbes to plants. *Journal of Bacteriology* **176**, 4795–4802.

Arnoux, P., Morosinotto, T., Saga, G., Bassi, R. and Pignol, D. (2009). A structural basis for the pH-dependent xanthophyll cycle in Arabidopsis thaliana. *The Plant Cell* **21**, 2036–2044.

Atienza, S. G., Ballesteros, J., Martin, A. and Hornero-Méndez, D. (2007). Genetic variability of carotenoid concentration and degree of esterification among tritordeum (× Tritordeum Ascherson et Graebner) and durum wheat accessions. *Journal of Agricultural and Food Chemistry* **55**, 4244–4251.

Auldridge, M. E., McCarty, D. R. and Klee, H. J. (2006). Plant carotenoid cleavage oxygenases and their apocarotenoid products. *Current Opinion in Plant Biology* **9**, 315–321.

Bai, L., Kim, E. H., DellaPenna, D. and Brutnell, T. P. (2009). Novel lycopene epsilon cyclase activities in maize revealed through perturbation of carotenoid biosynthesis. *The Plant Journal* **59**, 588–599.

Bailey, S. and Grossman, A. (2008). Photoprotection in cyanobacteria: Regulation of light harvesting. *Photochemistry and Photobiology* **84**, 1410–1420.

Barrero, J. M., Rodriguez, P. L., Quesada, V., Alabadi, D., Blazquez, M. A., Boutin, J. P., Marion-Poll, A., Ponce, M. R. and Micol, J. L. (2008). The ABA1 gene and carotenoid biosynthesis are required for late skotomorphogenic growth in Arabidopsis thaliana. *Plant, Cell & Environment* **31**, 227–234.

Bartley, G. and Scolnik, P. (1995). Plant carotenoids: Pigments for photoprotection, visual attraction, and human health. *The Plant Cell* **7**, 1027–1038.

Bechoff, A., Westby, A., Owori, C., Menya, G., Dhuique-Mayer, C., Dufour, D. and Tomlins, K. (2010). Effect of drying and storage on the degradation of total carotenoids in orange-fleshed sweetpotato cultivars. *Journal of the Science of Food and Agriculture* **90**, 622–629.

Beisel, K. G., Jahnke, S., Hofmann, D., Köppchen, S., Schurr, U. and Matsubara, S. (2010). Continuous turnover of carotenes and chlorophyll a in mature leaves of Arabidopsis thaliana revealed by 14CO2 pulse-chase labeling. *Plant Physiology* **152**, 2188–2199.

Beveridge, C. A., Ross, J. J. and Murfet, I. C. (1996). Branching in pea—Action of genes rms3 and rms4. *Plant Physiology* **110**, 859–865.

Beyer, P. (1989). Carotene biosynthesis in daffodil chromoplasts: On the membrane-integral desaturation and cyclization reactions. *In* Physiology, Biochemistry, and Genetics of Nongreen Plastids, (C. D. Boyer, J. C. Shannon and R. C. Hardison, eds.), pp. 157–170. The American Society of Plant Physiologists, Rockville, MD.

Beyer, P., Mayer, M. and Kleinig, H. (1989). Molecular oxygen and the state of geometric isomerism of intermediates are essential in the carotene desaturation and cyclization reactions in daffodil chromoplasts. *European Journal of Biochemistry* **184**, 141–150.

Booker, J., Auldridge, M., Wills, S., McCarty, D., Klee, H. and Leyser, C. (2004). MAX3/CCD7 is a carotenoid cleavage dioxygenase required for the synthesis of a novel plant signaling molecule. *Current Biology* **14**, 1232–1238.

Booker, J., Sieberer, T., Wright, W., Williamson, L., Willett, B., Stirnberg, P., Turnbull, C., Srinivasan, M., Goddard, P. and Leyser, O. (2005). MAX1 encodes a cytochrome P450 family member that acts downstream of MAX3/4 to produce a carotenoid-derived branch-inhibiting hormone. *Developmental Cell* **8**, 443–449.

Bouvier, F., Backhaus, R. A. and Camara, B. (1998). Induction and control of chromoplast-specific carotenoid genes by oxidative stress. *The Journal of Biological Chemistry* **273**, 30651–30659.

Bouvier, F., Dogbo, O. and Camara, B. (2003a). Biosynthesis of the food and cosmetic plant pigment bixin (annatto). *Science* **300**, 2089–2091.

Bouvier, F., Suire, C., Mutterer, J. and Camara, B. (2003b). Oxidative remodeling of chromoplast carotenoids: Identification of the carotenoid dioxygenase CsCCD and CsZCD genes involved in Crocus secondary metabolite biogenesis. *The Plant Cell* **15**, 47–62.

Britton, G., Weesie, R. J., Askin, D., Warburton, J. D., GallardoGuerrero, L., Jansen, F. J., de Groot, H. J. M., Lugtenburg, J., Cornard, J. P. and Merlin, J. C. (1997). Carotenoid blues: Structural studies on carotenoproteins. *Pure and Applied Chemistry* **69**, 2075–2084.

Burkhardt, P. K., Beyer, P., Wünn, J., Klöti, A., Armstrong, G. A., Schledz, M., von Lintig, J. and Potrykus, I. (1997). Transgenic rice (*Oryza sativa*) endosperm expressing daffodil (*Narcissus pseudonarcissus*) phytoene synthase accumulates phytoene, a key intermediate of provitamin A biosynthesis. *The Plant Journal* **11**, 1071–1078.

Cazzonelli, C. I. and Pogson, B. J. (2010). Source to sink: Regulation of carotenoid biosynthesis in plants. *Trends in Plant Science* **15**, 266–274.

Cazzonelli, C. I., Cuttriss, A. J., Cossetto, S. B., Pye, W., Crisp, P., Whelan, J., Finnegan, E. J., Turnbull, C. and Pogson, B. J. (2009). Regulation of carotenoid composition and shoot branching in Arabidopsis by a chromatin modifying histone methyltransferase, SDG8. *The Plant Cell* **21**, 39–53.

Chander, S., Guo, Y. Q., Yang, X. H., Zhang, J., Lu, X. Q., Yan, J. B., Song, T. M., Rocheford, T. R. and Li, J. S. (2008). Using molecular markers to identify two major loci controlling carotenoid contents in maize grain. *Theoretical and Applied Genetics* **116**, 223–233.

Chaudhary, N., Nijhawan, A., Khurana, J. P. and Khurana, P. (2010). Carotenoid biosynthesis genes in rice: Structural analysis, genome-wide expression profiling and phylogenetic analysis. *Molecular Genetics and Genomics* **283**, 13–33.

Chen, Y., Li, F. Q. and Wurtzel, E. T. (2010). Isolation and characterization of the Z-ISO gene encoding a missing component of carotenoid biosynthesis in plants. *Plant Physiology* **153**, 66–79.

Cong, L., Wang, C., Chen, L., Liu, H. J., Yang, G. X. and He, G. Y. (2009). Expression of phytoene synthase1 and carotene desaturase crtI genes result in an increase in the total carotenoids content in transgenic elite wheat (Triticum aestivum L.). *Journal of Agricultural and Food Chemistry* **57**, 8652–8660.

Cunningham, F. X. and Gantt, E. (2001). One ring or two? Determination of ring number in carotenoids by lycopene epsilon-cyclases. *Proceedings of the National Academy of Sciences of the United States of America* **98**, 2905.

Cunningham, F. X., Tice, A. B., Pham, C. and Gantt, E. (2010). Inactivation of genes encoding plastoglobulin-like proteins in Synechocystis sp PCC 6803 leads to a light-sensitive phenotype. *Journal of Bacteriology* **192**, 1700–1709.

Cuttriss, A. J., Chubb, A. C., Alawady, A., Grimm, B. and Pogson, B. J. (2007). Regulation of lutein biosynthesis and prolamellar body formation in Arabidopsis. *Functional Plant Biology* **34**, 663–672.

Dall'Osto, L., Cazzaniga, S., North, H., Marion-Poll, A. and Bassi, R. (2007). The Arabidopsis aba4-1 mutant reveals a specific function for neoxanthin in protection against photooxidative stress. *The Plant Cell* **19**, 1048–1064.

Datta, K., Rai, M., Parkhi, V., Oliva, N., Tan, J. and Datta, S. K. (2006). Improved 'golden' indica rice and post-transgeneration enhancement of metabolic target products of carotenoids (beta-carotene) in transgenic elite cultivars (IR64 and BR29). *Current Science* **91**, 935–939.

Datta, S. K., Datta, K., Parkhi, V., Rai, M., Baisakh, N., Sahoo, G., Rehana, S., Bandyopadhyay, A., Alamgir, M., Ali, M. S., Abrigo, E. Oliva, N. *et al.* (2007). Golden rice: Introgression, breeding, and field evaluation. *Euphytica* **154**, 271–278.

Davey, M. W., Van den Bergh, I., Markham, R., Swennen, R. and Keulemans, J. (2009). Genetic variability in Musa fruit provitamin A carotenoids, lutein and mineral micronutrient contents. *Food Chemistry* **115**, 806–813.

Davis, C. R., Howe, J. A., Rocheford, T. R. and Tanumihardjo, S. A. (2008). The xanthophyll composition of biofortified maize (Zea mays Sp.) does not influence the bioefficacy of provitamin A carotenoids in Mongolian gerbils (Meriones unguiculatus). *Journal of Agricultural and Food Chemistry* **56**, 6745–6750.

DellaPenna, D. and Pogson, B. J. (2006). Vitamin synthesis in plants: Tocopherols and carotenoids. *Annual Review of Plant Biology* **57**, 711–738.

Diretto, G., Al-Babili, S., Tavazza, R., Papacchioli, V., Beyer, P. and Giuliano, G. (2007). Metabolic engineering of potato carotenoid content through tuber-specific overexpression of a bacterial mini-pathway. *PloS One* **2**(4), e350.

Diretto, G., Al-Babili, S., Tavazza, R., Scossa, F., Papacchioli, V., Migliore, M., Beyer, P. and Giuliano, G. (2010). Transcriptional-metabolic networks in beta-carotene-enriched potato tubers: The long and winding road to the golden phenotype. *Plant Physiology* **154**, 899–912.

Dun, E. A., Brewer, P. B. and Beveridge, C. A. (2009). Strigolactones: Discovery of the elusive shoot branching hormone. *Trends in Plant Science* **14**, 364–372.

Estevez, J. M., Cantero, A., Romero, C., Kawaide, H., Jimenez, L. F., Kuzuyama, T., Seto, H., Kamiya, Y. and Leon, P. (2000). Analysis of the expression of *CLA1*, a gene that encodes the 1- deoxyxylulose 5-phosphate synthase of the 2-C-methyl-D-erythritol-4- phosphate pathway in Arabidopsis. *Plant Physiology* **124**, 95–104.

Fang, J., Chai, C. L., Qian, Q., Li, C. L., Tang, J. Y., Sun, L., Huang, Z. J., Guo, X. L., Sun, C. H., Liu, M., Zhang, Y. Lu, Q. T. *et al.* (2008). Mutations of genes in synthesis of the carotenoid precursors of ABA lead to pre-harvest sprouting and photo-oxidation in rice. *The Plant Journal* **54**, 177–189.

Fanning, K. J., Martin, I., Wong, L., Keating, V., Pun, S. and O'Hare, T. (2010). Screening sweetcorn for enhanced zeaxanthin concentration. *Journal of the Science of Food and Agriculture* **90**, 91–96.

Fernandez, M. G. S., Hamblin, M. T., Li, L., Rooney, W. L., Tuinstra, M. P. and Kresovich, S. (2008). Quantitative trait loci analysis of endosperm color and carotenoid content in sorghum grain. *Crop Science* **48**, 1732–1743.

Fernandez, M. G. S., Kapran, I., Souley, S., Abdou, M., Maiga, I. H., Acharya, C. B., Hamblin, M. T. and Kresovich, S. (2009). Collection and characterization of yellow endosperm sorghums from West Africa for biofortification. *Genetic Resources and Crop Evolution* **56**, 991–1000.

Galpaz, N., Wang, Q., Menda, N., Zamir, D. and Hirschberg, J. (2008). Abscisic acid deficiency in the tomato mutant high-pigment 3 leading to increased plastid number and higher fruit lycopene content. *The Plant Journal* **53**, 717–730.

Garcia-Limones, C., Schnabele, K., Blanco-Portales, R., Bellido, M. L., Caballero, J. L., Schwab, W. and Munoz-Blanco, J. (2008). Functional characterization of FaCCD1: A carotenoid cleavage dioxygenase from strawberry involved in lutein degradation during fruit ripening. *Journal of Agricultural and Food Chemistry* **56**, 9277–9285.

Gomez-Roldan, V., Fermas, S., Brewer, P. B., Puech-Pages, V., Dun, E. A., Pillot, J. P., Letisse, F., Matusova, R., Danoun, S., Portais, J. C., Bouwmeester, H. Becard, G. *et al.* (2008). Strigolactone inhibition of shoot branching. *Nature* **455**, 189–194.

Gunning, B. and Jagoe, M. (1967). The prolamellar body. *In* Biochemistry of Chloroplasts, (T. Goodwin, ed.), pp. 655–676. Academic Press, London.

Hagenimana, V., Anyango-Oyunga, M., Low, J., Njdroge, S. M., Gichuki, S. T. and Kabira, J. (1999). The effects of women farmers' adoption of orange-fleshed sweet potatoes: Raising vitamin A intake in Kenya. *In* Research Report Series, pp. 1–28. International Center for Research on Women, Washington, DC.

Hager, A. and Holocher, K. (1994). Localization of the xanthophyll-cycle enzyme violaxanthin de-epoxidase within the thylakoid lumen and abolition of its mobility by a (light-dependent) pH decrease. *Planta* **192**, 581–589.

Harjes, C. E., Rocheford, T. R., Bai, L., Brutnell, T. P., Kandianis, C. B., Sowinski, S. G., Stapleton, A. E., Vallabhaneni, R., Williams, M., Wurtzel, E. T., Yan, J. B. and Buckler, E. S. (2008). Natural genetic variation in lycopene epsilon cyclase tapped for maize biofortification. *Science* **319**, 330–333.

Havaux, M. and Niyogi, K. K. (1999). The violaxanthin cycle protects plants from photooxidative damage by more than one mechanism. *Proceedings of the National Academy of Sciences of the United States of America* **96**, 8762–8767.

He, X. Y., Zhang, Y. L., He, Z. H., Wu, Y. P., Xiao, Y. G., Ma, C. X. and Xia, X. C. (2008). Characterization of phytoene synthase 1 gene (Psy1) located on common wheat chromosome 7A and development of a functional marker. *Theoretical and Applied Genetics* **116**, 213–221.

Hentschel, V., Kranl, K., Hollmann, J., Lindhauer, M. G., Bohm, V. and Bitsch, R. (2002). Spectrophotometric determination of yellow pigment content and evaluation of carotenoids by high-performance liquid chromatography in durum wheat grain. *Journal of Agricultural and Food Chemistry* **50**, 6663–6668.

Howe, J. A., Maziya-Dixon, B. and Tanumihardjo, S. A. (2009). Cassava with enhanced beta-carotene maintains adequate vitamin A status in Mongolian gerbils (Meriones unguiculatus) despite substantial cis-isomer content. *The British Journal of Nutrition* **102**, 342–349.

Howitt, C. A., Cavanagh, C. R., Bowerman, A. F., Cazzonelli, C., Rampling, L., Mimica, J. L. and Pogson, B. J. (2009). Alternative splicing, activation of cryptic exons and amino acid substitutions in carotenoid biosynthetic genes

are associated with lutein accumulation in wheat endosperm. *Functional & Integrative Genomics* **9**, 363–376.
Ilg, A., Yu, Q. J., Schaub, P., Beyer, P. and Al-Babili, S. (2010). Overexpression of the rice carotenoid cleavage dioxygenase 1 gene in Golden Rice endosperm suggests apocarotenoids as substrates in planta. *Planta* **232**, 691–699.
Isaacson, T., Ronen, G., Zamir, D. and Hirschberg, J. (2002). Cloning of *tangerine* from tomato reveals a carotenoid isomerase essential for the production of b-carotene and xanthophylls in plants. *The Plant Cell* **14**, 333–342.
Isaacson, T., Ohad, I., Beyer, P. and Hirschberg, J. (2004). Analysis in vitro of the enzyme CRTISO establishes a poly-cis-carotenoid biosynthesis pathway in plants. *Plant Physiology* **136**, 4246–4255.
Johnson, M. P., Davidson, P. A., Ruban, A. V. and Horton, P. (2008). The xanthophyll pool size controls the kinetics of nonphotochemical chlorophyll quenching in Arabidopsis thaliana. *FEBS Letters* **582**, 262–266.
Joyard, J., Teyssier, E., Miege, C., Berny-Seigneurin, D., Marechal, E., Block, M. A., Dorne, A. J., Rolland, N., Ajlani, G. and Douce, R. (1998). The biochemical machinery of plastid envelope membranes. *Plant Physiology* **118**, 715–723.
Joyard, J., Ferro, M., Masselon, C., Seigneurin-Berny, D., Salvi, D., Garin, J. and Rolland, N. (2009). Chloroplast proteomics and the compartmentation of plastidial isoprenoid biosynthetic pathways. *Molecular Plant* **2**, 1154–1180.
Kang, Y. W., Kim, R. N., Cho, H. S., Kim, W. T., Choi, D. and Pai, H. S. (2008). Silencing of a BYPASS1 homolog results in root-independent pleiotrophic developmental defects in Nicotiana benthamiana. *Plant Molecular Biology* **68**, 423–437.
Kirk, J. T. O. and Tiliney-Bassett, R. A. E. (1978). Proplastids, etioplasts, amyloplasts, chromoplasts and other plastids. *In* The Plastids: Their Chemistry, Structure, Growth and Inheritance, (J. T. O. Kirk and R. A. E. Tinley-Bassett, eds.), pp. 217–239. Elsevier/North Holland Biomedical Press, Amsterdam.
Kohlen, W., Charnikhova, T., Liu, Q., Bours, R., Domagalska, M. A., Beguerie, S., Verstappen, F., Leyser, O., Bouwmeester, H. J. and Ruyter-Spira, C. (2010). Strigolactones are transported through the xylem and play a key role in shoot architectural response to phosphate deficiency in non-AM host Arabidopsis thaliana. *Plant Physiology* **55**, 974–987.
Kolotilin, I., Koltai, H., Tadmor, Y., Bar-Or, C., Reuveni, M., Meir, A., Nahon, S., Shlomo, H., Chen, L. and Levin, I. (2007). Transcriptional profiling of high pigment-2dg tomato mutant links early fruit plastid biogenesis with its overproduction of phytonutrients. *Plant Physiology* **145**, 389–401.
Lange, B. M., Rujan, T., Martin, W. and Croteau, R. (2000). Isoprenoid biosynthesis: The evolution of two ancient and distinct pathways across genomes. *Proceedings of the National Academy of Sciences of the United States of America* **97**, 13172–13177.
Lefebvre, V., Kuntz, M., Camara, B. and Palloix, A. (1998). The capsanthin-capsorubin synthase gene: A candidate gene for the y locus controlling the red fruit colour in pepper. *Plant Molecular Biology* **36**, 785–789.
Li, L. and Van Eck, J. (2007). Metabolic engineering of carotenoid accumulation by creating a metabolic sink. *Transgenic Research* **16**, 581–585.
Li, L., Paolillo, D. J., Parthasarathy, M. V., Dimuzio, E. M. and Garvin, D. F. (2001). A novel gene mutation that confers abnormal patterns of beta-carotene accumulation in cauliflower (*Brassica oleracea* var. botrytis). *The Plant Journal* **26**, 59–67.
Li, F. Q., Murillo, C. and Wurtzel, E. T. (2007). Maize Y9 encodes a product essential for 15-cis-zeta-carotene isomerization. *Plant Physiology* **144**, 1181–1189.

Li, F. Q., Vallabhaneni, R. and Wurtzel, E. T. (2008a). PSY3, a new member of the phytoene synthase gene family conserved in the poaceae and regulator of abiotic stress-induced root carotenogenesis. *Plant Physiology* **146**, 1333–1345.

Li, F. Q., Vallabhaneni, R., Yu, J., Rocheford, T. and Wurtzel, E. T. (2008b). The maize phytoene synthase gene family: Overlapping roles for carotenogenesis in endosperm, photomorphogenesis, and thermal stress tolerance. *Plant Physiology* **147**, 1334–1346.

Li, Z. R., Ahn, T. K., Avenson, T. J., Ballottari, M., Cruz, J. A., Kramer, D. M., Bassi, R., Fleming, G. R., Keasling, J. D. and Niyogi, K. K. (2009). Lutein accumulation in the absence of zeaxanthin restores nonphotochemical quenching in the Arabidopsis thaliana npq1 mutant. *The Plant Cell* **21**, 1798–1812.

Lin, H., Wang, R. X., Qian, Q., Yan, M. X., Meng, X. B., Fu, Z. M., Yan, C. Y., Jiang, B., Su, Z., Li, J. Y. and Wang, Y. H. (2009). DWARF27, an iron-containing protein required for the biosynthesis of strigolactones, regulates rice tiller bud outgrowth. *The Plant Cell* **21**, 1512–1525.

Lindgren, L., Stahlberg, K. G. and Hoglund, A.-S. (2003). Seed-specific overexpression of an endogenous Arabidopsis phytoene synthase gene results in delayed germination and increased levels of carotenoids, chlorophyll, and abscisic acid. *Plant Physiology* **132**, 779–785.

Lopez-Raez, J. A., Kohlen, W., Charnikhova, T., Mulder, P., Undas, A. K., Sergeant, M. J., Verstappen, F., Bugg, T. D. H., Thompson, A. J., Ruyter-Spira, C. and Bouwmeester, H. (2010). Does abscisic acid affect strigolactone biosynthesis? *The New Phytologist* **187**, 343–354.

Lu, S., Van Eck, J., Zhou, X., Lopez, A., O'Halloran, D. M., Cosman, K. M., Conlin, B. J., Paolillo, D. J., Garvin, D. F., Vrebalov, J., Kochian, L. V. Küpper, H. *et al.* (2006). The cauliflower Or gene encodes a DnaJ cysteinerich domain-containing protein that mediates high levels of betacarotene accumulation. *The Plant Cell* **18**, 3594–3605.

Ma, Y., Szostkiewicz, I., Korte, A., Moes, D., Yang, Y., Christmann, A. and Grill, E. (2009). Regulators of PP2C phosphatase activity function as abscisic acid sensors. *Science* **324**, 1064–1068.

Maass, D., Arango, J., Wuest, F., Beyer, P. and Welsch, R. (2009). Carotenoid crystal formation in Arabidopsis and carrot roots caused by increased phytoene synthase protein levels. *PloS One* **4,** Article No. e6373.

Matusova, R., Rani, K., Verstappen, F. W. A., Franssen, M. C. R., Beale, M. H. and Bouwmeester, H. J. (2005). The strigolactone germination stimulants of the plant-parasitic *Striga* and *Orobanche* spp. are derived from the carotenoid pathway. *Plant Physiology* **139**, 920–934.

Menkir, A., Liu, W. P., White, W. S., Mazlya-Dixon, B. and Rocheford, T. (2008). Carotenoid diversity in tropical-adapted yellow maize inbred lines. *Food Chemistry* **109**, 521–529.

Merzlyak, M. N. and Solovchenko, A. E. (2002). Photostability of pigments in ripening apple fruit: A possible photoprotective role of carotenoids during plant senescence. *Plant Science* **163**, 881–888.

Mialoundama, A. S., Heintz, D., Jadid, N., Nkeng, P., Rahier, A., Deli, J., Camara, B. and Bouvier, F. (2010). Characterization of plant carotenoid cyclases as members of the flavoprotein family functioning with no net redox change. *Plant Physiology* **153**, 970–979.

Misawa, N., Masamoto, K., Hori, T., Ohtani, T., Boger, P. and Sandmann, G. (1994). Expression of an Erwinia phytoene desaturase gene not only confers multiple resistance to herbicides interfering with carotenoid biosynthesis but also alters xanthophyll metabolism in transgenic plants. *The Plant Journal* **6**, 481.

Moran, N. A. and Jarvik, T. (2010). Lateral transfer of genes from fungi underlies carotenoid production in aphids. *Science* **328**, 624–627.

Nassar, N., Vizzotto, C. S., Schwartz, C. A. and Pires, O. R. (2007). Cassava diversity in Brazil: The case of carotenoid-rich landraces. *Genetics and Molecular Research* **6**, 116–121.

Neill, S. J., Horgan, R. and Parry, A. D. (1986). The carotenoid and abscisic acid content of viviparous kernels and seedlings of *Zea mays* L. *Planta* **169**, 87–96.

Nievelstein, V., Vandekerchove, J., Tadros, M., Lintig, J., Nitschke, W. and Beyer, P. (1995). Carotene desaturation is linked to a respiratory redox pathway in Narcissus pseudonarcissus chromoplast membranes. Involvement of a 23-kDa oxygene-evolving-complex-like protein. *European Journal of Biochemistry* **233**, 864–872.

Niyogi, K. K. (1999). Photoprotection revisited: Genetic and molecular approaches. *Annual Review of Plant Physiology and Plant Molecular Biology* **50**, 333–359.

Niyogi, K. K., Bjorkman, O. and Grossman, A. R. (1997). The roles of specific xanthophylls in photoprotection. *Proceedings of the National Academy of Sciences of the United States of America* **94**, 14162–14167.

Norris, S. R., Barrette, T. R. and DellaPenna, D. (1995). Genetic dissection of carotenoid synthesis in *Arabidopsis* defines plastoquinone as an essential component of phytoene desaturation. *The Plant Cell* **7**, 2139–2149.

North, H. M., De Almeida, A., Boutin, J. P., Frey, A., To, A., Botran, L., Sotta, B. and Marion-Poll, A. (2007). The Arabidopsis ABA-deficient mutant aba4 demonstrates that the major route for stress-induced ABA accumulation is via neoxanthin isomers. *The Plant Journal* **50**, 810–824.

Ohmiya, A., Kishimoto, S., Aida, R., Yoshioka, S. and Sumitomo, K. (2006). Carotenoid cleavage dioxygenase (CmCCD4a) contributes to white color formation in Chrysanthemum petals. *Plant Physiology* **142**, 1193–1201.

Paine, J. A., Shipton, C. A., Chaggar, S., Howells, R. M., Kennedy, M. J., Vernon, G., Wright, S. Y., Hinchliffe, E., Adams, J. L., Silverstone, A. L. and Drake, R. (2005). Improving the nutritional value of Golden Rice through increased pro-vitamin A content. *Nature Biotechnology* **23**, 482–487.

Panel on Micronutrients (2001). Dietary Reference Intakes for Vitamin A, Vitamin K, Arsenic, Boron, Chromium, Copper, Iodine, Iron, Manganese, Molybdenum, Nickel, Silicon, Vanadium, and Zinc. Food and Nutrition Board. Institute of Medicine, Washington, DC, pp. 82–69.

Park, H., Kreunen, S. S., Cuttriss, A. J., DellaPenna, D. and Pogson, B. J. (2002). Identification of the carotenoid isomerase provides insight into carotenoid biosynthesis, prolamellar body formation, and photomorphogenesis. *The Plant Cell* **14**, 321–332.

Park, S., Fung, P., Nishimura, N., Jensen, D., Fujii, H., Zhao, Y., Lumba, S., Santiago, J., Rodrigues, A., Chow, T., Alfred, S. Bonetta, D. *et al.* (2009). Abscisic acid inhibits type 2 C protein phosphatases via the PYR/PYL family of START proteins. *Science* **324**, 1068–1071.

Parry, A. D. and Horgan, R. (1992). Abscisic acid biosynthesis in roots I. The identification of potential abscisic acid precursors, and other carotenoids. *Planta* **187**, 185–191.

Pfundel, E. E., Renganathan, M., Gilmore, A. M., Yamamoto, H. Y. and Dilley, R. A. (1994). Intrathylakoid Ph in isolated pea-chloroplasts as probed by violaxanthin deepoxidation. *Plant Physiology* **106**, 1647–1658.

Pogson, B. J. and Rissler, H. M. (2000). Genetic manipulation of carotenoid biosynthesis and photoprotection. *Philosophical Transactions of the Royal Society London B: Biological Sciences* **355**, 1395–1403.

Pogson, B., McDonald, K., Truong, M., Britton, G. and DellaPenna, D. (1996). *Arabidopsis* carotenoid mutants demonstrate that lutein is not essential for photosynthesis in higher plants. *The Plant Cell* **8**, 1627–1639.

Pogson, B. J., Niyogi, K. K., Bjorkman, O. and DellaPenna, D. (1998). Altered xanthophyll compositions adversely affect chlorophyll accumulation and nonphotochemical quenching in Arabidopsis mutants. *Proceedings of the National Academy of Sciences of the United States of America* **95**, 13324–13329.

Randolph, L. F. and Hand, D. B. (1940). Relation between carotenoid content and number of genes per cell in diploid and tetraploid corn. *Journal of Agricultural Research* **60**, 51–64.

Robertson, D., Bachmann, M. and Anderson, I. (1966). Role of carotenoids in protecting chlorophyll from photodestruction-II. Studies on the effect of four modifiers of the albino *cl1* mutant of maize. *Photochemistry and Photobiology* **5**, 797–805.

Rodriguez-Concepcion, M. (2010). Supply of precursors for carotenoid biosynthesis in plants. *Archives of Biochemistry and Biophysics* **504**, 118–122.

Rodriguez-Villalon, A., Gas, E. and Rodriguez-Concepcion, M. (2009). Phytoene synthase activity controls the biosynthesis of carotenoids and the supply of their metabolic precursors in dark-grown Arabidopsis seedlings. *The Plant Journal* **60**, 424–435.

Rubio, A., Rambla, J. L., Santaella, M., Gomez, M. D., Orzaez, D., Granell, A. and Gomez-Gomez, L. (2008). Cytosolic and plastoglobule-targeted carotenoid dioxygenases from Crocus sativus are both involved in beta-ionone release. *The Journal of Biological Chemistry* **283**, 24816–24825.

Saga, G., Giorgetti, A., Fufezan, C., Giacometti, G. M., Bassi, R. and Morosinotto, T. (2010). Mutation analysis of violaxanthin de-epoxidase identifies substrate-binding sites and residues involved in catalysis. *The Journal of Biological Chemistry* **285**, 23763–23778.

Schaller, S., Latowski, D., Jemiola-Rzeminska, M., Wilhelm, C., Strzalka, K. and Goss, R. (2010). The main thylakoid membrane lipid monogalactosyldiacylglycerol (MGDG) promotes the de-epoxidation of violaxanthin associated with the light-harvesting complex of photosystem II (LHCII). *Biochimica et Biophysica Acta-Bioenergetics* **1797**, 414–424.

Schwartz, S. H., Tan, B. C., Cage, D. A., Zeevaart, J. A. D. and McCarty, D. R. (1997). Specific oxidative cleavage of carotenoids by VP14 of maize. *Science* **276**, 1872–1874.

Schwartz, S. H., Qin, X. Q. and Zeevaart, J. A. D. (2001). Characterization of a novel carotenoid cleavage dioxygenase from plants. *The Journal of Biological Chemistry* **276**, 25208–25211.

Schwartz, S. H., Tan, B. C., McCarty, D. R., Welch, W. and Zeevaart, J. A. (2003). Substrate specificity and kinetics for VP14, a carotenoid cleavage dioxygenase in the ABA biosynthetic pathway. *Biochimica et Biophysica Acta* **1619**, 9–14.

Schwartz, S. H., Qin, X. and Loewen, M. C. (2004). The biochemical characterization of two carotenoid cleavage enzymes from Arabidopsis indicates that a carotenoid-derived compound Inhibits lateral branching. *The Journal of Biological Chemistry* **279**, 46940–46945.

Seo, M. and Koshiba, T. (2002). Complex regulation of ABA biosynthesis in plants. *Trends in Plant Science* **7**, 41–48.

Shewmaker, C. K., Sheehy, J. A., Daley, M., Colburn, S. and Ke, D. Y. (1999). Seed-specific overexpression of phytoene synthase: Increase in carotenoids and other metabolic effects. *The Plant Journal* **20**, 401–412.

Sieburth, L. E. and Lee, D. K. (2010). BYPASS1: How a tiny mutant tells a big story about root-to-shoot signaling. *Journal of Integrative Plant Biology* **52**, 77–85.

Sommer, A. and Davidson, F. R. (2002). Assessment and control of vitamin A deficiency: The annecy accords. *The Journal of Nutrition* **132**, S2845.

Sorefan, K., Booker, J., Haurogne, K., Goussot, M. and Bainbridge, K. (2003). MAX4 and RMS1 are orthologous dioxygenase-like genes that regulate shoot branching in Arabidopsis and pea. *Genes & Development* **17**, 1469.

Steinmuller, D. and Tevini, M. (1985). Composition and function of plastoglobuli. 1. Isolation and purification from chloroplasts and chromoplasts. *Planta* **163**, 201–207.

Stirnberg, P., van de Sande, K. and Leyser, H. M. O. (2002). MAX1 and MAX2 control shoot lateral branching in Arabidopsis. *Development* **129**, 1131–1141.

Sun, Z., Gantt, E. and Cunningham, F. X., Jr. (1996). Cloning and functional analysis of the beta-carotene hydroxylase of *Arabidopsis thaliana. The Journal of Biological Chemistry* **271**, 24349–24352.

Tan, B. C. (1997). Genetic control of abscisic acid biosynthesis in maize. *Proceedings of the National Academy of Sciences of the United States of America* **94**, 12235–12240.

Tan, B. C., Joseph, L. M., Deng, W. T., Liu, L., Li, Q. B., Cline, K. and McCarty, D. R. (2003). Molecular characterization of the Arabidopsis 9-cis epoxycarotenoid dioxygenase gene family. *The Plant Journal* **35**, 44–56.

Tang, G. W., Qin, J., Dolnikowski, G. G., Russell, R. M. and Grusak, M. A. (2009). Golden Rice is an effective source of vitamin A. *The American Journal of Clinical Nutrition* **89**, 1776–1783.

Taylor, H. F. and Smith, T. A. (1967). Production of plant growth inhibitors from xanthophylls: A possible source of dormin. *Nature* **215**, 1513–1514.

Tevini, M. and Steinmuller, D. (1985). Composition and function of plastoglobuli. II. Lipid composition of leaves and plastoglobuli during senescence. *Planta* **163**, 91.

Tian, L., Musetti, V., Kim, J., Magallanes-Lundback, M. and DellaPenna, D. (2004). The Arabidopsis *LUT1* locus encodes a member of the cytochrome P450 family that is required for carotenoid e-ring hydroxylation activity. *Proceedings of the National Academy of Sciences of the United States of America* **101**, 402–407.

Toledo-Ortiz, G., Huq, E. and Rodriguez-Concepcion, M. (2010). Direct regulation of phytoene synthase gene expression and carotenoid biosynthesis by phytochrome-interacting factors. *Proceedings of the National Academy of Sciences of the United States of America* **107**, 11626–11631.

Tonhosolo, R., D'Alexandri, F. L., de Rosso, V. V., Gazarini, M. L., Matsumura, M. Y., Peres, V. J., Merino, E. F., Carlton, J. M., Wunderlich, G., Mercadante, A. Z., Kimura, E. A. and Katzin, A. M. (2009). Carotenoid biosynthesis in intraerythrocytic stages of Plasmodium falciparum. *The Journal of Biological Chemistry* **284**, 9974–9985.

Treharne, K. J., Mercer, E. I. and Goodwin, T. W. (1966). Carotenoid biosynthesis in some maize mutants. *Phytochemistry* **5**, 581–587.

Umehara, M., Hanada, A., Yoshida, S., Akiyama, K., Arite, T., Takeda-Kamiya, N., Magome, H., Kamiya, Y., Shirasu, K., Yoneyama, K., Kyozuka, J. and

Yamaguchi, S. (2008). Inhibition of shoot branching by new terpenoid plant hormones. *Nature* **455**, 195–200.

Underwood, B. A. (2004). Vitamin A deficiency disorders: International efforts to control a preventable "pox". *The Journal of Nutrition* **134**, 231S–236S.

Vallabhaneni, R. and Wurtzel, E. T. (2009). Timing and biosynthetic potential for carotenoid accumulation in genetically diverse germplasm of maize. *Plant Physiology* **150**, 562–572.

Vallabhaneni, R., Gallagher, C. E., Licciardello, N., Cuttriss, A. J., Quinlan, R. F. and Wurtzel, E. T. (2009). Metabolite sorting of a germplasm collection reveals the Hydroxylase3 locus as a new target for maize provitamin A biofortification. *Plant Physiology* **151**, 1635–1645.

Vallabhaneni, R., Bradbury, L. and Wurtzel, E. T. (2010). The carotenoid dioxygenase gene family in maize, sorghum, and rice. *Archives in Biochemistry and Biophysics* **504**, 104–111, (Carotenoids Highlight edition).

van Jaarsveld, P., Faber, M., Tanumihardjo, S., Nestel, P., Lombard, C. and Benade, A. (2005). Beta-carotene-rich orange-fleshed sweet potato improves the vitamin A status of primary school children assessed with the modified-relative-dose–response test. *The American Journal of Clinical Nutrition* **81**, 1080–1087.

Van Norman, J. M., Frederick, R. L. and Sieburth, L. E. (2004). BYPASS1 negatively regulates a root-derived signal that controls plant architecture. *Current Biology* **14**, 1739–1746.

Verhoeven, A. S., Bugos, R. C. and Yamamoto, H. Y. (2001). Transgenic tobacco with suppressed zeaxanthin formation is susceptible to stress-induced photoinhibition. *Photosynthesis Research* **67**, 27–39.

Vishnevetsky, M., Ovadis, M., Zuker, A. and Vainstein, A. (1999). Molecular mechanisms underlying carotenogenesis in the chromoplast: Multilevel regulation of carotenoid-associated genes. *The Plant Journal* **20**, 423–431.

Vogel, J. T., Walter, M. H., Giavalisco, P., Lytovchenko, A., Kohlen, W., Charnikhova, T., Simkin, A. J., Goulet, C., Strack, D., Bouwmeester, H. J., Fernie, A. R. and Klee, H. J. (2010). SlCCD7 controls strigolactone biosynthesis, shoot branching and mycorrhiza-induced apocarotenoid formation in tomato. *The Plant Journal* **61**, 300–311.

von Lintig, J. (2010). Colors with functions: Elucidating the biochemical and molecular basis of carotenoid metabolism. *Annual Review of Nutrition* **30**, 35–56.

von Lintig, J. and Vogt, K. (2000). Filling the gap in vitamin a research: Molecular identification of an enzyme cleaving beta-carotene to retinal. *The Journal of Biological Chemistry* **275**, 11915–11920.

Welsch, R., Beyer, P., Hugueney, P., Kleinig, H. and von Lintig, J. (2000). Regulation and activation of phytoene synthase, a key enzyme in carotenoid biosynthesis, during photomorphogenesis. *Planta* **211**, 846–854.

Welsch, R., Medina, J., Giuliano, G., Beyer, P. and Von Lintig, J. (2003). Structural and functional characterization of the phytoene synthase promoter from Arabidopsis thaliana. *Planta* **216**, 523–534.

Welsch, R., Maass, D., Voegel, T., DellaPenna, D. and Beyer, P. (2007). Transcription factor RAP2.2 and its interacting partner SINAT2: Stable elements in the carotenogenesis of Arabidopsis leaves. *Plant Physiology* **145**, 1073–1085.

Welsch, R., Wust, F., Bar, C., Al-Babili, S. and Beyer, P. (2008). A third phytoene synthase is devoted to abiotic stress-induced abscisic acid formation in rice and defines functional diversification of phytoene synthase genes. *Plant Physiology* **147**, 367–380.

Welsch, R., Arango, J., Bar, C., Salazar, B., Al-Babili, S., Beltran, J., Chavarriaga, P., Ceballos, H., Tohme, J. and Beyer, P. (2010). Provitamin A accumulation in cassava (Manihot esculenta) roots driven by a single nucleotide polymorphism in a phytoene synthase gene. *The Plant Cell* **22**, 3348–3356.

Wurtzel, E. T. (2004). Genomics, genetics, and biochemistry of maize carotenoid biosynthesis. *In* Recent Advances in Phytochemistry, (J. Romeo, ed.), **38**, pp. 85–110. Elsevier Ltd., Oxford, UK.

Wyss, A., Wirtz, G., Woggon, W., Brugger, R., Wyss, M., Friedlein, A., Bachmann, H. and Hunziker, W. (2000). Cloning and expression of beta, beta-carotene 15,15′-dioxygenase. *Biochemical and Biophysical Research Communications* **271**, 334–336.

Yamamoto, H. Y. (1979). Biochemistry of the xanthophyll cycle in higher plants. *Pure and Applied Chemistry* **51**, 639–648.

Yan, J. B., Kandianis, C. B., Harjes, C. E., Bai, L., Kim, E. H., Yang, X. H., Skinner, D. J., Fu, Z. Y., Mitchell, S., Li, Q., Fernandez, M. G. S. Zaharieva, M. *et al.* (2010). Rare genetic variation at Zea mays crtRB1 increases beta-carotene in maize grain. *Nature Genetics* **42**, 322–327.

Ye, X., Al-Babili, S., Kloti, A., Zhang, J., Lucca, P., Beyer, P. and Potrykus, I. (2000). Engineering the provitamin A (beta-carotene) biosynthetic pathway into (carotenoid-free) rice endosperm. *Science* **287**, 303–305.

Zhu, C. F., Naqvi, S., Breitenbach, J., Sandmann, G., Christou, P. and Capell, T. (2008). Combinatorial genetic transformation generates a library of metabolic phenotypes for the carotenoid pathway in maize. *Proceedings of the National Academy of Sciences of the United States of America* **105**, 18232–18237.

Vitamin B_1 (Thiamine): A Cofactor for Enzymes Involved in the Main Metabolic Pathways and an Environmental Stress Protectant

MARIA RAPALA-KOZIK

Department of Analytical Biochemistry, Faculty of Biochemistry, Biophysics and Biotechnology, Jagiellonian University, Krakow, Poland

Advances in Botanical Research, Vol. 58

0065-2296/11 $35.00
DOI: 10.1016/B978-0-12-386479-6.00004-4

ABSTRACT

Thiamine (vitamin B_1) is essential for human metabolism and is particularly important for proper brain functioning. Plants, which are the best source of this vitamin for human nutrition, synthesize thiamine in three stages. The first of these involves the independent formation of thiazole and pyrimidine moieties. In the next phase, these are coupled together to form thiamine monophosphate. The final step results in the formation of the active form of vitamin B_1, thiamine diphosphate, which functions as a major enzymatic cofactor. The biosynthesis of thiamine is regulated through feedback inhibition by the end product of the pathway, that is, thiamine diphosphate. This regulatory mechanism involves the binding of thiamine diphosphate by mRNA elements, riboswitches (THI-BOXes). The transport of thiamine and thiamine diphosphate between plant tissues and into cell compartments determines the proper functioning of major metabolic pathways such as the acetyl-CoA synthesis, the tricarboxylic acid cycle, the pentose phosphate pathway, Calvin–Benson cycle and isoprenoid biosynthesis pathway. The recently reported activation of thiamine production in plant cells under biotic or abiotic stress conditions also suggests a non-cofactor role of this vitamin as a stress alarmone or stress protectant to enable plants to survive in unfavourable environments.

I. INTRODUCTION

The discovery of vitamin B_1 was made from studies of plants, with the finding by Umetaro Suzuki in 1910 that unpolished rice could cure patients with a nutritional deficiency-based disease, beriberi. Two years later, Casimir Funk isolated the compound from rice bran (Funk, 1912) and its biosynthesis was accomplished in 1935 by Robert R. Williams who first coined the name “thiamine” for this vitamin (Williams and Cline, 1936). Thiamine is essential for the normal growth and development of all living organisms. It plays a crucial role in carbohydrate metabolism, NADPH and ATP biosynthesis and in the production of nucleic acid pentoses. In mammals, thiamine is also essential for the proper functioning of the heart, muscles and nervous system. The biologically active form of vitamin B_1 is thiamine diphosphate (TDP), which in most organisms is formed from free thiamine in a one-step process catalysed by thiamine pyrophosphokinase (TPK). Thiamine can be synthesized *de novo*, that is, from simple precursors, in bacteria, yeast and plants. However, humans and other mammals are dependent on its dietary uptake.

A. STRUCTURE AND BIOLOGICAL FUNCTIONS OF PHOSPHORYLATED THIAMINE ANALOGUES

The thiamine molecule is composed of two heterocyclic moieties, a substituted pyrimidine (4-amino-2-methyl-5-pyrimidyl) and substituted thiazole (4-methyl-5-(2-hydroxyethyl)-thiazolium) rings which are linked by a methylene bridge (Fig. 1).

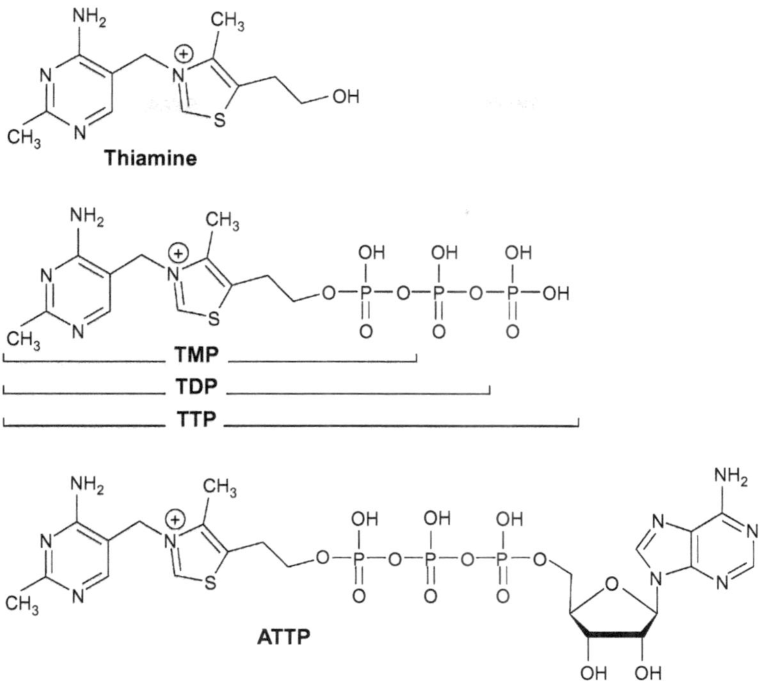

Fig. 1. Chemical structure of thiamine (vitamin B_1) and its biologically occurring phosphorylated derivatives.

In living organisms, thiamine is present in its free form and also as four phosphorylated derivatives, thiamine monophosphate (TMP), TDP, thiamine triphosphate (TTP) and adenosine thiamine triphosphate (ATTP). TMP is a product of thiamine biosynthetic pathways in bacteria, plants and yeast and is a reservoir for further transformations to thiamine or TDP. However, no other physiological function has been proposed for this compound. TDP is the main thiamine compound that functions as a cofactor for a number of enzymes involved in major metabolic pathways. These critical TDP-dependent enzymes include pyruvate dehydrogenase (PDH), α-ketoglutarate dehydrogenase (KGDH), branched-chain α-ketoacid dehydrogenase (BCKDH), transketolase (TK) and pyruvate decarboxylase (PDC; Frank *et al.*, 2007). TTP represents the smallest fraction of the total thiamine pool in humans, but it has been proven to play an important role in the physiology of the nervous system (Gangolf *et al.*, 2010b) owing to its involvement in the phosphorylation of key regulatory proteins (Nghiêm *et al.*, 2000) and in the activation of high-conductance anion channels in nerve cells (Bettendorff *et al.*, 1993). Recent reports of the common occurrence of this compound in bacteria, fungi and plants under specific metabolic conditions (e.g. amino acid starvation) suggest its more general cellular function as an "alarmone" (Bettendorff and Wins, 2009; Makarchikov *et al.*, 2003). ATTP has only recently been identified (Bettendorff *et al.*, 2007) and has been since detected at a minimal level in mammalian tissues and in some cell lines (Frédérich *et al.*, 2009; Gangolf *et al.*, 2010a). However, the levels of ATTP are dramatically increased in *Escherichia coli* in response to carbon starvation (Gigliobianco *et al.*, 2010).

In human, the main fraction of total thiamine contains TDP (72–80%) which exists mostly in a form that is bound to TDP-dependent enzymes. Free thiamine and TMP constitute about 20–26% of the total thiamine content and appear to be a flexible fraction of this vitamin pool that is easily transferable and transformable into TDP or TTP, depending on the requirements at the time. TTP constitutes only 1–2% of the total thiamine (Gangolf *et al.*, 2010a; Lonsdale, 2006). The best sources of vitamin B_1 for human consumption are cereals, whole grains (especially wheat germ), fortified bread, beans, peas, soybeans, nuts, fish, eggs and lean meats (especially pork). The average physiologic requirement for thiamine is about 1.5 mg per day for humans but this value may vary with age, gender and living conditions (e.g. physical activity, stress or pregnancy; Linus Pauling Institute Recommendation; Rakel, 2007). A content of thiamine in various plants of nutritional interest is presented in Table I.

B. THIAMINE DEFICIENCY SYMPTOMS IN MAMMALS

The clinical symptoms of thiamine deficiency in humans manifest in the cardiovascular system (such as "wet" beriberi, which is associated with vasodilatation, myocardial failure, edema and fulminant cardiovascular collapse) and the nervous system ("dry" beriberi which is related to mental confusion, a disordered ocular motility and ataxia; also Wernicke–Korsakoff syndrome, a neuropathy). However, a thiamine deficiency may not only be the result of a low-vitamin diet problem but may also arise due to metabolic dysfunction. The available thiamine levels in cells depend on (i) the absorption of thiamine from gastrointestinal tract, (ii) the effective phosphorylation to TDP, (iii) active TDP transport to organelles and (iv) the incorporation of TDP into properly functioning TDP-dependent enzymes. For each of these events, some disturbances may evoke thiamine deficiency symptoms. Additionally, in some countries, seasonal ataxia is observed due to the consumption of local special meals (shellfish, raw fermented fish or pupae of the African silkworm) that contain heat labile enzyme, thiaminase, which effectively degrades thiamine molecules (Bos and Kozik, 2000;

TABLE I
Average Thiamine Content in Plant Foods

Thiamine (mg/100 g)		Thiamine (mg/100 g)	
Cereal:		*Vegetables:*	
Cornmeal	0.18	Broccoli	0.10
Oatmeal	0.62	Cabbage	0.06
Rice (brown)	0.33	Carrots	0.06
Sorghum	0.15	Cassava leaves	0.16
Wheat (whole grain)	0.41	Cauliflower	0.11
		Spinach	0.11
		Tomatoes	0.06
Pulses, nuts and seeds:			
Beans	0.46–0.63	*Fruits:*	
Chickpeas	0.40	Bananas	0.05
Groundnuts	0.84	Breadfruit	0.09
Lentils	0.50	Grapes	0.06
Peas	0.72	Mangoes	0.05
Soybeans (dry whole seeds)	1.03	Oranges	0.08
		Pineapples	0.08
Tubers/starchy roots:			
Cassava	0.06		
Potato	0.10		
Yam	0.09		

Source: WHO (1999). Thiamine deficiency and its prevention and control in major emergencies.

Jenkins *et al.*, 2007). In addition, heat-stable polyphenolic compounds which are produced in plants (ferns, tea, coffee, betel nuts) may react with thiamine to yield non-absorbable forms of this molecule (Hilker and Somogyi, 1982).

1. Damage to the uptake or transport of thiamine and TDP

Thiamine is distributed between tissues via the bloodstream and the blood thiamine level is critically dependent on the intestinal thiamine absorption which requires thiamine to cross the brush border and basolateral membranes of the enterocytes (Ricci and Rindi, 1992; Rindi and Laforenza, 2000). To be available for uptake by neuronal cells, thiamine must additionally cross the blood–brain barrier to reach the cerebrospinal fluid (Tallaksen *et al.*, 1993). The active intestinal transport of thiamine at its physiological concentrations involves two types of membrane transporters, THTR1 (the product of *SLC19A2* gene in humans) and THTR2 (*SLC19A3* gene; Fig. 2) which probably function through a thiamine/H^+ antiport mechanism.

THTR1 operates at the brush border membrane and undergoes saturation at micromolar thiamine concentrations, whilst THTR2 becomes saturated at nanomolar levels (Said *et al.*, 2004). The entry of thiamine into the

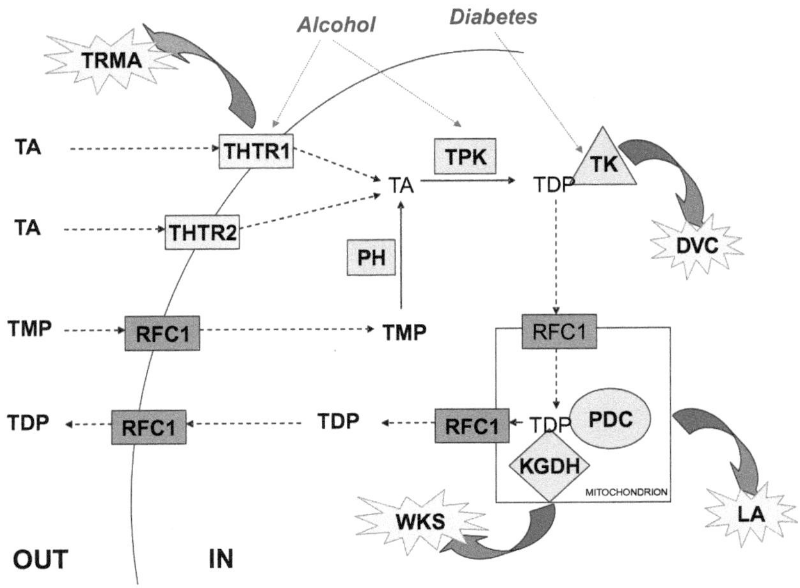

Fig. 2. Diseases caused by damages to thiamine transport or functionality of TDP-dependent enzymes in mammalian cells. Thiamine (TA) enters the cell via two types of specific thiamine/H^+ antiporters, THTR1 and THTR2. A mutation in *slc19a2* gene which codes for THTR1 causes a thiamine-responsive megaloblastic anaemia (TRMA). In the cytosol, thiamine is converted to TDP by the action of thiamine pyrophosphokinase (TPK). The cytosolic TDP pool can be enriched by the uptake of external TMP via a cell membrane-bound reduced folate carrier (RFC1); this relatively non-specific anion transport is coupled with H^+ symport into the cell and OH^- antiport out of the cell. Intracellular TMP must be dephosphorylated by unspecific phosphatase (PH) to become the TPK substrate. TDP can be (i) exported from the cell via plasma membrane-bound RFC1, (ii) bound by cytosolic TDP-dependent enzymes such as transketolase (TK) or (iii) cross the inner mitochondrial membrane via the same anion transporter RFC1 to be used by intramitochondrial TDP-dependent enzymes such as pyruvate dehydrogenase (PDH) or ketoglutarate dehydrogenase (KGDH). The thiamine transporters as well as TPK can be damaged by alcohol to cause the thiamine deficiency and Wernicke–Korsakoff syndrome (WKS). Neurological effects of thiamine deficiency are associated with the impairment of PDH or KGDH which ensure the biosynthesis of several neurotransmitters such as acetylcholine, glutamate and GABA. Reduced activity of these enzymes can also lead to lactic acid accumulation within the brain (lactic acidosis, LA). In diabetics, a low plasma thiamine level is accompanied with low transketolase activity, decreasing the utilization of high carbohydrate levels and resulting in a raised level of advanced glycation end products which promotes a diabetic vascular complication (DVC).

bloodstream across the basolateral membrane is dependent on the Na^+ concentration and upon ATP hydrolysis by the universal Na^+/K^+ ATP-ase (Rindi and Laforenza, 2000). Most cell types actively take up thiamine from the blood via THTR1 or THTR2 located on their plasma membranes. A reduced folate transporter RFC1 (*SCL19A1*) seems also to be involved in cellular TMP import (Zhao *et al.*, 2002) and TDP export (Zhao *et al.*, 2001). Mutations in the *SLC19A2* gene that encodes THTR1 cause thiamine deficiency and thiamine-responsive megaloblastic anaemia syndrome (Fleming *et al.*, 1999).

In the cytosol, thiamine is rapidly phosphorylated to TDP by TPK, and TDP is taken up by mitochondria to be bound by the main TDP-dependent dehydrogenases. TDP transfer across the inner mitochondrial membrane probably occurs via a TDP/TMP antiport mechanism with the engagement of the RFC1 transporter (Barile *et al.*, 1990; Song and Singleton, 2002). A large body of evidence also suggests that intestinal thiamine absorption and further thiamine phosphorylation in the peripheral tissues and brain are impaired by alcohol (Martin *et al.*, 2003).

2. Functional disorders in mitochondria

Three dehydrogenase complexes involved in mitochondrial energy production, PDH, KGDH and BCKDH, utilize TDP as their cofactor. One of the best recognized thiamine deficiency-based disorders is the Wernicke–Korsakoff syndrome in which the selective damage of mammillary bodies, the thalamus and pons has been commonly observed. Analyses at the cellular level have shown that this disorder is associated with neuronal loss, microglial activation and astrocyte proliferation (Hazell, 2009; Hazell *et al.*, 1998; Wang and Hazell, 2010). It has been demonstrated also in Wernicke–Korsakoff syndrome that the activity of all TDP-dependent enzymes is reduced, but KGDH is principally affected. The treatment of experimental animals with pyrithiamine, a known thiamine antagonist (Fig. 3), has confirmed that KGDH depletion leads to a decrease in glutamate, aspartate and γ-aminobutyric acid (GABA) production (Héroux and Butterworth, 1995), as well as mitochondria disintegration and chromatin clumping (Zhang *et al.*, 1995). This impairment of cerebral energy metabolism causes lactate accumulation and acidosis, which results in neuronal cell loss (Hakim, 1984; Navarro *et al.*, 2005).

Recent brain studies in rats treated with pyrithiamine have provided evidence that in thiamine deficiency, it is oxidative stress that causes cellular energy depletion and neuronal damage. The increase in hemooxygenase and ICAM-1 levels, as well as microglial activation and the induction of neuronal peroxidase, have also been observed during a thiamine deficiency (Gibson and Zhang, 2002). A high NOS expression level, NO production as well as nitrotyrosine immunodetection have indicated that the formation of peroxynitrites is likely to be responsible for KGDH

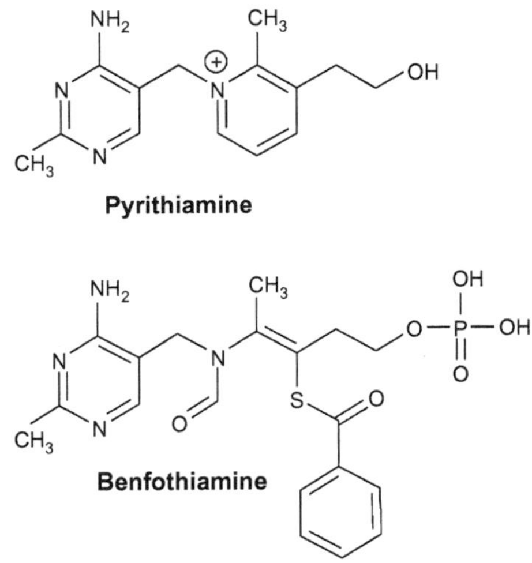

Fig. 3. Chemical structure of pyrithiamine and benfothiamine. Pyrithiamine is a thiamine antagonist commonly used in model studies of thiamine deficiency in animals. Benfothiamine, a lipophilic thiamine analogue which easily crosses biological membranes, can be used for treatment of thiamine deficiency-related diseases.

deactivation. These observations have led to the hypothesis that thiamine may act as an antioxidant (Gibson and Blass, 2007; Huang *et al.*, 2010) but the chemical mechanism underlying this putative activity remains unknown. Analogical changes have been detected also in Alzheimer's disease and Parkinson's disease (Hazell and Butterworth, 2009).

3. Diabetes and diabetic complications

The high glucose cytosolic concentration associated with hyperglycemia leads to triosephosphate accumulation and the development of diabetic complications such as diabetic nephropathy, neuropathy and retinopathy. Decreased thiamine concentrations and TK activity in whole blood samples are often observed in diabetic patients. Supplementation with thiamine or its lipophilic analogue, benfothiamine (Fig. 3), applied in cell culture or diabetic rat models restore the disposal of excess triosephosphate by the pentose phosphate pathway (Thornalley, 2005). Benfothiamine, a lipid-soluble compound from the allithiamine family, was originally described in onions and leeks (Fujiwara, 1976). Its high cellular bioavailability depends on a thiazole ring-open structure that facilitates cell membrane crossing more readily. After oral administration, benfothiamine is dephosphorylated by alkaline phosphatase in the brush border of intestinal mucosal cells. The product of this reaction, *S*-benzoylthiamine, enters the cells by passive diffusion and is further converted to thiamine, mostly in erythrocytes. Equivalent doses of thiamine have a fivefold lower bioavailability (Balakumar *et al.*, 2010).

II. THIAMINE BIOSYNTHESIS

Thiamine biosynthesis pathways in bacteria, yeast and plants (Figs. 4 and 5) consist of three general stages: (i) the independent formation of phosphorylated pyrimidine (4-amino-2-methyl-5-hydroxmethylypyrimidine diphosphate, HMP-PP) and thiazole (4-methyl-5-(2-hydroxyethyl)thiazole phosphate, HET-P) precursors, (ii) their condensation into TMP molecules and (iii) the formation of biologically active TDP (Goyer, 2010; Jurgenson *et al.*, 2009; Kowalska and Kozik, 2008).

In spite of this common general scheme, however, the biosynthetic pathways in the main groups of thiamine-synthesizing organisms differ in many details. These differences are highest between prokaryotic and eukaryotic organisms and, among the latter, plants seem to have a combination of the synthetic systems of bacteria and yeast (Begley *et al.*, 2008; Nosaka, 2006; Rapala-Kozik *et al.*, 2009). The reaction rate of the entire pathway is tightly regulated by the final product, TDP, albeit via different mechanisms in bacteria, yeast and plants (Bocobza and Aharoni, 2008; Miranda-Ríos, 2007; Nosaka, 2006).

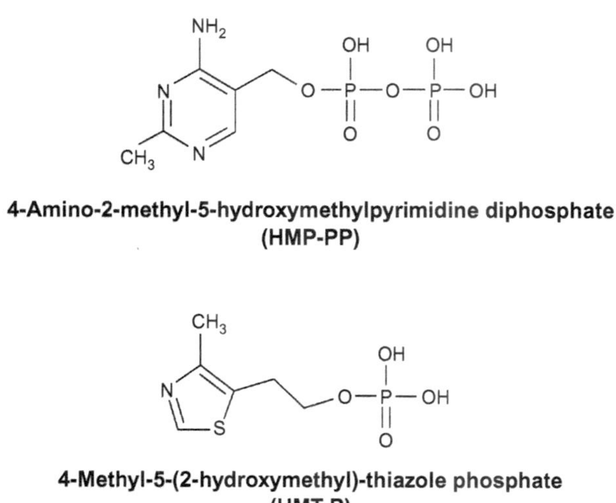

Fig. 4. Biosynthetic precursors of pyrimidine and thiazole moieties of thiamine.

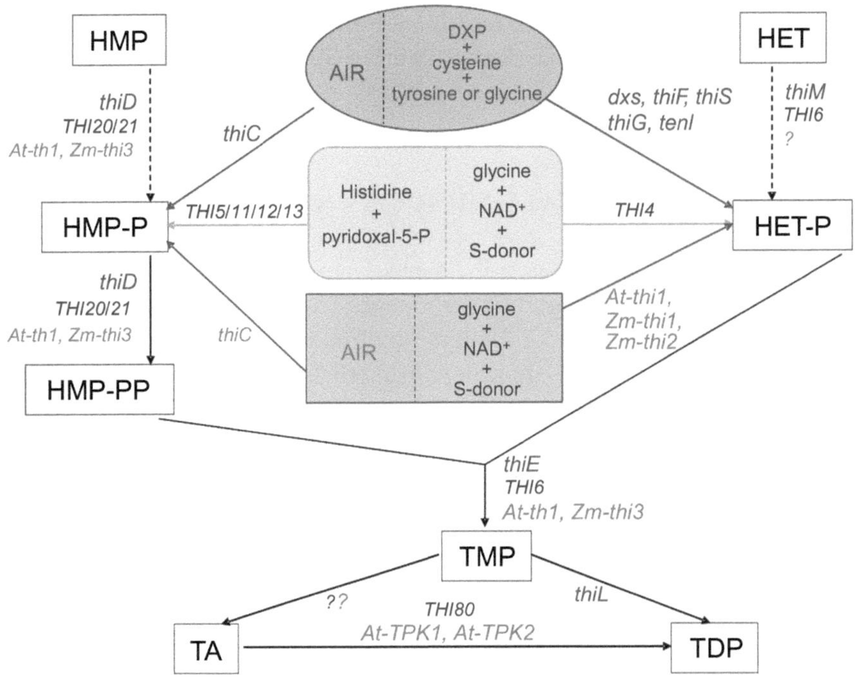

Fig. 5. A comparative scheme of thiamine biosynthesis in bacteria, yeast and plants. Thiamine biosynthesis pathways use different sets of substrates in bacteria (red), baker's yeast *Saccharomyces cerevisiae* (blue) and plants (green), but the late steps are common to all thiamine-synthesizing organisms and include the independent formation of pyrimidine (HMP-PP) and thiazole (HET-P) precursors, followed by their condensation into TMP. The symbols of genes coding for the proteins involved in these late steps of thiamine synthesis are specified on the scheme with correspondingly coloured fonts (*At—Arabidopsis thaliana, Zm—Zea mays*). These enzymes include HMP-P synthase, HMP-P kinase, HET-P synthase and TMP synthase. Common to bacteria, yeast and plants utilize also the salvage pathways that engage HMP kinase and HET kinase. Only in bacteria, the metabolically active coenzyme, TDP, can be formed through a direct phosphorylation of TMP by TMP kinase. In yeast and plants, TMP is first dephosphorylated by non-specific phosphatases to give free thiamine which is used as a substrate by TPK. (For interpretation of the references to colour in this figure legend, the reader is referred to the Web version of this chapter.)

A. THIAMINE BIOSYNTHESIS IN BACTERIA AND YEAST

1. *Pyrimidine component synthesis*

Extensive genetic and biochemical studies in *E. coli* and *Bacillus subtilis* have revealed that 4-amino-2-methyl-5-hydroxymethylpyrimidine phosphate (HMP-P) is formed by a rearrangement of 5-aminoimidazole ribonucleotide (AIR), an intermediate in the purine nucleotide biosynthesis pathway.

This process is catalysed by the product of the *thiC* gene (Begley *et al.*, 1999; Zhang *et al.*, 1997). HMP-P synthase (ThiC) activity is dependent on *S*-adenosyl methionine (SAM) and a functional iron–sulphur cluster localized on the ThiC C-terminal domain. It has been suggested that this Fe–S cluster is available to bind SAM and that a reductive cleavage may generate the 5′-deoxyadenosyl radical (Martinez-Gomez *et al.*, 2009). This step may enable the further rearrangement of AIR to HMP-P. The presence of the Fe–S cluster within the structure of ThiC and possible free radical reaction chemistry has confirmed this enzyme as a member of the radical SAM superfamily (Chatterjee *et al.*, 2008a).

The next HMP-P phosphorylation event, resulting in HMP-PP formation, is performed by a *thiD* gene product. This kinase is bifunctional as it also can take part in a salvage pathway through which external 4-amino-2-methyl-5-hydroxymethylpyrimidine (HMP) may be phosphorylated to HMP-P (Mizote *et al.*, 1999).

In *Saccharomyces cerevisiae*, the pyrimidine moiety of thiamine is derived from histidine and pyridoxal 5′-phosphate, with the involvement of the *THI5* gene family (*THI5/THI11/THI13*) but the exact mechanism underlying HMP-P formation remains insufficiently understood (Nosaka, 2006; Zeidler *et al.*, 2003). In the final HMP-P phosphorylation step, another multigene family (*THI20/THI21*) is engaged. Again, the latter kinases perform the salvage HMP phosphorylation reactions (Kawasaki *et al.*, 2005).

2. Thiazole component synthesis

For the biosynthesis of the thiazole moiety, bacteria utilize 1-deoxy-D-xylulose-5-phosphate (DXP), glycine or tyrosine and a sulphur carrier protein (ThiS). This reaction is initiated by thiazole phosphate synthase (ThiG), an enzyme that performs DXP tautomerization and further oxidative condensation with glycine and cysteine, in cooperation with a Ten1 protein (Dorrestein *et al.*, 2004; Kriek *et al.*, 2007). Early isotopic labelling studies have identified cysteine, glycine and D-pentulose-5-phosphate (D-ribulose-5-phosphate or D-xylulose-5-phosphate) as primary precursors of the thiazole moiety in *S. cerevisiae*. However, a recent study of the structure and mechanism underlying enzymatic thiazole formation by thiazole synthase (THI4) and analysis of a thiazole derivative tightly bound to this protein has revealed that NAD^+ is the most likely source of the carbohydrate (ribose) required for thiazole synthesis (Chatterjee *et al.*, 2007).

The final product of these pathways, HET-P, may be also regenerated from HET by a salvage enzyme, HET kinase, encoded by the *thiM* gene in *E. coli* (Mizote and Nakayama, 1989) and *THI6* gene in *S. cerevisiae* (Nosaka *et al.*, 1994).

3. Condensation of pyrimidine and thiazole components into TMP

A TMP synthase (also known as thiamine-phosphate diphosphorylase), encoded by the *thiE* gene in bacteria, couples the diphosphorylated pyrimidine compound and the phosphorylated thiazole compound to form TMP. The structure of this enzyme and its catalytic mechanism suggests a dissociative mechanism of TMP formation (Chiu *et al.* 1999; Peapus *et al.*, 2001; Reddick *et al.*, 2001), with a pyrimidine carbocation as an intermediate (Hanes *et al.*, 2007). The same condensation reaction in *S. cerevisiae* is performed by the bifunctional TMP synthase/HET kinase encoded by the *THI6* gene (Nosaka *et al.*, 1994).

4. Formation of TDP, a biologically relevant cofactor

The last step in the TDP biosynthesis pathway differs between bacteria and yeast. In most bacteria, TMP is simply further phosphorylated to TDP by a kinase encoded by the *thiL* gene (Webb and Downs, 1997). The active site structure of ThiL suggests a direct, inline transfer of the γ-phosphate of ATP to TMP (McCulloch *et al.*, 2008). Yeast utilize a more complex mechanism to produce TDP. First, TMP is dephosphorylated to thiamine, probably by an unspecific but not yet identified phosphatase. The free thiamine is then diphosphorylated by TPK, encoded by a single *THI80* gene (Nosaka *et al.*, 1993). The structure of this enzyme is well documented but its mechanism of catalysis is still under debate (Baker *et al.*, 2001; Voskoboyev and Ostrovsky, 1982).

5. Synthesis of TTP and ATTP

The biosynthetic pathways for the TTP and ATTP in bacteria and yeast have not yet been identified. However, in the rat brain, TTP synthesis was recently suggested to occur in mitochondria and to be coupled to the respiratory chain (Gangolf *et al.*, 2010a,b).

B. GENES AND PROTEINS INVOLVED IN PLANT THIAMINE BIOSYNTHESIS AND THE CELLULAR DISTRIBUTION OF THE BIOSYNTHETIC PATHWAYS

A large number of auxotrophic mutants have been used to elucidate the specific steps involved in thiamine biosynthesis in plants. Thiamine auxotrophs identified in the model plant, *Arabidopsis thaliana*, manifest seedling lethal phenotypes that can be complemented by exogenous thiamine. The first identified group of auxotrophs, *th1* (chromosome I), *th2* (chromosome V) and *th3* (chromosome IV) were rescued only using thiamine supplementation, indicating that the respective genes are involved in the latest steps of TDP formation. The second group of auxotrophs, *py* (chromosome II) and *tz* (chromosome V), require thiamine or HMP (*py*) and thiamine or HET (*tz*) for growth, suggesting the involvement of these genes in the biosynthesis of the pyrimidine and thiazole moiety, respectively (Koornneef and Hanhart, 1981).

1. Pyrimidine component biosynthesis

Plants appear to take advantage of both bacterial and yeast thiamine biosynthetic processes (Fig. 5). A search for a HMP-P synthase candidate in *A. thaliana* revealed only one homologue of the *thiC* gene, which encodes a protein with a 60% sequence identity to ThiC from *B. subtilis* and *E. coli*. A partial confirmation of the involvement of plant *THIC* in thiamine biosynthesis came from a previous finding that seedlings of *THIC* knockdown mutants possess a significantly decreased thiamine level, present a chlorotic phenotype and are unable to develop beyond the cotyledon stage. However, supplementation with an external dose of thiamine was found to rescue this phenotype (Kong *et al.*, 2008; Raschke *et al.*, 2007). Further analyses of the *THIC* mutant for metabolites originating from the reactions which engage TDP as a cofactor (the tricarboxylic acid cycle, the Calvin–Benson cycle and the oxidative pentose phosphate pathway) have suggested that THIC is essential for plant viability (Raschke *et al.*, 2007). This hypothesis was further supported by findings that *THIC*-overexpressing plants possess a higher thiamine content in their tissues and that the *A. thaliana THIC* gene can complement a bacterial *thiC* mutant (Kong *et al.*, 2008). Similarly to bacteria, plant THIC requires a reducing agent and a Fe–S cluster for the catalytic conversion of AIR into HMP-P. Raschke *et al.* (2007) have demonstrated in *A. thaliana* that a cysteine desulfurylase (NifS) may be the sulphur source for the Fe–S cluster and speculated that the thioredoxin system could be involved in the activation of this enzyme. However, the mechanism of this reaction has not yet been elucidated.

YFP-fusion protein analysis has indicated that THIC is localized in chloroplasts. This finding confirmed early suggestions that plant thiamine synthesis occurs in plastids (Faith *et al.*, 1995; Julliard and Douce, 1991). The *THIC* transcript was found to be expressed in leaves, flowers and siliques, and at small amounts in roots. The expression levels are dependent on the stage of seedling development (commencing on the fifth day after imbibition) and also the thiamine levels in the medium. The *THIC* transcript levels increase also under light exposure (Kong *et al.*, 2008; Raschke *et al.*, 2007). The *THIC* gene is tightly regulated by a riboswitch-dependent mechanism (see Section II.C) in which TDP plays a role of a feedback inhibitor whilst thiamine, available in the medium, may be easily converted to TDP inside the cells (Fig. 6).

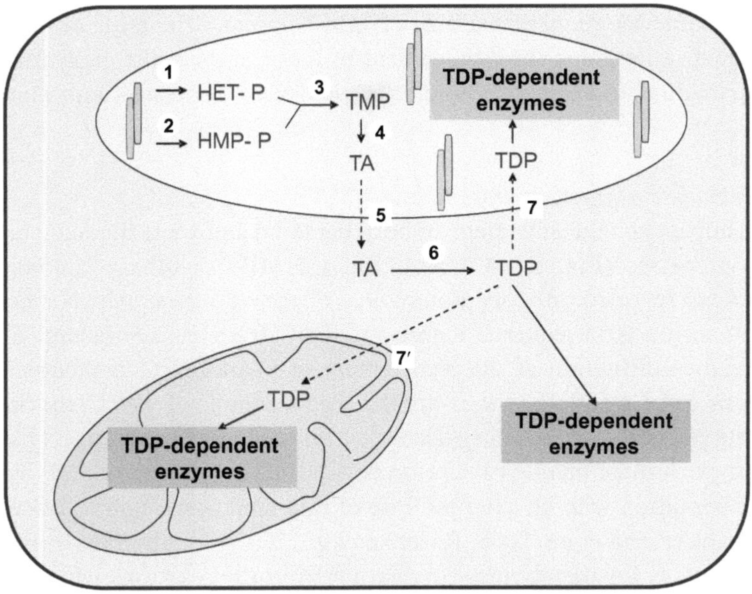

Fig. 6. Compartment localization of biosynthetic pathways and intracellular traffic of thiamine diphosphate (TDP) in the plant cell. The plant biosynthesis of thiamine monophosphate (TMP) is localized in chloroplasts where the thiazole precursor (HET-P) and pyrimidine precursor (HMP-P) are formed on independent pathways catalysed by HET-P synthase (1) and HMP-P synthase (2), respectively. Coupling of these two moieties (after additional phosphorylation of HMP-P) is performed by TMP synthase (3). Since thiamine (TA) is present in the cytosol, the product of its biosynthesis (TMP) must first be dephosphorylated by yet unidentified, probably non-specifc phosphatases (4). Hence, this process is proposed to proceed in the chloroplast but its occurrence in the cytosol is also possible. After dephosphorylation, TA is transported to cytosol by a yet unidentified transporter (5) and is further converted to TDP by cytosolic thiamine pyrophosphokinase (6). TDP plays the cofactor function for the cytosol-, mitochondrion- or chloroplast-localized enzymes. TDP must be transported into mitochondria and chloroplasts by high effective specific transporter(s) (7, 7′) that have not yet been identified.

2. Thiazole component biosynthesis

For the building of the thiazole component, plants use the same pathway developed in yeast (Fig. 5), in which NAD^+ and glycine are converted to HET-P by thiazole synthase (THI4) in cooperation with a protein sulphur donor (Chatterjee *et al.*, 2006, 2008b). Numerous genes with high sequence similarity to *THI4* have been identified in the genomes of *Zea mays* (*thi1* and *thi2*; Belanger *et al.*, 1995), *Alnus glutinosa* (*agthi1*; Ribeiro *et al.*, 1996), *A. thaliana* (*thi1*; Machado *et al.*, 1996) and *Oryza sativa* (*OsDR8*; Wang *et al.*, 2006). Complementation studies using Arabidopsis *THI1* in *E. coli* mutant strains defective in DNA repair pathways or *THI4*-defective yeast mutant strains further supported a role for *THI1* in thiamine biosynthesis and its possible involvement also in plant tolerance to mitochondrial DNA damage (Machado *et al.*, 1996, 1997).

Sequence analysis of the THI1 protein encoded by a single gene in Arabidopsis (*tz* locus) has identified an N-terminal chloroplast transit peptide and a mitochondria targeting-like presequence just downstream, suggesting the dual targeting of this gene product to both plastids and mitochondria. The resolved crystal structure of Arabidopsis THI1, heterologously expressed and overproduced in *E. coli* (Godoi *et al.*, 2006) revealed that the protein (244 kDa) is an octamer containing dinucleotide binding domains adapted to NAD^+ binding. To date, this is the only plant thiamine biosynthetic enzyme whose three-dimensional structure has been elucidated at an atomic resolution (Fig. 7).

Similarly to the yeast THI4 protein, the tightly bound 2-carboxylate-4-methyl-5-(β-ethyl adenosine 5′-diphosphate) thiazole was identified within the THI1 structure and was suggested to be a late intermediate on the thiazole biosynthetic route, additionally supporting the hypothesis that yeast-like biosynthetic pathways are utilized by plants, with NAD^+ as the substrate. The dual function of this gene was confirmed by the observation that some site-directed mutations of THI1 prevent thiazole biosynthetic

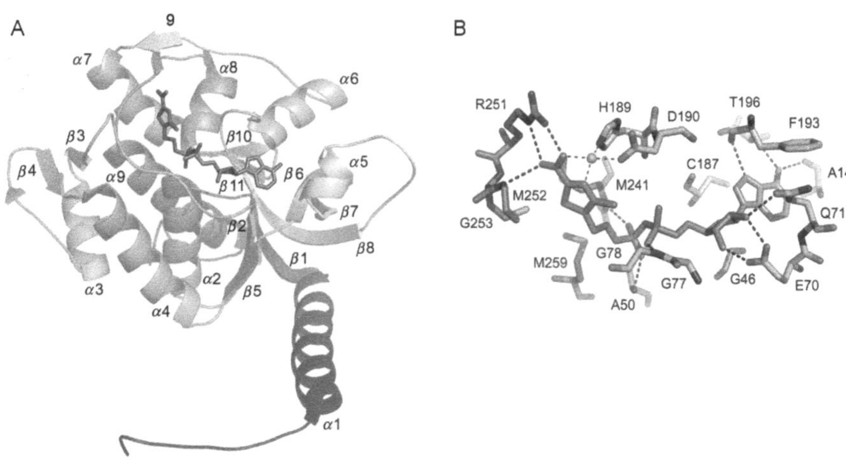

Fig. 7. Structure of thiazole-synthesizing protein THI1 from *Arabidopsis thaliana*. (A) The structure of THI1 monomer with a visualized molecule of tightly bound 2-carboxylate-4-methyl-5-(β-ethyl adenosine 5′-diphosphate) thiazole (ADT), an apparent product of the catalysed reaction. (B) Amino acid residues which surround the ADT molecule bound in the active centre of THI1 protein. The structure was imported from UniProt KB (access No Q38814) and drawn with PyMol program (ExPASy server).

activity but do not affect mitochondrial DNA stability (Godoi *et al.*, 2006), the latter being controlled by the same gene through a yet unidentified mechanism.

Thiazole synthesis was found to be localized to chloroplasts in spinach (Julliard and Douce, 1991) and maize (Belanger *et al.*, 1995). The dominant accumulation of transcripts of *thi2*, the maize paralog of *thi1*, was observed in young, rapidly dividing tissues, whilst *thi1* is detectable in mature green leaves. This may reflect a subfunctionalization of both encoded proteins. The *thi2-blk1* mutant is a thiamine auxotroph which shows defects in shoot meristem maintenance and a novel leaf blade reduction phenotype (Woodward *et al.*, 2010).

In Arabidopsis, an analysis of the organelle localization of a β-glucuronidase-fused THI1 protein (GUS-THI1) confirmed chloroplasts and mitochondria as the targets of THI1 localization and provided evidence that two isoforms of THI1 are produced from a single nuclear transcript. Hence, this targeting occurs through a post-transcriptional mechanism (Chabregas *et al.*, 2001, 2003; Ribeiro *et al.*, 2005). The intensive expression of *THI1* was observed in all organs at different plant development stages, for example, during nodule differentiation (Ribeiro *et al.*, 1996) and ethylene-induced fruit maturation (Jacob-Wilk *et al.*, 1997). *THI1* expression was also found to predominate in shoot tissues as compared with roots (Ribeiro *et al.*, 2005) and is twofold higher in plants grown under light (Papini-Terzi *et al.*, 2003). The presence of thiamine in the medium did not affect the *THI1* expression level, in sharp contrast to the strong repression of the yeast orthologous gene (*THI4*) by external thiamine.

3. Coupling the pyrimidine and thiazole compounds

The condensation of pyrimidine and thiazole components to form TMP is the common step in thiamine biosynthesis in all autotrophic organisms. Similarly to *S. cerevisiae*, plants use a bifunctional enzyme for this reaction (Fig. 5), although the additional activity (HMP/HMP-P kinase) combined with TMP synthase activity in one molecule is different from that in the yeast THI6 protein (HET kinase). The occurrence of TMP synthase in plants was demonstrated in studies on the functional complementation of thiamine-requiring mutants in bacteria (Ajjawi *et al.*, 2007b; Kim *et al.*, 1998). The protein identified in *Z. mays* (THI3) shows a 39% sequence similarity to *B. subtilis* ThiD and a 60–80% similarity to several plant orthologues in *O. sativa, Medicago truncatula, A. thaliana* and *Brassica napus* (Rapala-Kozik *et al.*, 2007). This analysis also indicated that THI3, similarly to all of its plant orthologues, possesses two putative conserved domains, an N-terminal domain with a high sequence similarity to bacterial HMP-P kinases, and a C-terminal domain highly similar in sequence to the bacterial TMP synthases. In contrast, the yeast bifunctional TMP synthase (THI6) is associated with HET kinase activity localized in a C-terminal domain, that is, downstream from the TMP synthase domain (Kawasaki, 1993). On the basis of sequence similarity of THI3 to structurally characterized bacterial TMP synthase (*B. subtilis*) and HMP-P kinase (*Salmonella typhimurium*) (Cheng *et al.*, 2002; Chiu *et al.*, 1999), the overall structures of the THI3 domains as well as the arrangements of conserved amino acid residues within the active centres have been modelled (Rapala-Kozik *et al.*, 2007). The kinase domain reveals a ribokinase-like fold, whilst the synthase domain harbours a triose phosphate isomerase fold.

THI3 was heterologously expressed in *E. coli* and yielded as a soluble dimer of 55 kDa subunits which possessed the expected enzymatic activities (Rapala-Kozik *et al.*, 2007). These included TMP synthesis and two successive steps of HMP phosphorylation, with the production of HMP-P and HMP-PP, the latter serving as the substrate for TMP synthase. HMP phosphorylation to HMP-P also appears to be a salvage pathway, as in bacteria. The predicted arrangements of the active centre amino acid residues were confirmed by site-directed mutagenesis experiments. Detailed kinetic analysis showed that TMP formation was strongly inhibited by an excess of one of TMP synthase substrates (HMP-PP) and uncompetitively inhibited by ATP. Both compounds are involved in the reaction catalysed by the HMP-P kinase domain of THI3, one as a substrate (ATP) and the other as a product (HMP-PP). It was suggested that this unique fusion of both enzyme activities in one protein molecule may provide a regulatory mechanism for TMP biosynthesis in plants.

All members of the plant TMP synthase family contain the N-terminal signal sequence responsible for chloroplast targeting (Rapala-Kozik *et al.*, 2007). The detection of fluorescent protein-fused TMP synthase in Arabidopsis mesophyll protoplasts also indicated the chloroplasts as the location of TMP biosynthesis (Ajjawi *et al.*, 2007b). During seedling development, most plants on the early stage of growth utilize thiamine reserves accumulated in the seeds and *de novo* thiamine biosynthesis, manifested by the induction of TMP synthase activity, which starts between days 3 and 6 after imbibition (Golda *et al.*, 2004).

4. TDP synthesis

In the final step of thiamine coenzyme formation, plants, unlike bacteria but similar to yeast, do not perform a direct phosphorylation of *de novo* synthesized TMP. Instead, TMP is first dephosphorylated to free thiamine which is then pyrophosphorylated to TDP, in a reaction catalysed by TPK (Fig. 5).

It is generally assumed that the production of free thiamine from TMP in plants involves numerous broad-specificity phosphatases. However, this does not exclude the possibility that some phosphatases may be more important than others in this process. Recently, a homogeneous acid phosphatase (a dimer of 24 kDa subunits) with a broad specificity was isolated from *Z. mays* seedlings on the basis of its ability to dephosphorylate TDP and TMP (Rapala-Kozik *et al.*, 2009). The purified enzyme showed some preference for thiamine phosphates (TDP > TMP) over other organic phosphate esters. Purified TPK preparations have been obtained from parsley leaves (Mitsuda *et al.*, 1979), soybean seedlings (Molin and Fites, 1980) and maize seedlings (Rapala-Kozik *et al.*, 2009). They differed slightly in terms of subunit size, subunit association states and basic kinetic parameters. For example, the maize TPK is a 29-kDa monomeric protein. In Arabidopsis, this enzyme is encoded by two genes, *At-TPK1* and *At-TPK2*, and the predicted amino acid sequence of their protein products show a significant similarity with the structurally characterized fungal and animal TPKs (Ajjawi *et al.*, 2007a). Both genes are expressed at comparable levels, predominantly in leaves but also in the stems, siliques and flowers. However, in the roots, their expression levels differ, with a clear preference for *At-TPK1*. An analysis of a TPK double knockout mutant in Arabidopsis further showed that the seedling had a lethal phenotype and survived only in the presence of external doses of TDP (Ajjawi *et al.*, 2007a).

Negative regulation by light was suggested for TPK activity in *Z. mays* seedlings, whilst a presence of thiamine in the culture medium exerted only minor effects upon TPK expression (Rapala-Kozik *et al.*, 2009). TPK is involved in TDP biosynthesis from the very early stages of seed germination when thiamine reserves stored in seeds serve as the substrate for TPK-catalysed pyrophosphorylation (Molin *et al.*, 1980; Golda *et al.*, 2004). The *de novo* formation of TDP was found to be localized to the plant cell cytosol, as in yeast and mammals (Barile *et al.*, 1990; Bettendorff, 1995; Hohmann and Meacock, 1998). As TDP is necessary for many biochemical processes in different cell compartments, effective systems for its transport must exist in plants, but the underlying mechanisms have not yet been characterized.

5. *Thiamine triphosphate and adenosine thiamine triphosphate*

The presence of the highly phosphorylated thiamine compound, TTP, in the germ axes of higher plants was reported many years ago (Kochibe *et al.*, 1963; Yusa, 1961) and recently, its presence in withering plants was also confirmed (Makarchikov *et al.*, 2003). Whereas TTP accumulation was detected in *E. coli* in response to amino acid starvation in carbon containing medium, leading to the hypothesis of a general alarmone function of this compound (Bettendorff and Wins, 2009; Lakaye *et al.*, 2004), its actual significance and functions in plants remain unknown. The newly discovered thiamine compound, ATTP, first detected in bacteria upon carbon starvation, has also been found in the roots of higher plants (Bettendorff *et al.*, 2007). It has been further proposed that ATTP may serve as a TTP precursor deposited in a less reactive storage form (Jordan, 2007) or as a source of TDP. To date, however, nothing is known about the biosynthetic routes for TTP and ATTP in plants.

C. REGULATION OF PLANT THIAMINE BIOSYNTHESIS

For half a century, it has been well established that thiamine synthesis and transport in bacteria and yeast are strongly repressed by the presence of exogenous thiamine in the culture media (Begley *et al.*, 2008; Kowalska and Kozik, 2008; Nosaka, 2006). It is actually the intracellular TDP concentration that provides this regulatory signal. Although less frequent, a similar system of feedback regulation has been reported in plants (Kim *et al.*, 1998; Rapala-Kozik *et al.*, 2009). Only recently, however, have the molecular mechanisms underlying the regulation of plant thiamine biosynthesis been characterized at the molecular level with the discovery of plant TDP-dependent riboswitches that regulate the expression of the *THIC* pyrimidine-synthesizing gene and, albeit not in the entire plant kingdom, of the *THI1* thiazole-synthesizing gene (Bocobza *et al.*, 2007; Sudarsan *et al.*, 2003). Other reports have also suggested that the expression of thiamine-synthetic enzymes may depend on some tissue-specific transcription factors (Ribeiro *et al.*, 2005), light (Kong *et al.*, 2008; Rapala-Kozik *et al.*, 2009; Raschke *et al.*, 2007; Ribeiro *et al.*, 1996, 2005), the thioredoxin system (Balmer *et al.*, 2003; Lemaire *et al.*, 2004; Raschke *et al.*, 2007) and elements responding to abiotic and biotic stress signalling (see Section VI). Additionally, the allosteric inhibition of plant TMP synthase activity by ATP and HMP-PP has been reported (Rapala-Kozik *et al.*, 2007).

1. Riboswitch-dependent regulation of HMP-P synthase (THIC) and HET-P synthase (THI1)

The precise mechanism of plant *THIC* gene regulation by accessible TDP has recently been identified and shown to engage a TDP-binding riboswitch (THI-BOX) (Bocobza *et al.*, 2007; Sudarsan *et al.*, 2003; Wachter *et al.*, 2007). Riboswitches are non-coding mRNA domains that can selectively bind some metabolites and subsequently affect the expression of adjacent coding sequences (Breaker, 2010; Serganov, 2010; Smith *et al.*, 2010; Wachter, 2010). They are believed to be the modern descendents of an ancient sensory and regulatory system which may have functioned in the "RNA world" (Breaker, 2010). The THI-BOXes represent the most abundant class of riboswitches, and are found in prokaryotes, archea and eukaryotes (Bocobza and Aharoni, 2008). Like other riboswitches, they are composed of a highly conserved TDP-binding domain ("aptamer") responsible for coenzyme sensing and a more variable expression platform which, when forced to rearrange by the ligand-induced conformational change in the aptamer, affects gene expression (Rodionov *et al.*, 2002; Winkler *et al.*, 2002). The high-resolution crystal structure of a complex between TDP and the Arabidopsis THI-BOX that controls *THIC* gene expression has been reported (Thore *et al.*, 2006) and is schematically presented in Fig. 8.

Two helical domains are involved in TDP binding. The first of these forms a deep pocket for the pyrimidine moiety of TDP and the other is responsible for the binding of the diphosphate tail, bridged by an Mg^{2+} ion (Miranda-Ríos, 2007; Serganov *et al.*, 2006; Thore *et al.*, 2006). The TDP molecule lies in a perpendicular orientation against the two parallel helices and adopts an extended conformation, in contrast to its V-conformation observed in

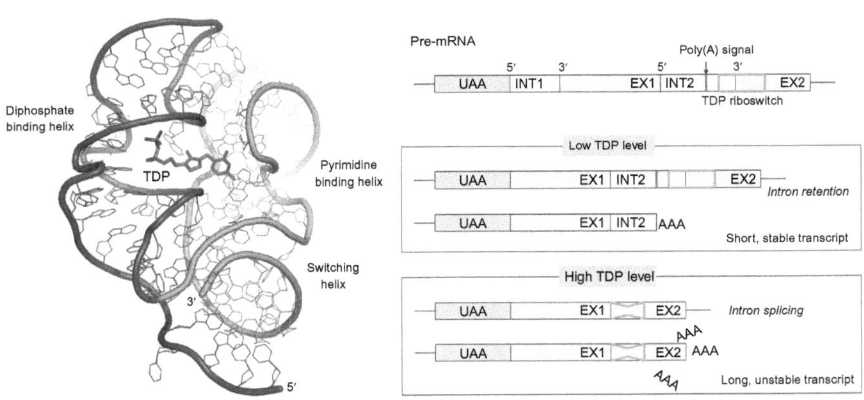

Fig. 8. Structure and action mechanism of TDP-binding riboswitch (THI-BOX) which regulates the expression of the *thiC* gene in *Arabidopsis thaliana.* (Left panel) The three-dimensional structure of THI-BOX (Thore *et al.*, 2006; Protein Data Bank access No 3D2G, drawn with PyMol program). (Right panel) A suggested mechanism of THI-BOX-dependent regulation of plant *thiC* expression (Bocobza and Aharoni, 2008). Different modes of *thiC* pre-mRNA processing are dependent on the intracellular TDP level. At low TDP concentration, the riboswitch folding enables its interaction with 5′ splice site and prevents splicing. The major processing site is retained (the intron retention variant), resulting in the formation of short transcript that permits a high THIC expression. At high TDP level, it binds to riboswitch and induces its conformational changes that prevent the riboswitch interaction with 5′ splice site. The splicing takes place (the intron splaced variant), leading to the removal of poly(A) signal and the formation of long unstable transcripts.

TDP-dependent enzymes. The apparent dissociation constant of the complex formed (500 nM) is consistent with the TDP level detected in plants after they initiate its biosynthesis (0.25–5 μM; Winkler *et al.*, 2002; Bocobza and Aharoni, 2008). The accommodation of a pyrithiamine diphosphate by the THI-BOX aptamer suggests that the thiazole ring is not essential for fixing TDP in the binding pocket. However, the binding of TMP by a bacterial THI-BOX is weaker by 3 orders of magnitude, suggesting incomplete stabilization of both helical domains and presenting additional evidence for the preferential regulation of *THIC* expression by TDP (Agyei-Owusu and Leeper, 2009; Winkler *et al.*, 2002).

The binding of TDP generates a parallel localization of sensor helices and alters the expression platform (Bocobza and Aharoni, 2008; Thore *et al.*, 2006; Winkler *et al.*, 2002). This ancient-origin mechanism is widespread in both prokaryotes and eukaryotes, although differs in terms of gene expression alteration (Bocobza and Aharoni, 2008; Sudarsan *et al.*, 2003). In Gram-positive bacteria, TDP binding by THI-BOX, such as that involved in the *tenA* regulation in *B. subtilis*, causes structural rearrangements that lead to the formation of a transcription termination hairpin. In Gram-negative bacteria (e.g. *thiM* in *E. coli*), the presence of TDP leads to translation repression via the sequestration of the Shine–Delgarno sequence and the prevention of ribosome binding (Mironov *et al.*, 2002; Ontiveros-Palacios *et al.*, 2008; Winkler *et al.*, 2002). In some non-yeast fungi (e.g. *Aspergillus oryzae* or *Neurospora crassa*), THI-BOX is located within an intron in the 5′-untranslated region (5′-UTR) of *THI4* (an orthologue of plant *THI1*) mRNA and TDP binding alters the mRNA splicing so that it does not occur at the 5′ splice site (distal) as normal but at a more proximal site. This alternative splicing leads to upstream open-reading frame (ORF) expression and premature termination (Cheah *et al.*, 2007; Sudarsan *et al.*, 2003).

In *THIC* transcripts in flowering plants, the TDP riboswitch element is located in the 3′-UTR and controls the splicing toward an alternative 3′ end processing of precursor *THIC* mRNA (Bocobza *et al.*, 2007, Sudarsan *et al.*, 2003; Wachter *et al.*, 2007). The pre-mRNA 3′-UTR consists of a constitutively spliced intron just after the ORF stop codon, followed by a sequence which contains a polyadenylation signal and a potential splice site (Fig. 7B). The TDP riboswitch is located 70 bp downstream of the polyadenylation signal and is followed by the last variable-length exon. The 3′ splice site of the second intron is located within the riboswitch ("P2 box"). At a low TDP concentration, the riboswitch interacts with 5′ splice site and splicing of the second intron is prevented. In this situation, mRNA processing leads to variant transcripts with the second intron retained and harbouring the major processing site that permits transcript cleavage and polyadenylation.

This variant is short and can be translated. At a high TDP concentration, the conformational change of the riboswitch exposes the 5′ splice site of the second intron and an alternative splicing reaction proceeds and removes the normal processing site to form an intron-spliced variant which is both long and unstable (Bocobza and Aharoni, 2008). Analysis of the effects of TDP supplementation upon the *THIC* transcript levels in Arabidopsis and tomato auxotrophic mutants with low endogenous TDP contents directly supports the hypothesis of TDP involvement in thiamine biosynthesis regulation (Bocobza *et al.*, 2007; Wachter *et al.*, 2007).

The THI-BOX-dependent regulation of pyrimidine-synthesizing *THIC* gene expression has been confirmed in all major plant taxa, from species of moss (bryophytes) to flowering plants (angiosperms). A similar type of riboswitch-dependent regulation of the expression of the thiazole-synthesizing *THI1* gene was lost during gymnosperm evolution. *Cycas revoluta* is the plant of the highest evolutionary order for which a TDP-dependent riboswitch in the 3′-UTR of *THI1* mRNA can be detected (Bocobza and Aharoni, 2008).

III. TDP-DEPENDENT ENZYMES IN PLANTS

TDP functions as the cofactor for enzymes involved in key metabolic pathways such as ethanolic fermentation, acetyl-CoA formation, the tricarboxylic acid cycle, the oxidative pentose phosphate pathway, the Calvin–Benson cycle, the mevalonate-independent isoprenoid synthesis pathway and branched-chain amino acid biosynthesis. In all of these processes, the first step depends on the special structure and charge of the TDP molecule in the enzyme active centres. A comparison of the crystal structures of the major TDP-dependent enzymes reveals that this cofactor is accommodated by the protein in the V-conformation in which the amino group of the pyrimidine ring is closely positioned with the C2 atom of the thiazole ring. This orientation influences the mechanism of enzyme catalysis (Frank *et al.*, 2007).

Although the sequence similarity between TDP-dependent enzymes is low (less than 20% amino acid identity), the tertiary structures of these proteins show high similarities in terms of the TDP-binding folds, particularly at the geometric positions of the conserved residues which are involved in TDP binding. It has been demonstrated that TDP-dependent enzymes contain at least two conserved domains: (i) a phosphate-binding domain in which the TDP cofactor is bound primarily through its diphosphate group coordinated by a divalent cation and (ii) a pyrimidine-binding domain containing the conserved glutamic acid residue which plays a crucial role in the catalytic mechanism (Duggleby, 2006; Widmann *et al.*, 2010). TDP-dependent enzymes form either dimers or tetramers, in which TDP is bound at the dimer interface, with the pyrimidine moiety associated with one monomer and the diphosphate residue bound to the other monomer (Duggleby, 2006; Lindqvist *et al.*, 1992).

A. CATALYTIC MECHANISMS OF TDP-DEPENDENT ENZYMES

The close proximity of the 4′-amino group to the C2 atom of thiazolium ring in the enzyme-bound TDP molecule permits an intramolecular proton abstraction which leads to the formation of a nucleophilic C2-"ylide". This is the initiating step of all reactions catalysed by TDP-dependent enzymes (Fig. 9).

It has been demonstrated in many earlier reports that TDP first undergoes tautomerization into an imino-form and that the nitrogen atom of the imine is responsible for abstracting the C2 proton in the thiazolium ring of TDP (Jordan *et al.*, 2003; Nemeria *et al.*, 2007; Tittmann *et al.*, 2003). This process is rendered possible by the charge of the N1′ atom, which forms a hydrogen bond with the conserved glutamate residue located in the catalytic centres of many TDP-dependent enzymes (Shaanan and Chipman, 2009). Ylide stabilization and further catalytic steps are also favoured by the effective polarity of the binding site (Zhang *et al.*, 2005).

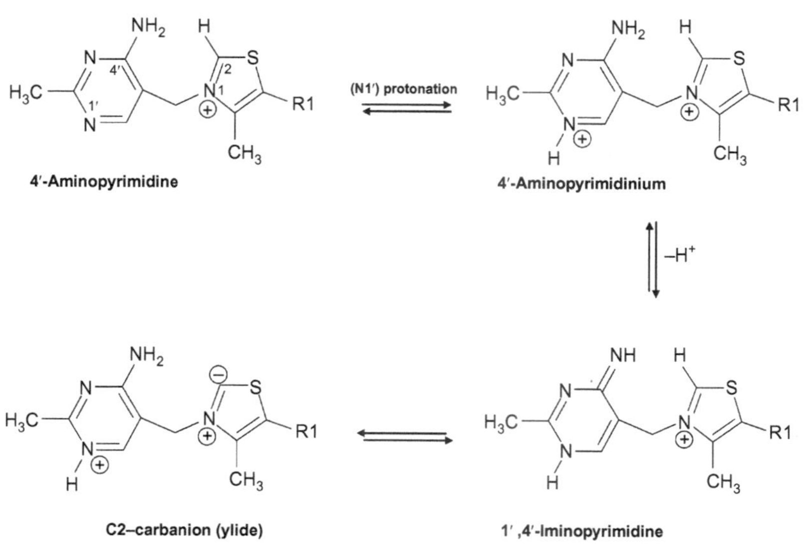

Fig. 9. Generation of active ylide-like carbanion as an initiating step of all reactions catalysed by TDP-dependent enzymes.

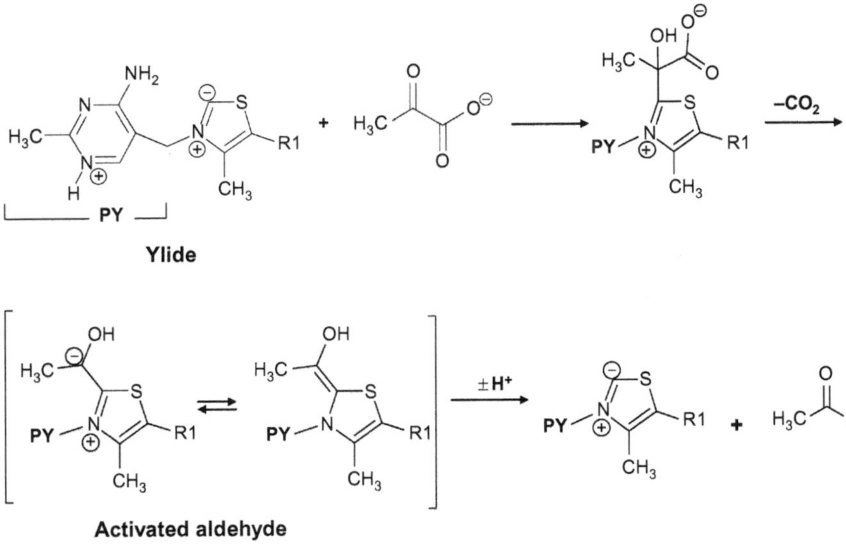

Fig. 10. Generation of activated aldehyde intermediate in the pyruvate decarboxylase reaction. The activated aldehyde, also known as the enamine-carbanion intermediate, is a common early stage in catalytic mechanisms of all TDP-dependent enzymes, in spite of very different first substrate and downstream reaction steps.

The reactions catalysed by TDP-dependent enzymes can be generally divided into decarboxylation and transferase-type reactions. All share certain mechanistic similarities, that is, the first part of the reaction involves the breaking of a C—C or C—H bond adjacent to a carbonyl group of the substrate and the formation of a metastable enamine intermediate (Fig. 10). In the next step of the catalysis reaction, the second substrate is bound and the final product is eventually released with ylide regeneration.

B. CLASSIFICATION OF TDP-DEPENDENT ENZYMES AND THEIR LOCALIZATION WITHIN THE PLANT CELL

The TDP-dependent enzymes involved in plant vital functions belong to three of the main enzyme classes, namely oxidoreductases, transferases and lyases. Each enzyme has an important function in major metabolic pathways and their localizations within the plant cell are presented in Fig.11.

1. *Oxidoreductases*

The plant TDP-dependent oxidoreductases are α-ketoacid dehydrogenase complexes that have crucial functions in all aerobic organisms. These enzymes link glycolysis with the tricarboxylic acid cycle, drive the further

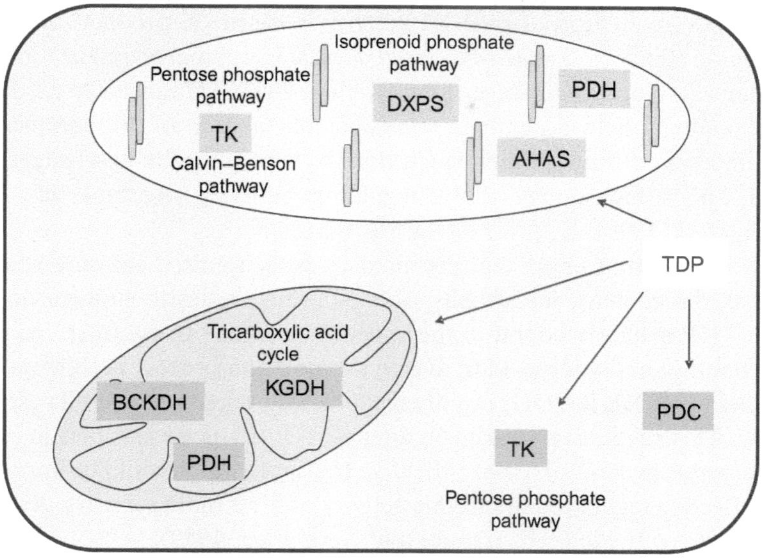

Fig. 11. Localization of TDP-dependent pathways in the plant cell. The major mitochondrial TDP-dependent enzymes include pyruvate dehydrogenase (PDH) involved in the acetyl-CoA synthesis, α-ketoglutarate dehydrogenase (KGDH) of the tricarboxylic acid cycle and branched-chain α-ketoacid dehydrogenase (BCKDH). The chloroplast contains PDH, acetohydroxyacid synthase (AHAS), 2-deoxy-D-xylulose-5- phosphate synthase (DXPS) of non-mevalonate isoprenoid synthesis pathway and transketolase (TK), involved in the pentose phosphate pathway and the Calvin–Benson cycle. Pyruvate decarboxylase (PDC) and, at least in some species, an additional pool of TK are present in the cytosol.

oxidation of glucose in this cycle and are involved in substrate production for fatty acid biosynthesis (Randall *et al.*, 1996; Money *et al.*, 2002). In general, these dehydrogenase complexes consist of multiple copies of three components: (E1) a specific α-ketoacid dehydrogenase that also decarboxylates α-ketoacid with the participation of TDP (E2) dihydrolipoyl acyltransferase that transfers the acyl group to CoA and (E3) dihydrolipoyl dehydrogenase that regenerates the E2 prosthetic group and produces NADH. The E2 component forms the core of the complex to which E1 and E3 are noncovalently attached. Plants possess three types of α-ketoacid dehydrogenase complex as described in further detail below.

a. Mitochondrial pyruvate dehydrogenase complex. Mitochondrial pyruvate dehydrogenase complex (mtPDH) converts pyruvate to acetyl-CoA. Localized in the mitochondrial matrix, mtPDH is a linker between cytoplasmic glycolysis and the mitochondrial tricarboxylic acid cycle (Fig. 10).

The precise regulation of mtPDH activity is based on product inhibition, metabolite effectors (Tovar-Méndez *et al.*, 2003), compartmentalization and the plant developmental stage (Luethy *et al.*, 2001; Thompson *et al.*, 1998). Additionally, a light-dependent, reversible inactivation of this complex has been observed during its phosphorylation by a bound E2 kinase, suggesting that E2 phosphatase may play a regulatory function (Roche *et al.*, 2001; Thelen *et al.*, 1998).

It is interesting to note that mammalian mitochondrial enzymes that use TDP as the cofactor are usually isolated from tissues as holoenzymes in which TDP is tightly bound to the apoenzyme forms. In contrast, the plant mitochondria easily loose TDP during the isolation process but the purified enzymes (mtPDH, KGDH) rapidly recapture the coenzyme after its external supply. This suggests a weaker binding of TDP by these enzymes in plants with possible benefits of a more effective transport which could be important for a effective regulation of enzyme activity or for a more sensitive detection of TDP biosynthetic needs (Douce and Neuburger, 1989).

b. Plastidial pyruvate dehydrogenase complex. The plastidial pyruvate dehydrogenase complex (ptPDH) supplies the acetyl-CoA and NADH for *de novo* fatty acid biosynthesis in the stroma (Camp and Randall, 1985; Ke *et al.*, 2000). Unlike mtPDH, ptPDH is upregulated under photosynthetic conditions by an increase in the stromal pH and Mg^{2+} concentrations (Miernyk *et al.*, 1985; Williams and Randall, 1979), and is not regulated by reversible phosphorylation (Miernyk *et al.*, 1998).

c. α-Ketoglutarate dehydrogenase. As a component of the tricarboxylic acid cycle, KGDH catalyses the oxidative decarboxylation of α-ketoglutarate to succinyl-CoA and NADH and is localized at the inner mitochondrial membrane (Millar *et al.*, 1999). Analyses of KGDH activity in the presence of some inhibitors (Araújo *et al.*, 2008; Bunik and Fernie, 2009) have shown that it is the limiting enzyme for cellular respiration and plays a role in nitrogen assimilation and amino acid (glutamate, glutamine and GABA) metabolism. It has also been proposed that at low levels of NAD^+, KGDH may be involved in a side reaction of reactive oxygen species (ROS) production, thus being a signal of a metabolic disorder (Bunik and Fernie, 2009).

d. Branched-chain α-ketoacid dehydrogenase. BCKDH is a mitochondrial enzyme (Anderson *et al.*, 1998) which catalyses the irreversible oxidative decarboxylation of the branched-chain α-ketoacids derived from valine, leucine and isoleucine (Paxton *et al.*, 1986; Wynn *et al.*, 1996; Yeaman, 1989).

Its regulation by light is probably dependent on a mechanism similar to that of mtPDH (Fujiki *et al.*, 2000).

2. Transferases

The transketolases (TKs) belong to the class of transferases and catalyse the reversible transfer of a keto group from a ketose to an aldose via a non-oxidative mechanism (Schenk *et al.*, 1998).

a. Transketolase. Plant TKs operate mostly in chloroplasts (Debnam and Emes, 1999; Schnarrenberger *et al.*, 1995). The spinach *TK* gene harbours a plastid-targeting sequence (Flechner *et al.*, 1996; Teige *et al.*, 1995) and is expressed in both photosynthesizing and non-photosynthesizing tissues (Bernacchia *et al.*, 1995; Teige *et al.*, 1998). In chloroplast stroma, TK takes part in the photosynthesis-associated carbon fixation that occurs in the Calvin–Benson cycle (Raines, 2003). Its activity is a limiting factor for the maximum rate of photosynthesis. In the Calvin–Benson cycle, TK catalyses the conversion of glyceraldehyde-3-P and fructose-6-P to xylulose-5-P and erythrose-4-P, as well as that of glyceraldehyde-3-P and sedoheptulose-7-P to ribose-5-P and xylulose-5-P. Although TK is a non-regulated enzyme, its decreased level can suppress sucrose production and the photosynthesis rate (Henkes *et al.*, 2001). TK is also universally required for the pentose phosphate pathway. Most of the enzymes involved in NADPH generation in the oxidative part of this pathway are present in both plastids and the cytosol. However, the plant cell localization of the non-oxidative part of pentose phosphate pathway, where TK is responsible for the carbon skeleton production for nucleotide biosynthesis, is still under debate (Bernacchia *et al.*, 1995). Previous TK activity analyses (Hong and Copeland, 1990; Journet and Douce, 1985; Nishimura and Beevers, 1979) and isotopic carbohydrate labelling studies (Krook *et al.*, 1998; Rontein *et al.*, 2002) have indicated that TK catalysis can vary between species, tissues and different stages of plant development, and may also depend on the environmental conditions (Kruger and von Schaewen, 2003).

3. Lyases

Among the well-characterized plant TDP-dependent lyases are (i) PDC, the key enzyme in ethanolic fermentation; (ii) acetolactate synthase (AHAS) which is involved in branched-chain amino acid synthesis; and (iii) 1-deoxy-D-xylulose-5-phosphate synthase (DXPS), the enzyme for isoprenoid formation.

a. Pyruvate decarboxylase. PDC catalyses the irreversible, non-oxidative decarboxylation of pyruvate to acetaldehyde with CO_2 liberation (Fig. 9). This enzyme is predominant in seeds and has been detected in *O. sativa* (Hossain *et al.*, 1994; Rivoal *et al.*, 1990), *Z. mays* (Forlani *et al.*, 1999) and *Pisum sativum* (Mücke *et al.*, 1995). During ethanolic fermentation, acetaldehyde is reduced to ethanol by alcohol dehydrogenase. The activation of PDC, resulting in ethanol production, has mostly been observed under stress conditions, for example, in the adaptation of rice plants to low temperature, probably owing to alterations in the physical properties of membrane lipids (Kato-Noguchi and Yasuda, 2007) or in changes in plant growth under anoxia and hypoxia (Ismail *et al.*, 2009; Ismond *et al.*, 2003; Kürsteiner *et al.*, 2003). The induction of fermentative metabolism was also observed previously under aerobic conditions in the roots of pea plants as a result of the inhibition of branched-chain amino acid biosynthesis (Zabalza *et al.*, 2005). PDC was also shown to be critically involved in the growth of pollen tubes in *Petunia hybrida* (Gass *et al.*, 2005).

b. Acetohydroxyacid synthase. AHAS catalyses the first step in the biosynthesis of branched-chain amino acids (Duggleby *et al.*, 2008), the condensation of two pyruvate molecules during the synthesis of Val and Leu, or that of pyruvate and α-ketobutyrate for the synthesis of Ile. This enzyme is unstable during purification, but its activities have been demonstrated in maize (Muhitch *et al.*, 1987), barley (Durner and Boger, 1988) and wheat (Southan and Copeland, 1996) and, using a heterologous expression system in bacteria, also in cocklebur (Bernasconi *et al.*, 1995), Arabidopsis (Chang and Duggleby, 1997; Dumas *et al.*, 1997; Ott *et al.*, 1996) and tobacco (Chang *et al.* 1997). Plant AHASs are composed of a catalytic subunit with a TDP-binding site and a regulatory subunit necessary for feedback inhibition by branched-chain amino acids (Lee and Duggleby, 2001; McCourt and Duggleby, 2006). The identified N-terminal signal sequences suggest the translocation of this protein to chloroplasts (Ott *et al.*, 1996). The AHAS enzymes are also involved in the binding of several herbicide classes (McCourt *et al.*, 2006). However, some herbicide-resistant mutations in the *AHAS* gene have been reported in rice, tobacco and Arabidopsis (Chang and Duggleby, 1998; Kawai *et al.*, 2007; Okuzaki *et al.*, 2007; Shimizu *et al.*, 2002; Tan *et al.*, 2005). These observations have prompted a number of attempts to produce transgenic, herbicide-resistant crop plants (Ott *et al.*, 1996).

*c. 1-Deoxy-*D*-xylulose-5-phosphate synthase.* DXPS catalyses the first reaction in an alternative, non-mevalonate pathway of isoprenoid biosynthesis, in which glyceraldehyde 3-phosphate is condensed with pyruvate

(Lichtenthaler, 1999; Sprenger *et al.*, 1997). The product, DXP, was also identified as a precursor in the thiamine and pyridoxol (a form of vitamin B_6) biosynthesis pathways in plants and in *E. coli* (Begley *et al.*, 1999; Hill *et al.*, 1996). Multiple *DXPS* genes have been found in several plant species that encode isoforms involved in the biosynthesis of different classes of isoprenoids (Cordoba *et al.*, 2009). *DXPS* expression has also been detected in all photosynthetic tissues, with an unequivocal plastidial localization of this enzyme (Zhang *et al.*, 2009). *DXPS* overexpression in tomato, Arabidopsis and tobacco correlates with the accumulation of chlorophyll, carotenoids, tocopherols and abscisic acid (ABA), indicating that this enzyme catalyses the rate-limiting reaction in the isoprenoid phosphate pathway (Estevez *et al.*, 2001; Lois *et al.*, 2000, Zhang *et al.*, 2009). Some growth conditions, for example, light exposure (Kim *et al.*, 2005), mechanical wounding or fungal elicitors (Phillips *et al.*, 2007), also modulate *DXPS* transcript accumulation.

IV. THIAMINE TRANSPORT, DISTRIBUTION AND STORAGE IN PLANT TISSUES

Depending on the development stage, plants use different sources for thiamine acquisition. These include seed storage tissues, biosynthetic processes and soils. During seed maturation, thiamine accumulates in the germ in parallel with the increase in the total soluble protein content (Shimizu *et al.*, 1990). Thiamine is stored in the unphosphorylated form and even in mature seeds, the phosphate esters represent only 5% of the total thiamine content. The long-term thiamine storage in seeds depends on specific thiamine-binding proteins (TBPs) which are present in many plant species (Adamek-Swierczynska and Kozik, 2002; Adamek-Swierczynska *et al.*, 2000; Kozik and Rapala-Kozik, 1995; Mitsunaga *et al.*, 1986a,b, 1987; Nishimura *et al.*, 1984; Nishino *et al.*, 1983, Rapala-Kozik and Kozik, 1998; Shimizu *et al.*, 1995). The chemical mechanism of thiamine binding by these proteins has been extensively studied (Kozik and Rapala-Kozik, 1996; Rapala-Kozik and Kozik, 1992, 1996; Rapala-Kozik *et al.*, 1999). TBPs are suggested to represent specific variants of the major seed storage globulins (Adamek-Swierczynska and Kozik, 2002; Rapala-Kozik *et al.*, 2003).

Developing seedlings first utilize the thiamine that is stored in seeds, as demonstrated from previous analyses of the total seed thiamine content which does not change (Kylen and McCready, 1975; Mitsunaga *et al.*, 1987) or decrease (Golda *et al.*, 2004) during seed germination. Depending on the species, this takes 2–4 days after imbibition, before the seedlings commence thiamine biosynthesis (Golda *et al.*, 2004). At least in cereal seeds, TPK activity progressively increases during seed germination and seedling growth (Golda *et al.*, 2004; Mitsunaga *et al.*, 1987).

As discussed in the preceding sections, the biosynthesis of thiamine takes place in chloroplasts but TDP is formed in the cytosol (Fig. 6). As different compartments utilize TDP as an enzyme cofactor, it is highly probable that plants possess thiamine-, TMP- or TDP-specific cellular transporters, but none have yet been identified. It has been shown that TMP, synthesized *de novo* in chloroplasts, readily undergoes dephosphorylation by relatively non-specific phosphatases (Rapala-Kozik *et al.*, 2009), but the actual subcellular localization of this process remains unknown. Free thiamine is pyrophosphorylated in the cytosol but the TDP produced is also important for fundamental mitochondrial functions. This suggests that a TDP transporter should exist in the inner mitochondrial membrane. Mitochondrial TDP transporters were previously identified in human, yeast and *Drosophila melanogaster* (Iacopetta *et al.*, 2010; Lindhurst *et al.*, 2006; Marobbio *et al.*, 2002) and belong to a broad mitochondrial carrier family, the members of which have also been detected in Arabidopsis (Millar and Heazlewood, 2003). Similar hypothetical transporters may also be useful for thiamine uptake from seed storage tissues or soil.

Owing to the chloroplastic localization of the entire pathway of TMP *de novo* synthesis, green tissues are the primary location where thiamine is formed and from which it is transported to thiamine-requiring tissues such as the roots. Accordingly, the genes encoding HMP-P synthase, HET-P synthase and HMP-P kinase/TMP synthase are predominantly or sometimes exclusively detected in leaves (Belanger *et al.*, 1995; Kim *et al.*, 1998; Kong *et al.*, 2008; Papini-Terzi *et al.*, 2003; Raschke *et al.*, 2007; Ribeiro *et al.*, 1996). In contrast, TPK is expressed in all plant tissues, albeit at variable levels (Ajjawi *et al.*, 2007a,b), to ensure that both endogenous and exogenous thiamine sources will be equally useful for the synthesis of TDP.

An alternative way to acquire thiamine is via absorption from the soil by the roots (Mozafar and Oertli, 1992, 1993), which in most plant species have no thiamine-synthetic capacity. The transport of external thiamine appears to be independent of the level of metabolic energy and probably represents a passive transpiration-mediated process. Root-absorbed thiamine flows to other plant parts via the xylem (Mozafar and Oertli, 1992, 1993). Thiamine and its phosphate esters can also be introduced into plant seedlings through the leaves (Mozafar and Oertli, 1992, 1993). After a sufficient period of time from its application, thiamine appears to be uniformly distributed between different parts of the plant. This transport probably occurs via the phloem and may be strictly polarized (basopetal), as seen in the tomato petiole (Kruszewski and Jakobs, 1974) or may proceed in both the acropetal and basopetal directions (Mozafar and Oertli, 1992, 1993). More recently, it has been reported that the foliar application of TMP and TDP can trigger plant disease resistance (Ahn *et al.*, 2005, 2007) and complement the Arabidopsis TPK double mutant (Ajjawi *et al.*, 2007a,b). These findings provide good evidence for thiamine-phosphate transport by the plant vascular system via an apoplastic route. Leaves which develop after thiamine application can concentrate this vitamin, suggesting its possible re-mobilization from older parts of the plant (Mozafar and Oertli, 1993). Thiamine transport via the phloem from leaves to the kernels in maize, wheat and rice was reported many years ago (Kondo *et al.*, 1951; Shimamoto and Nelson, 1981). The thiamine levels decrease in the glumes, leaves and stem and increase in the kernels towards the end of kernel-filling process (Geddes and Levine, 1942). In maize, the concentration of thiamine in the embryo is more than 10-fold greater than that in the endosperm (Shimamoto and Nelson, 1981).

In summary, the current knowledge of thiamine transport in plant tissues and cells is not well advanced and further research, paying particular attention to the identification of the TDP- and/or thiamine transporters, is necessary.

V. ROLE OF THIAMINE IN THE SENSING, RESPONSE AND ADAPTATION TO PLANT STRESS

The environmental conditions which exert abiotic stress in plants (drought, high salinity, heavy metals, drastic changes in temperature or light intensity) can significantly alter plant metabolism, growth and development. However, the mechanisms underlying the responses or even perception of these environmental stresses by plants are not well understood. The current evidence with regards to the pathways by which plants sense or adapt to stress is based on transcript changes (genetic analyses), protein induction or suppression (proteomics) or protein activity determination. However, an increase in the mRNA levels could be interpreted in terms of increased requirement for the translated protein product during stress conditions but it may also indicate that this protein is susceptible to damage during stress and its resulting degradation requires an increase in transcription to maintain its normal cellular levels. These possibilities must be taken into account in future data analysis.

A. ABIOTIC STRESS RESPONSES

As plants are unable to avoid exposure to extreme environmental conditions, they have developed many types of specific responses in order to survive. Most metabolic analyses in this regard have been concerned with changes in various pathways of carbon metabolism including glycolysis, the tricarboxylic acid cycle and photosynthesis, which probably represent the primary responses of plants to stress, mediated by chemical reactions and enzymatic components. However, the changes observed in thiamine biosynthesis processes should be considered as a second line of defence, once the stress stimulus has been sensed by the plant and transcriptional, translational or post-translational responses have been initiated.

Previous studies that have focused on the activity of the main metabolic pathways that operate during abiotic stress conditions have shown that primary anabolic metabolism is largely downregulated in favour of catabolic and antioxidant metabolism. For example, in Arabidopsis roots or in the cells of other organs subjected to oxidative stress, an impairment of the tricarboxylic acid cycle and of amino acid metabolism has been observed and this was followed by the initiation of a backup system for glycolysis comprising a redirection of carbon metabolism to the oxidative pentose phosphate pathway for NADPH production (Baxter *et al.*, 2007; Lehmann *et al.*, 2009).

As many enzymes which operate in the sensing, response activation and adaptation to plant stress require TDP as a cofactor (Fig. 11), it is not surprising that the *de novo* biosynthesis of this compound is upregulated in plants under stress conditions. The upregulation in the transcript levels (three- to fourfold) of two initial thiamine biosynthetic genes, *THI1* and *THIC*, was observed during the adaptation of Arabidopsis seedlings to growth under paraquat-induced oxidative stress (Tunc-Ozdemir *et al.*, 2009). Additionally, a twofold increase in β-GUS activity was observed under salt stress or flooding conditions in transgenic plants carrying the *GUS* promoter gene fused to *THI1* promoter fragments (Ribeiro *et al.*, 2005). These results confirmed earlier findings from proteomic and DNA microarray studies of plant responses to cold, heat and drought (Ferreira *et al.*, 2006; Wong *et al.*, 2006). The *THI1* gene may be precisely regulated under stress conditions since its promoter possesses an ABA-responsive element (Ribeiro *et al.*, 2005). It has also been suggested that the *THIC* promoter possesses a stress–response element (Tunc-Ozdemir *et al.*, 2009). However, in both cases, there is no evidence of the actual functioning of these putative regulatory elements. A three- to sixfold increase of the levels of TMP synthase and TPK transcripts was also observed in Arabidopsis seedlings under oxidative stress conditions and these results correlated with a detectable increase of the levels of thiamine and its phosphate esters (Tunc-Ozdemir *et al.*, 2009). Analogical responses were observed in *Z. mays* seedlings under salt, water and oxidative stress conditions under which the activities of both TMP synthase and TPK, as well as total thiamine levels, significantly increased

(Rapala-Kozik *et al.*, 2008). Interestingly, the latter effect in stressed *Z. mays* seedlings was predominantly due to an increase of free thiamine, whilst in Arabidopsis, a TDP increase was more predominantly detected under similar stress conditions. This could be explained by the different plant response phases analysed in these two studies. In the study of *Z. mays* seedlings, an overproduction of thiamine, ready to be transported into the appropriate organelles, was detected and in the Arabidopsis model, the response may be shifted to the production of the functional coenzyme form of thiamine. A drop in the steady-state TDP levels may be important as TDP is the major regulatory factor for thiamine biosynthesis (Nosaka, 2006), and is known to operate via a riboswitch which is present in the 3′-UTR of the *THIC* gene (Bocobza *et al.*, 2007; Raschke *et al.*, 2007; Wachter *et al.*, 2007).

After the regeneration of a significant source of TDP, damaged pathways can be restarted, probably at a higher rate to compensate for any stress-induced deficiencies and to support adaptive responses (Fig. 12).

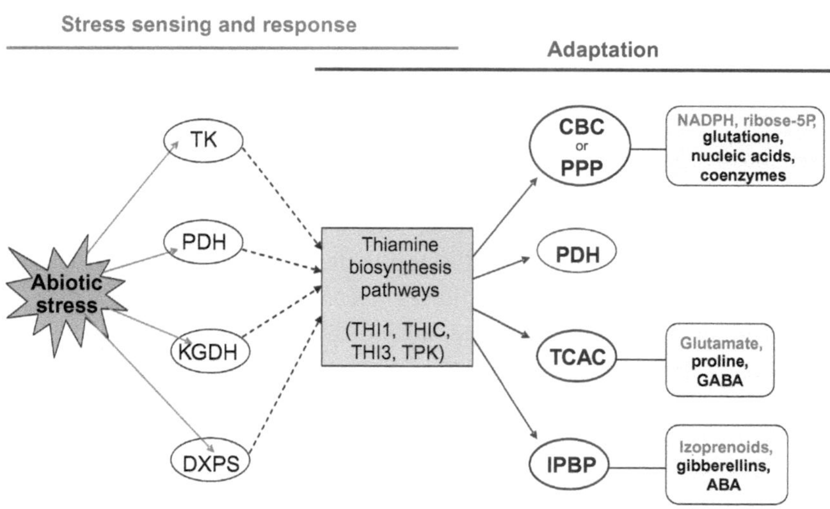

Fig. 12. Thiamine biosynthesis and TDP-dependent pathways in the sensing, response and adaptation to plant stress. A sensing of environment stress factors by the plant involves damages to the main TDP-dependent enzymes (TK, PDH, KGDH, DXPS). In a response, the activities of thiamine biosynthetic enzymes (THI1, THIC, THI3,TPK) increase and subsequently a regeneration of the main metabolic pathways occurs. In an adaptation phase, some of the TDP-dependent pathways such as the Calvin–Benson cycle (CBC), the pentose phosphate pathway (PPP), the tricarboxylic acid cycle (TCAC) and isoprenoid phosphate biosynthesis pathway (IPBC) can be upregulated to compensate for the previous damages and to provide important defence molecules (e.g. antioxidants) and stress protectants.

It has been reported that the oxidative pentose phosphate pathway (Baxter *et al.*, 2007; Couée *et al.*, 2006), isoprenoid biosynthesis pathway (Paterami and Kanellis, 2010; Schroeder and Nambara, 2006), the tricarboxylic acid cycle (Lehmann *et al.*, 2009) and ethanolic fermentation (Conley *et al.*, 1999; Drew, 1997; Kürsteiner *et al.*, 2003) are accelerated or induced under different abiotic stress conditions, in which an intensive increase in ROS production was observed in all plant cell compartments in most cases (Zhu, 2002).

The cytosolic enzymes involved in the early stages of glycolysis, triosephosphate isomerase and glyceraldehyde-3-phosphate dehydrogenase, may be partly inhibited by excessive ROS, causing a rerouting of the main carbohydrate-metabolic flux from the glycolytic to the pentose phosphate pathway (Ralser *et al.*, 2007). This pathway is activated by the upregulation of regulatory enzymes involved in the oxidative steps (Couée *et al.*, 2006; Debnam *et al.*, 2004; Valderrama *et al.*, 2006) and produces more NADPH, which is recycled via numerous antioxidant systems, such as the ascorbate-glutathione cycle, to quickly restore the cytoplasmic redox equilibrium (Valderrama *et al.*, 2006).

TK is one of the major TDP-dependent enzymes for which the increased transcript and protein levels, as well as a higher enzymatic activity, has been shown in several plant species under different stress conditions (Bernacchia *et al.*, 1995; Ferreira *et al.*, 2006; Rapala-Kozik *et al.*, 2008; Wolak *et al.*, 2010). TK operates in chloroplasts and probably, at least in some species, also in the cytoplasm, and is involved in the Calvin–Benson cycle and pentose phosphate pathway. These two processes produce NADPH which feeds a variety of ROS-scavenging systems such as the plastidial Asada–Halliwell pathway that engages two powerful antioxidants, reduced glutathione and ascorbate (Arora *et al.*, 2002). Although TK is not a regulatory enzyme, its levels need to be suitably adjusted during the response to environmental stresses to assure a balanced flow of all intermediates of the NADPH producing pathways (Henkes *et al.*, 2001).

Another stress defence system which operates in chloroplast is dependent on the non-mevalonate isoprenoid synthesis pathway which engages another TDP-dependent enzyme, DXPS (Lange *et al.*, 1998). This pathway provides precursors for the synthesis of carotenoids, terpenes, tocopherols and is also a source of chlorophyll, plastoquinone, gibberellins and ABA (Lichtenthaler *et al.*, 1997). Carotenoids are powerful antioxidants (Hix *et al.*, 2004; Vallabhaneni and Wurtzel, 2010) and ABA participates in the signal transduction pathways required for plant adaptation to several types of abiotic stress. DXPS transcript accumulation is induced in *Cistus creticus* in response to heat, drought, wounding and elicitors including salicylic acid and methyl jasmonate (Paterami and Kanellis, 2010). These results are consistent with previous finding that isoprenoids participate also in thermotolerance-related activities involved in plant adaptation (Peñuelas and Munné-Bosch, 2005).

The activation of the ethanolic fermentation pathway in plants which grow at low temperatures, or under water deficiency or hypoxia, is well documented (Ismond *et al.*, 2003; Kato-Noguchi and Yasuda, 2007). A cytosolic TDP-dependent enzyme, PDC, is the main regulatory enzyme in this pathway (Kürsteiner *et al.*, 2003) and its overexpression in Arabidopsis improves the plant tolerance to hypoxia (Ismond *et al.*, 2003). This finding suggests that mitochondrial dysfunction and the inhibition of pyruvate conversion to acetyl-CoA cause a redirection of the main glycolytic pathway to cytoplasmic ethanolic fermentation. Ethanol production prevents lipid degradation in the plant membrane and enables the maintenance of energy production until the more effective aerobic respiration processes are recovered (Kürsteiner *et al.*, 2003; Tadege *et al.*, 1999).

The major stress sensing pathway in plants seems to be the tricarboxylic acid cycle and mitochondrial production of acetyl-CoA (Baxter *et al.*, 2007; Sweetlove *et al.*, 2002; Taylor *et al.*, 2004a). Both pathways engage the TDP-activated complex enzymes PDH and KGDH which are readily inactivated by oxidative damage of their lipoic acid-dependent components (Taylor *et al.*, 2004b). After antioxidant stress responses are activated, these pathways are restored during the adaptation phase (Taylor *et al.*, 2004a).

B. THIAMINE FUNCTION IN BIOTIC STRESS

The improved growth of plants in the presence of thiamine was observed some years ago, but it has only been recently that a better understanding of this effect of thiamine has come to light, particularly under biotic stress conditions. The thiamine supplementation of plants undergoing bacterial, fungal or viral infection triggers systemic acquired resistance (SAR) to these pathogens (Ahn *et al.*, 2005; Malamy *et al.*, 1996). It was further found that in the presence of thiamine, the expression of defence-related (*PDF1.2*) and SAR-related (*PR1*) genes is induced more rapidly compared with pathogen inoculation. The expression of these genes was found to be even higher when pathogen inoculation was followed by thiamine treatment (Ahn *et al.*, 2005). Interestingly, thiamine-phosphate esters were also found to rescue infected plants, and at even lower concentrations than thiamine. This could be due to either a higher effectiveness of TDP/TMP or a slower effect of thiamine due to restrictions in its transport.

The signalling processes that are affected by thiamine during pathogen infections involve the salicylic acid-dependent and mobilized calcium-related signalling pathways and also the priming of plant defence responses that suppress pathogen propagation (Ahn *et al.*, 2005). SAR activation by thiamine is accompanied by hydrogen peroxide accumulation which can trigger a stress response (Ahn *et al.*, 2007). This suggestion is supported by the observation that rice plants with a repressed disease resistance-responsive gene (*OsDR8*) produce a lower level of thiamine (Wang *et al.*, 2006). This effect can be related to the high sequence similarity between the OsDR8 protein and maize thiazole synthase (THI1, THI2). Additionally, *OsDR8*-silenced plants express lower levels of several defence-responsive genes suggesting the involvement of *OsDR8* in the regulation of signal transduction pathways that function in the defence response.

C. RESCUE OF STRESSED PLANTS BY THIAMINE SUPPLEMENTATION—IS THIAMINE A REAL ANTIOXIDANT?

In many types of plant stress, ROS production in the cells is considered to be a secondary stress event and the prime activator of antioxidative response pathways. Some reports have suggested that thiamine can function directly as an antioxidant. The products of thiamine oxidation *in vitro* are thiochrome or thiamine disulphide analogues (Lukienko *et al.*, 2000; Stepuro *et al.*, 2005). Thiochrome can be easily detected owing to its strong fluorescence but for its formation, a non-physiological highly basic environment is necessary. Thiamine disulphide-related compounds are difficult to analyse in cell extracts. A similar hypothesis for the antioxidant activity of thiamine has come from analyses of human nerve cells with a thiamine deficiency (Hazell and Butterworth, 2009). It has been documented that thiamine can normalize the lipid peroxidation levels and elevate the activity of glutathione reductase, and that thiamine deficiency leads to peroxynitrite accumulation. However, thiamine was found not to exert a phytotoxic effect at any concentration tested. In addition, the participation of thiamine in DNA repair in bacteria, yeast and plants has been proposed (Machado *et al.*, 1996, 1997). However, the mechanism underlying the role of thiamine action as an antioxidant defence trigger remains obscure.

VI. PRACTICAL ASPECTS AND FUTURE PERSPECTIVES

The current knowledge of the physiological functions of thiamine compounds in human, including the crucial role of TDP as cofactor in cellular metabolism and the non-cofactor neurophysiological role of TTP, is well advanced. Modern medicine has taken advantage of this knowledge in the development of treatments for numerous pathophysiological conditions which are the results of a low vitamin B_1 content in the diet, inefficient intestinal thiamine absorption, an impaired uptake of thiamine by individual tissues and cells or thiamine-dependent metabolic malfunctions. At least some of these pathogenic influences can be prevented or eliminated nutritionally through the enrichment of foods with vitamin B_1, supplementation with more easily absorbable derivatives (e.g. benfothiamine) and the elimination of antithiamine factors (thiaminases, polyphenolic compounds) among others. In developed countries, thiamine imbalances in the diet are usually overcome by industrial fortification of foods such as bread with this vitamin. However, in developing countries, crops are the major source of thiamine. Unfortunately, the limited advances thus far in thiamine-focused plant science seriously hinder the potential for improving plant constituents as a strategy to lower the rate of thiamine deficiency-related diseases.

As noted in this review, considerable progress has been made in our understanding of thiamine biosynthesis and metabolism in higher plants in recent years. One of the highlights in this regard has been the discovery of the riboswitch-dependent feedback inhibition of thiamine synthesis. This and other regulatory mechanisms must be further elucidated to the point where it is possible to engineer plant cultivars with a higher thiamine content in the consumable tissues. However, thiamine produced by microorganisms in the soil can be absorbed by roots, transported to plant cells and converted to TDP, but our current understanding of these transport processes is still in its infancy. It is possible that cytoplasmic TPK, which is probably less tightly regulated than chloroplastic enzymes of the main thiamine biosynthetic route, may be a viable target for genetic manipulation (overexpression) to increase the production of TDP and, after its quick dephosphorylation, augment the total thiamine content in plant tissues. It is likely, however, that the best material to increase the nutrition value with respect to vitamin B_1 will prove to be the seeds in which specific globulins are deposited together with captured thiamine, to provide the necessary reserves for the growing seedling. An increase in the expression of these TBPs by genetic engineering should be possible and thereby provide an enriched store of this vitamin.

The recent unequivocal establishment of the critical role of thiamine in the plant response and adaptation to biotic and abiotic stresses should have a practical impact, for example, in developing plant cultivars with higher stress resistance. Once our general understanding of the mechanism of thiamine transport in plants is improved, methods for a more effective supplementation of plants may be developed to increase plant resistance to stress factors such as high temperature, drought or environmental pollution. The development of plant cultivars with high stress tolerance should improve global plant production levels, which would represent another approach in contemporary agriculture to overcoming thiamine deficiency problems. Recently, a strategy for engineering herbicide-tolerant crops has been proposed which utilizes the known properties of a specific TDP-dependent enzyme, AHAS. This enzyme, which is critical for the biosynthesis of branched-chain amino acids in plants, is a potential target for herbicide action (Duggleby *et al.*, 2008; McCourt *et al.*, 2006). It was found that the imidazolinone herbicides bind to the AHAS regulatory subunit, blocking the active centre of this enzyme (Trucco *et al.*, 2006). In mutagenesis analysis of the herbicide-binding pocket in AHAS, some amino acids were selected whose substitution resulted in the resistance of this enzyme to these herbicides (Jung *et al.*, 2004; Kolkman *et al.*, 2004). Mutagenesis or selection approaches, that utilize conventional plant breeding techniques, have created many imidazolinone-resistant crops including maize, rice, wheat, sunflower and oilseed rape (Tan *et al.*, 2005). The application of imidazolinone herbicides in the cultivation of resistant crops has facilitated the control of a broad spectrum of grasses and broadleaf weeds. However, effectiveness at low doses, low mammalian toxicity as well as a favourable environmental profile have made imidazolinone herbicides attractive agents for efficient crop production. In addition, DXPS, the key enzyme in mevalonate-independent isoprenoid biosynthesis, has been suggested to be a promising target for new herbicide development as well as for improving the nutritional value of crop plants (Cordoba *et al.*, 2009; Müller *et al.*, 2000).

In summary, the recent strong progress in the biochemical and physiological study of thiamine in plants, albeit less advanced than analogous research in animals and microorganisms, is expected to continue in the near future and to have an important impact in modern agriculture for improving the nutritional value of plant crops, thereby reducing the rate of chronic disease states that are dependent on the impaired uptake and metabolism of vitamin B_1.

ACKNOWLEDGEMENTS

The author thanks prof. Andrzej Kozik for helpful discussion and critical chapter reading. This work was supported in part by the Ministry of Science and Higher Education, Poland (the grant No. NN303 320937) and the Jagiellonian University (statutory funds No. DS/15/WBBiB).

REFERENCES

Adamek-Swierczynska, S. and Kozik, A. (2002). Multiple thiamine-binding proteins of legume seeds. Thiamine-binding vicilin of Vicia faba versus thiamine-binding albumin of Pisum sativum. *Plant Physiology and Biochemistry* **40**, 735–741.

Adamek-Swierczynska, S., Rapala-Kozik, M. and Kozik, A. (2000). Purification and preliminary characterisation of a thiamine-binding protein from maize seeds. *Journal of Plant Physiology* **156**, 635–639.

Agyei-Owusu, K. and Leeper, F. J. (2009). Thiamin diphosphate in biological chemistry: Analogues of thiamin diphosphate in studies of enzymes and riboswitches. *The FEBS Journal* **276**, 2905–2916.

Ahn, I. P., Kim, S. and Lee, Y. H. (2005). Vitamin B1 functions as an activator of plant disease resistance. *Plant Physiology* **138**, 1505–1515.

Ahn, I. P., Kim, S., Lee, Y. H. and Suh, S. C. (2007). Vitamin B1-induced priming is dependent on hydrogen peroxide and the NPR1 gene in Arabidopsis. *Plant Physiology* **143**, 838–848.

Ajjawi, I., Rodriguez Milla, M. A., Cushman, J. and Shintani, D. K. (2007a). Thiamin pyrophosphokinase is required for thiamin cofactor activation in Arabidopsis. *Plant Molecular Biology* **65**, 151–162.

Ajjawi, I., Tsegaye, Y. and Shintani, D. (2007b). Determination of the genetic, molecular, and biochemical basis of the *Arabidopsis thaliana* thiamin auxotroph *th1*. *Archives of Biochemistry and Biophysics* **459**, 107–114.

Anderson, M. D., Che, P., Song, J., Nikolau, B. J. and Wurtele, E. S. (1998). 3-Methylcrotonyl-coenzyme A carboxylase is a component of the mitochondrial leucine catabolic pathway in plants. *Plant Physiology* **118**, 1127–1138.

Araújo, W. L., Nunes-Nesi, A., Trenkamp, S., Bunik, V. I. and Fernie, A. R. (2008). Inhibition of 2-oxoglutarate dehydrogenase in potato tuber suggests the enzyme is limiting for respiration and confirms its importance in nitrogen assimilation. *Plant Physiology* **148**, 1782–1796.

Arora, A., Sairam, R. K. and Srivastava, G. C. (2002). Oxidative stress and antioxidative system in plants. *Current Science* **82**, 1227–1238.

Baker, L. J., Dorocke, J. A., Harris, R. A. and Timm, D. E. (2001). The crystal structure of yeast thiamin pyrophosphokinase. *Structure* **9**, 539–546.

Balakumar, P., Rohilla, A., Krishan, P., Solairaj, P. and Thangathirupathi, A. (2010). The multifaceted therapeutic potential of benfotiamine. *Pharmacological Research* **61**, 482–488.

Balmer, Y., Koller, A., del Val, G., Manieri, W., Schurmann, P. and Buchanan, B. B. (2003). Proteomics gives insight into the regulatory function of chloroplast thioredoxins. *Proceedings of the National Academy of Sciences of the United States of America* **100**, 370–375.

Barile, M., Passarella, S. and Quagliariello, E. (1990). Thiamine pyrophosphate uptake into isolated rat liver mitochondria. *Archives of Biochemistry and Biophysics* **280**, 352–357.

Baxter, C. J., Redestig, H., Schauer, N., Repsilber, D., Patil, K. R., Nielsen, J., Selbig, J., Liu, J., Fernie, A. R. and Sweetlove, L. J. (2007). The metabolic response of heterotrophic Arabidopsis cells to oxidative stress. *Plant Physiology* **143**, 312–325.

Begley, T. P., Downs, D. M., Ealick, S. E., Mc Lafferty, F. W., Van Loon, A. P., Taylor, S., Campobasso, N., Chiu, H. J., Kinsland, C., Reddick, J. J. and Xi, J. (1999). Thiamin biosynthesis in prokaryotes. *Archives of Microbiology* **171**, 293–300.

Begley, T. P., Chatterjee, A., Hanes, J. W., Hazra, A. and Ealick, S. E. (2008). Cofactor biosynthesis—Still yielding fascinating new biological chemistry. *Current Opinion in Chemical Biology* **12**, 118–125.

Belanger, F., Leustek, T., Chu, B. and Kirz, A. (1995). Evidence for the thiamine biosynthetic pathway in higher plant plastids and its developmental regulation. *Plant Molecular Biology* **29**, 809–821.

Bernacchia, G., Schwall, G., Lottspeich, F., Salamini, F. and Bartels, D. (1995). The transketolase gene family of the resurrection plant *Craterostigma plantagineum*: Differential expression during the rehydration phase. *The EMBO Journal* **14**, 610–618.

Bernasconi, P., Woodworth, A. R., Rosen, B. A., Subramanian, M. V. and Siehl, D. L. (1995). A naturally occurring point mutation confers broad range tolerance toherbicides that target acetolactate synthase. *The Journal of Biological Chemistry* **270**, 17381–17385.

Bettendorff, L. (1995). Thiamine homeostasis in neuroblastoma cells. *Neurochemistry International* **26**, 295–302.

Bettendorff, L. and Wins, P. (2009). Thiamin diphosphate in biological chemistry: New aspects of thiamin metabolism, especially triphosphate derivatives acting other than as cofactors. *The FEBS Journal* **276**, 2917–2925.

Bettendorff, L., Kolb, H. A. and Schoffeniels, E. (1993). Thiamine triphosphate activates an anion channel of large unit conductance in neuroblastoma cells. *The Journal of Membrane Biology* **136**, 281–288.

Bettendorff, L., Wirtzfeld, B., Makarchikov, A. F., Mazzucchelli, G., Frédérich, M., Gigliobianco, T., Gangolf, M., De Pauw, E., Angenot, L. and Wins, P. (2007). Discovery of a natural thiamine adenine nucleotide. *Nature Chemical Biology* **3**, 211–212.

Bocobza, S. and Aharoni, A. (2008). Switching the light on plant riboswitches. *Trends in Plant Science* **13**, 526–533.

Bocobza, S., Adato, A., Mandel, T., Shapira, M., Nudler, E. and Aharoni, A. (2007). Riboswitch-dependent gene regulation and its evolution in the plant kingdom. *Genes & Development* **21**, 2874–2879.

Bos, M. and Kozik, A. (2000). Some molecular and enzymatic properties of a homogeneous preparation of thiaminase I purified from carp liver. *Journal of Protein Chemistry* **19**, 75–84.

Breaker, R. R. (2010). Riboswitches and the RNA World. *Cold Spring Harbor Perspectives in Biology* 10.1101/cshperspect.a003566.

Bunik, V. I. and Fernie, A. R. (2009). Metabolic control exerted by the 2-oxoglutarate dehydrogenase reaction: A cross-kingdom comparison of the crossroad between energy production and nitrogen assimilation. *The Biochemical Journal* **422**, 405–421.

Camp, P. J. and Randall, D. D. (1985). Purification and characterization of the pea chloroplast pyruvate dehydrogenase complex: A source of acetyl-CoA and NADH for fatty acid biosynthesis. *Plant Physiology* **77**, 571–577.

Chabregas, S. M., Luche, D. D., Farias, L. P., Ribeiro, A. F., van Sluys, M. A., Menck, C. F. and Silva-Filho, M. C. (2001). Dual targeting properties of the N-terminal signal sequence of *Arabidopsis thaliana* THI1 protein to mitochondria and chloroplasts. *Plant Molecular Biology* **46**, 639–650.

Chabregas, S. M., Luche, D. D., Van Sluys, M. A., Menck, C. F. and Silva-Filho, M. C. (2003). Differential usage of two in-frame translational start codons regulates subcellular localization of *Arabidopsis thaliana* THI1. *Journal of Cell Science* **116**, 285–291.

Chang, A. K. and Duggleby, R. G. (1997). Expression, purification and characterization of *Arabidopsis thaliana* acetohydroxyacid synthase. *The Biochemical Journal* **327**, 161–169.

Chang, A. K. and Duggleby, R. G. (1998). Herbicide resistant forms of *Arabidopsis thaliana* acetohydroxyacid synthase: Characterization of the catalytic properties and sensitivity to inhibitors of four defined mutants. *The Biochemical Journal* **333**, 765–777.

Chang, S. I., Kang, M. K., Choi, J. D. and Namgoong, S. K. (1997). Soluble overexpression in *Escherichia coli*, and purification and characterization of wild-type recombinant tobacco acetolactate synthase. *Biochemical and Biophysical Research Communications* **234**, 549–553.

Chatterjee, A., Jurgenson, C. T., Schroeder, F. C., Ealick, S. E. and Begley, T. P. (2006). Thiamin biosynthesis in eukaryotes: Characterization of the enzyme-bound product of thiazole synthase from *Saccharomyces cerevisiae* and its implications in thiazole biosynthesis. *Journal of the American Chemical Society* **128**, 7158–7159.

Chatterjee, A., Jurgenson, C. T., Schroeder, F. C., Ealick, S. E. and Begley, T. P. (2007). Biosynthesis of thiamin thiazole in eukaryotes: Conversion of NAD to an advanced intermediate. *Journal of the American Chemical Society* **129**, 2914–2922.

Chatterjee, A., Li, Y., Zhang, Y., Grove, T. L., Lee, M., Krebs, C., Booker, S. J., Begley, T. P. and Ealick, S. E. (2008a). Reconstitution of ThiC in thiamine pyrimidine biosynthesis. *Nature Chemical Biology* **4**, 758–765.

Chatterjee, A., Schroeder, F. C., Jurgenson, C. T., Ealick, S. E. and Begley, T. P. (2008b). Biosynthesis of the thiamin-thiazole in eukaryotes: Identification of a thiazole tautomer intermediate. *Journal of the American Chemical Society* **130**, 11394–11398.

Cheah, M. T., Wachter, A., Sudarsan, N. and Breaker, R. R. (2007). Control of alternative RNA splicing and gene expression by eukaryotic riboswitches. *Nature* **447**, 497–500.

Cheng, G., Bennett, E. M., Begley, T. P. and Ealick, S. E. (2002). Crystal structure of 4-amino-5-hydroxymethyl-2-methylpyrimidine phosphate kinase from Salmonella typhimurium at 2.3 A resolution. *Structure* **10**, 225–235.

Chiu, H. J., Reddick, J. J., Begley, T. P. and Ealick, S. E. (1999). Crystal structure of thiamin phosphate synthase from *Bacillus subtilis* at 1.25 A resolution. *Biochemistry* **38**, 6460–6470.

Conley, T. R., Peng, H. P. and Shih, M. C. (1999). Mutations affecting induction of glycolytic and fermentative genes during germination and environmental stress in Arabidopsis. *Plant Physiology* **119**, 599–607.

Cordoba, E., Salmi, M. and León, P. (2009). Unravelling the regulatory mechanisms that modulate the MEP pathway in higher plants. *Journal of Experimental Botany* **60**, 2933–2943.

Couée, I., Sulmon, C., Gouesbet, G. and El Amrani, A. (2006). Involvement of soluble sugars in reactive oxygen species balance and responses to oxidative stress in plants. *Journal of Experimental Botany* **57**, 449–459.

Debnam, P. M. and Emes, M. J. (1999). Subcellular distribution of enzymes of the oxidative pentose phosphate pathway in root and leaf tissues. *Journal of Experimental Botany* **50**, 1653–1661.

Debnam, P. M., Fernie, A. R., Leisse, A., Golding, A., Bowsher, C. G., Grimshaw, C., Knight, J. S. and Emes, M. J. (2004). Altered activity of the P2 isoform of plastidic glucose 6-phosphate dehydrogenase in tobacco (*Nicotiana tabacum* cv. Samsun) causes changes in carbohydrate metabolism and response to oxidative stress in leaves. *The Plant Journal* **38**, 49–59.

Dorrestein, P. C., Zhai, H., McLafferty, F. W. and Begley, T. P. (2004). The biosynthesis of the thiazole phosphate moiety of thiamin: The sulfur transfer mediated by the sulfur carrier protein ThiS. *Chemistry & Biology* **11**, 1373–1381.

Douce, R. and Neuburger, M. (1989). The uniqueness of plant mitochondria. *Annual Review of Plant Physiology and Plant Molecular Biology* **40**, 371–414.

Drew, M. C. (1997). Oxygen deficiency and root metabolism: Injury and acclimation under hypoxia and anoxia. *Annual Review of Plant Physiology and Plant Molecular Biology* **48**, 223–250.

Duggleby, R. G. (2006). Domain relationships in thiamine diphosphate-dependent enzymes. *Accounts of Chemical Research* **39**, 550–557.

Duggleby, R. G., McCourt, J. A. and Guddat, L. W. (2008). Structure and mechanism of inhibition of plant acetohydroxyacid synthase. *Plant Physiology and Biochemistry* **46**, 309–324.

Dumas, R., Biou, V. and Douce, R. (1997). Purification and characterization of a fusion protein of plant acetohydroxy acid synthase and acetohydroxy acid isomeroreductase. *FEBS Letter* **408**, 156–160.

Durner, J. and Boger, P. (1988). Acetolactate synthase from barley (*Hordeum vulgare* L.): Purification and partial characterization. *Zeitschrift für Naturforschung* **43C**, 850–856.

Estevez, J. M., Cantero, A., Reindl, A., Reichler, S. and Leon, P. (2001). 1-Deoxy-D-xylulose-5-phosphate synthase, a limiting enzyme for plastidic isoprenoid biosynthesis in plants. *The Journal of Biological Chemistry* **276**, 22901–22909.

Faith, C. B., Thomas, L., Boyang, C. and Alan, L. K. (1995). Evidence for the thiamine biosynthetic pathway in higher-plant plastids and its developmental regulation. *Plant Molecular Biology* **29**, 809–821.

Ferreira, S., Hjerno, K., Larsen, M., Wingsle, G., Larsen, P., Fey, S., Roepstorff, P. and Salome Pais, M. (2006). Proteome profiling of *Populus euphratica Oliv.* upon heat stress. *Annals of Botany* **98**, 361–377.

Flechner, A., Dressen, U., Westhoff, P., Henze, K., Schnarrenberger, C. and Martin, W. (1996). Molecular characterization of transketolase (EC 2.2.1.1) active in the Calvin cycle of spinach chloroplasts. *Plant Molecular Biology* **32**, 475–484.

Fleming, J. C., Tartaglini, E., Steinkami, M. P., Schorderet, D. F., Cohen, N. and Neufeldt, E. J. (1999). The gene mutated in thiamine-responsive anaemia with diabetes and deafness (TRMA) encodes a functional thiamine transporter. *Nature Genetics* **22**, 305–308.

Forlani, G., Mantelli, M. and Nielsen, E. (1999). Biochemical evidence for multiple acetoin-forming enzymes in cultured plant cells. *Phytochemistry* **50**, 255–262.

Frank, R. A., Leeper, F. J. and Luisi, B. F. (2007). Structure, mechanism and catalytic duality of thiamine-dependent enzymes. *Cellular and Molecular Life Science* **64**, 892–905.

Frédérich, M., Delvaux, D., Gigliobianco, T., Gangolf, M., Dive, G., Mazzucchelli, G., Elias, B., De Pauw, E., Angenot, L., Wins, P. and Bettendorff, L. (2009). Thiaminylated adenine nucleotides. Chemical synthesis, structural characterization and natural occurrence. *The FEBS Journal* **276**, 3256–3268.

Fujiki, Y., Sato, T., Ito, M. and Watanabe, A. (2000). Isolation and characterization of cDNA clones for the e1beta and E2 subunits of the branched-chain alpha-ketoacid dehydrogenase complex in Arabidopsis. *The Journal of Biological Chemistry* **275**, 6007–6013.

Fujiwara, M. (1976). Allithiamine and itsproperties. *Journal of Nutritional Science and Vitaminology (Tokyo)* **22**, 57–62.

Funk, C. (1912). The etiology of the deficiency diseases. Beri-beri, polyneuritis in birds, epidemic dropsy, scurvy, experimental scurvy in animals, infantile scurvy, ship beri-beri, pellagra. *Journal of State Medicine* **20**, 341–368.

Gangolf, M., Czerniecki, J., Radermecker, M., Detry, O., Nisolle, M., Jouan, C., Martin, D., Chantraine, F., Lakaye, B., Wins, P., Grisar, T. and Bettendorff, L. (2010a). Thiamine status in humans and content of phosphorylated thiamine derivatives in biopsies and cultured cells. *PloS ONE* **5**, 13616–13629.

Gangolf, M., Wins, P., Thiry, M., El Moualij, B. and Bettendorff, L. (2010b). Thiamine triphosphate synthesis in rat brain occurs in mitochondria and is coupled to the respiratory chain. *The Journal of Biological Chemistry* **285**, 583–594.

Gass, N., Glagotskaia, T., Mellema, S., Stuurman, J., Barone, M., Mandel, T., Roessner-Tunali, U. and Kuhlemeier, C. (2005). Pyruvate decarboxylase provides growing pollen tubes with a competitive advantage in petunia. *The Plant Cell* **17**, 2355–2368.

Geddes, W. F. and Levine, M. N. (1942). The distribution of thiamin in the wheat plant at successive stage of kernel development. *Cereal Chemistry* **19**, 547–552.

Gibson, G. E. and Blass, J. P. (2007). Thiamine-dependent processes and treatment strategies in neurodegeneration. *Antioxidants & Redox Signaling* **9**, 1605–1619.

Gibson, G. E. and Zhang, H. (2002). Interactions of oxidative stress with thiamine homeostasis promote neurodegeneration. *Neurochemistry International* **40**, 493–504.

Gigliobianco, T., Lakaye, B., Wins, P., El Moualij, B., Zorzi, W. and Bettendorff, L. (2010). Adenosine thiamine triphosphate accumulates in *Escherichia coli* cells in response to specific conditions of metabolic stress. *BMC Microbiology* **10**, 148–152.

Godoi, P. H., Galhardo, R. S., Luche, D. D., Van Sluys, M. A., Menck, C. F. and Oliva, G. (2006). Structure of the thiazole biosynthetic enzyme THI1 from *Arabidopsis thaliana*. *The Journal of Biological Chemistry* **281**, 30957–30966.

Golda, A., Szyniarowski, P., Ostrowska, K., Kozik, A. and Rapala-Kozik, M. (2004). Thiamine binding and metabolism in germinating seeds of selected cereals and legumes. *Plant Physiology and Biochemistry* **42**, 187–195.

Goyer, A. (2010). Thiamine in plants: Aspects of its metabolism and functions. *Phytochemistry* **71**, 1615–1624.

Hakim, A. M. (1984). The induction and reversibility of cerebral acidosis in thiamine deficiency. *Annals of Neurology* **16**, 673–679.

Hanes, J. W., Ealick, S. E. and Begley, T. P. (2007). Thiamin phosphate synthase: The rate of pyrimidine carbocation formation. *Journal of the American Chemical Society* **129**, 4860–4861.

Hazell, A. S. (2009). Astrocytes are a major target in thiamine deficiency and Wernicke's encephalopathy. *Neurochemistry International* **55**, 129–135.

Hazell, A. S. and Butterworth, R. F. (2009). Update of cell damage mechanisms in thiamine deficiency: Focus on oxidative stress, excitotoxicity and inflammation. *Alcohol and Alcoholism* **44**, 141–147.

Hazell, A. S., Todd, K. G. and Butterworth, R. F. (1998). Mechanisms of neuronal cell death in Wernicke's encephalopathy. *Metabolic Brain Disease* **13**, 97–122.

Henkes, S., Sonnewald, U., Badur, R., Flachmann, R. and Stitt, M. (2001). A small decrease of plastid transketolase activity in antisense tobacco transformants has dramatic effects on photosynthesis and phenylpropanoid metabolism. *The Plant Cell* **13**, 535–551.

Héroux, M. and Butterworth, R. F. (1995). Regional alterations of thiamine phosphate esters and of thiamine diphosphate-dependent enzymes in relation to function in experimental Wernicke's encephalopathy. *Neurochemical Research* **20**, 87–93.

Hilker, D. M. and Somogyi, J. C. (1982). Antithiamins of plant origin: Their chemical nature and mode of action. *Annals of the New York Academy of Sciences* **378**, 137–145.

Hill, R. E., Himmeldirk, K., Kennedy, I. A., Pauloski, R. M., Sayer, B. G., Wolf, E. and Spenser, I. D. (1996). The biogenetic anatomy of vitamin B6. A 13C NMR investigation of the biosynthesis of pyridoxol in *Escherichia coli*. *The Journal of Biological Chemistry* **271**, 30426–30435.

Hix, L. M., Lockwood, S. F. and Bertram, J. S. (2004). Bioactive carotenoids: Potent antioxidants and regulators of gene expression. *Redox Report* **9**, 181–191.

Hohmann, S. and Meacock, P. A. (1998). Thiamin metabolism and thiamin diphosphate-dependent enzymes in the yeast *Saccharomyces cerevisiae*: Genetic regulation. *Biochimica et Biophysica Acta* **1385**, 201–219.

Hong, Z. Q. and Copeland, L. (1990). Pentose phosphate pathway enzymes in nitrogen-fxing leguminous root nodules. *Phytochemistry* **29**, 2437–2440.

Hossain, M. A., Hug, E. and Hodges, T. K. (1994). Sequence of a cDNA from *Oryza sativa* (L.) encoding the pyruvate decarboxylase 1 gene. *Plant Physiology* **106**, 799–800.

Huang, H. M., Chen, H. L. and Gibson, G. E. (2010). Thiamine and oxidants interact to modify cellular calcium stores. *Neurochemical Research* **35**, 2107–2116.

Iacopetta, D., Carrisi, C., De Filippis, G., Calcagnile, V. M., Cappello, A. R., Chimento, A., Curcio, R., Santoro, A., Vozza, A., Dolce, V., Palmieri, F. and Capobianco, L. (2010). The biochemical properties of the mitochondrial thiamine pyrophosphate carrier from *Drosophila melanogaster*. *The FEBS Journal* **277**, 1172–1181.

Ismail, A. M., Ella, E. S., Vergara, G. V. and Mackill, D. J. (2009). Mechanisms associated with tolerance for flooding during germination and early seedling growth in rice *(Oryza sativa)*. *Annals of Botany* **103**, 197–209.

Ismond, K. P., Dolferus, R., de Pauw, M., Dennis, E. S. and Good, A. G. (2003). Enhanced low oxygen survival in Arabidopsis through increased metabolic flux in the fermentative pathway. *Plant Physiology* **132**, 1292–1302.

Jacob-Wilk, D., Goldschmidt, E. E., Riov, J., Sadka, A. and Holland, D. (1997). Induction of a Citrus gene highly homologous to plant and yeast *thi* genes involved in thiamine biosynthesis during natural and ethylene-induced fruit maturation. *Plant Molecular Biology* **35**, 661–666.

Jenkins, A. H., Schyns, G., Potot, S., Sun, G. and Begley, T. P. (2007). A new thiamin salvage pathway. *Nature Chemical Biology* **3**, 492–497.

Jordan, F. (2007). Adenosine triphosphate and thiamine cross paths. *Nature Chemical Biology* **3**, 202–203.

Jordan, F., Nemeria, N. S., Zhang, S., Yan, Y., Arjunan, P. and Furey, W. (2003). Dual catalytic apparatus of the thiamin diphosphate coenzyme: Acid-base via the 1′,4′-iminopyrimidine tautomer along with its electrophilic role. *Journal of the American Chemical Society* **125**, 12732–12738.

Journet, E. P. and Douce, R. (1985). Enzymic capacities of purified cauliflower bud plastids for lipid synthesis and carbohydrate metabolism. *Plant Physiology* **79**, 458–467.

Julliard, J. H. and Douce, R. (1991). Biosynthesis of the thiazole moiety of thiamin (vitamin B1) in higher plant chloroplasts. *Proceedings of the National Academy of Sciences of the United States of America* **88**, 2042–2045.

Jung, S. M., Le, D. T., Yoon, S. S., Yoon, M. Y., Kim, Y. T. and Choi, J. D. (2004). Amino acid residues conferring herbicide resistance in tobacco acetohydroxy acid synthase. *The Biochemical Journal* **383**, 53–61.

Jurgenson, C. T., Begley, T. P. and Ealick, S. E. (2009). The structural and biochemical foundations of thiamin biosynthesis. *Annual Review of Biochemistry* **78**, 569–603.

Kato-Noguchi, H. and Yasuda, Y. (2007). Effect of low temperature on ethanolic fermentation in rice seedlings. *Journal of Plant Physiology* **164**, 1013–1018.

Kawai, K., Kaku, K., Izawa, N., Shimizu, T., Fukuda, A. and Tanaka, Y. (2007). A novel mutant acetolactate synthase gene from rice cells, which confers resistance to ALS-inhibiting herbicides. *Journal of Pesticide Science* **32**, 89–98.

Kawasaki, Y. (1993). Copurification of hydroxyethylthiazole kinase and thiamine-phosphate pyrophosphorylase of *Saccharomyces cerevisiae*: Characterization of hydroxyethylthiazole kinase as a bifunctional enzyme in the thiamine biosynthetic pathway. *Journal of Bacteriology* **175**, 5153–5158.

Kawasaki, Y., Onozuka, M., Mizote, T. and Nosaka, K. (2005). Biosynthesis of hydroxymethylpyrimidine pyrophosphate in *Saccharomyces cerevisiae*. *Current Genetics* **47**, 156–162.

Ke, J., Behal, R. H., Back, S. L., Nikolau, B. J., Wurtele, E. S. and Oliver, D. J. (2000). The role of pyruvate dehydrogenase and acetyl-coenzyme A synthetase in fatty acid synthesis in developing *Arabidopsis seeds*. *Plant Physiology* **123**, 497–508.

Kim, Y. S., Nosaka, K., Downs, D. M., Kwak, J. M., Park, D., Chung, I. K. and Nam, H. G. (1998). A Brassica cDNA clone encoding a bifunctional hydroxymethylpyrimidine kinase/thiamin-phosphate pyrophosphorylase involved in thiamin biosynthesis. *Plant Molecular Biology* **37**, 955–966.

Kim, B. R., Kim, S. U. and Chang, Y. J. (2005). Differential expression of three 1-deoxy-D-xylulose-5-phosphate synthase genes in rice. *Biotechnology Letters* **27**, 997–1001.

Kochibe, N., Yusa, T. and Hyashi, K. (1963). Occurrence of thiamine triphosphate in higher plants. *Plant & Cell Physiology* **4**, 239–244.

Kolkman, J. M., Slabaugh, M. B., Bruniard, J. M., Berry, S., Bushman, B. S., Olungu, C., Maes, N., Abratti, G., Zambelli, A., Miller, J. F., Leon, A. and Knapp, S. J. (2004). Acetohydroxyacid synthase mutations conferring resistance to imidazolinone or sulfonylurea herbicides in sunflower. *Theoretical and Applied Genetics* **109**, 1147–1159.

Kondo, K., Mitsuda, H. and Iwai, K. (1951). Thiamine synthesis in leaves of cereal crops. *Journal of Agricultural Chemical Society of Japan* **24**, 128–132.

Kong, D., Zhu, Y., Wu, H., Cheng, X., Liang, H. and Ling, H. Q. (2008). *AtTHIC*, a gene involved in thiamine biosynthesis in *Arabidopsis thaliana*. *Cell Research* **18**, 566–576.

Koornneef, M. and Hanhart, C. (1981). A new thiamine locus in *Arabidopsis*. *Arabidopsis Information Service* **18**, 52–58.

Kowalska, E. and Kozik, A. (2008). The genes and enzymes involved in the biosynthesis of thiamin and thiamin diphosphate in yeasts. *Cellular & Molecular Biology Letters* **13**, 271–282.

Kozik, A. and Rapala-Kozik, M. (1995). Protein-attributable thiamine-binding activity in gymnosperm seeds. *Journal of Plant Physiology* **146**, 760–762.

Kozik, A. and Rapala-Kozik, M. (1996). Comparative study on binding of thiamine-related compounds by extracts of seeds of *Zea mays*, *Spinacia oleracea*, *Picea abies* and *Cycas revolute*. *Plant Physiology and Biochemistry* **34**, 779–786.

Kriek, M., Martins, F., Leonardi, R., Fairhurst, S. A., Lowe, D. J. and Roach, P. L. (2007). Thiazole synthase from *Escherichia coli*: An investigation of the substrates and purified proteins required for activity in vitro. *The Journal of Biological Chemistry* **282**, 17413–17423.

Krook, J., Vreugdenhil, D., Dijkema, C. and van der Plas, L. H. W. (1998). Sucrose and starch metabolism in carrot (*Daucuscarota L.*) cell suspensions analysed by C-labelling: Indications for a cytosol and a plastid-localized oxidative pentose phosphate pathway. *Journal of Experimental Botany* **49**, 1917–1924.

Kruger, N. J. and von Schaewen, A. (2003). The oxidative pentose phosphate pathway: Structure and organisation. *Current Opinion in Plant Biology* **6**, 236–246.

Kruszewski, S. P. and Jakobs, W. P. (1974). Polarity of thiamine movement through tomato petioles. *Plant Physiology* **54**, 310–311.

Kürsteiner, O., Dupuis, I. and Kuhlemeier, C. (2003). The pyruvate decaroboxylase1 gene of Arabidopsis is required during anoxia but not other environmental stresses. *Plant Physiology* **132**, 968–978.

Kylen, A. and McCready, R. M. (1975). Nutrients in seeds and sprouts of alfalfa, lentils, mung beans, and soybeans. *Journal of Food Science* **40**, 1008–1009.

Lakaye, B., Wirtzfeld, B., Wins, P., Grisar, T. and Bettendorff, L. (2004). Thiamine triphosphate, a new signal required for optimal growth of *Escherichia coli* during amino acid starvation. *The Journal of Biological Chemistry* **279**, 17142–17147.

Lange, B. M., Wildung, M. R., McCaskill, D. and Croteau, R. (1998). A family of transketolases that directs isoprenoid biosynthesis via a mevalonate-independent pathway. *Proceedings of the National Academy of Sciences of the United States of America* **95**, 2100–2104.

Lee, Y. T. and Duggleby, R. G. (2001). Identification of the regulatory subunit of *Arabidopsis thaliana* acetohydroxyacid synthase and reconstitution with its catalytic subunit. *Biochemistry* **40**, 6836–6844.

Lehmann, M., Schwarzländer, M., Obata, T., Sirikantaramas, S., Burow, M., Olsen, C. E., Tohge, T., Fricker, M. D., Møller, B. L., Fernie, A. R., Sweetlove, L. J. and Laxa, M. (2009). The metabolic response of Arabidopsis roots to oxidative stress is distinct from that of heterotrophic cells in culture and highlights a complex relationship between the levels of transcripts, metabolites, and flux. *Molecular Plant* **2**, 390–406.

Lemaire, S. D., Guillon, B., Le Marechal, P., Keryer, E., Miginiac-Maslow, M. and Decottignies, P. (2004). New thioredoxin targets in the unicellular photosynthetic eukaryote Chlamydomonas reinhardtii. *Proceedings of the National Academy of Sciences of the United States of America* **101**, 7475–7480.

Lichtenthaler, H. K. (1999). The 1-deoxy-D-phosphate pathway of isoprenoid biosynthesis in plants. *Annual Review of Plant Physiology and Plant Molecular Biology* **50**, 47–65.

Lichtenthaler, H. K., Schwender, J., Disch, A. and Rohmer, M. (1997). Biosynthesis of isoprenoids in higher plant chloroplasts proceeds via a mevalonate-independent pathway. *FEBS Letters* **400**, 271–274.

Lindhurst, M. J., Fiermonte, G., Song, S., Struys, E., De Leonardis, F., Schwartzberg, P. L., Chen, A., Castegna, A., Verhoeven, N., Mathews, C. K., Palmieri, F. and Biesecker, L. G. (2006). Knockout of Slc25a19 causes mitochondrial thiamine pyrophosphate depletion, embryonic lethality, CNS malformations, and anemia. *Proceedings of the National Academy of Sciences of the United States of America* **103**, 15927–15932.

Lindqvist, Y., Schneider, G., Ermler, U. and Sundström, M. (1992). Three-dimensional structure of transketolase, a thiamine diphosphate dependent enzyme, at 2.5 A resolution. *The EMBO Journal* **11**, 2373–2379.

Lois, L. M., Rodriguez-Concepcion, M., Gallego, F., Campos, N. and Boronat, A. (2000). Carotenoid biosynthesis during tomato fruit development: Regulatory role of 1-deoxy-D-xylulose5-phosphate synthase. *The Plant Journal* **22**, 503–513.

Lonsdale, D. (2006). A review of the biochemistry, metabolism and clinical benefits of thiamin(e) and its derivatives. *Evidence Based Complementary and Alternative Medicine* **3**, 49–59.

Luethy, M. H., Gemel, J., Johnston, M. L., Mooney, B. P., Miernyk, J. A. and Randall, D. D. (2001). Developmental expression of the mitochondrial pyruvate dehydrogenase complex in pea (*Pisum sativum*) seedlings. *Physiologia Plantarum* **112**, 559–566.

Lukienko, P. I., Mel'nichenko, N. G., Zverinskii, I. V. and Zabrodskaya, S. V. (2000). Antioxidant properties of thiamine. *Bulletin of Experimental Biology and Medicine* **130**, 874–876.

Machado, C. R., de Oliveira, R. L., Boiteux, S., Praekelt, U. M., Meacock, P. A. and Menck, C. F. (1996). Thi1, a thiamine biosynthetic gene in *Arabidopsis thaliana*, complements bacterial defects in DNA repair. *Plant Molecular Biology* **31**, 585–593.

Machado, C. R., Praekelt, U. M., de Oliveira, R. C., Barbosa, A. C., Byrne, K. L., Meacock, P. A. and Menck, C. F. (1997). Dual role for the yeast THI4 gene in thiamine biosynthesis and DNA damage tolerance. *Journal of Molecular Biology* **273**, 114–121.

Makarchikov, A. F., Lakaye, B., Gulyai, I. E., Czerniecki, J., Coumans, B., Wins, P., Grisar, T. and Bettendorff, L. (2003). Thiamine triphosphate and thiamine triphosphatase activities: From bacteria to mammals. *Cellular and Molecular Life Science* **60**, 1477–1488.

Malamy, J., SanchezCasas, P., Hennig, J., Guo, A. L. and Klessig, D. F. (1996). Dissection of the salicylic acid signaling pathway in tobacco. *Molecular Plant-Microbe Interaction* **9**, 474–482.

Marobbio, C. M., Vozza, A., Harding, M., Bisaccia, F., Palmieri, F. and Walker, J. E. (2002). Identification and reconstitution of the yeast mitochondrial transporter for thiamine pyrophosphate. *The EMBO Journal* **21**, 5653–5661.

Martin, P. R., Singleton, C. K. and Hiller-Sturmhöfel, S. (2003). The role of thiamine deficiency in alcoholic brain disease. *Alcohol Research and Health* **27**, 134–142.

Martinez-Gomez, N. C., Poyner, R. R., Mansoorabadi, S. O., Reed, G. H. and Downs, D. M. (2009). Reaction of AdoMet with ThiC generates a backbone free radical. *Biochemistry* **48**, 217–219.

McCourt, J. A. and Duggleby, R. G. (2006). Acetohydroxyacid synthase and its role in the biosynthetic pathway for branched-chain amino acids. *Amino Acids* **31**, 173–210.

McCourt, J. A., Nixon, P. F. and Duggleby, R. G. (2006). Thiamin nutrition and catalysis-induced instability of thiamin diphosphate. *The British Journal of Nutrition* **96**, 636–638.

McCulloch, K. M., Kinsland, C., Begley, T. P. and Ealick, S. E. (2008). Structural studies of thiamin monophosphate kinase in complex with substrates and products. *Biochemistry* **47**, 3810–3821.

Miernyk, J. A., Camp, P. J. and Randall, D. D. (1985). Regulation of plant pyruvate dehydrogenase complexes. *Current Topics in Plant Biochemistry and Physiology* **4**, 175–190.

Miernyk, J. A., Thelen, J. J. and Randall, D. D. (1998). Reversible phosphorylation as a mechanism for regulating activity of the mitochondrial PDC. *In* Plant Mitochondria: from Gene to Function (P. Gardestrom, I. M. Muller, K.

Glimelius and E. Glaser, eds.), pp. 321–327. Backhuys Publishers, Leiden, The Netherlands.

Millar, A. H. and Heazlewood, J. L. (2003). Genomic and proteomic analysis of mitochondrial carrier proteins in Arabidopsis. *Plant Physiology* **131**, 443–453.

Millar, A. H., Hill, S. A. and Leaver, C. J. (1999). Plant mitochondrial 2-oxoglutarate dehydrogenase complex: Purification and characterization in potato. *The Biochemical Journal* **43**, 327–334.

Miranda-Ríos, J. (2007). The THI-box riboswitch, or how RNA binds thiamin pyrophosphate. *Structure* **15**, 259–265.

Mironov, A. S., Gusarov, I., Rafikov, R., Lopez, L. E., Shatalin, K., Kreneva, R. A., Perumov, D. A. and Nudler, E. (2002). Sensing small molecules by nascent RNA: A mechanism to control transcription in bacteria. *Cell* **111**, 747–756.

Mitsuda, H., Takii, Y., Iwami, K., Yasumoto, K. and Nakajima, K. (1979). Enzymatic formation of thiamine pyrophosphate in plants. *Methods in Enzymology* **62**, 107–111.

Mitsunaga, T., Matsuda, M., Shimizu, M. and Iwashima, A. (1986a). Isolation and properties of a thiamine-binding protein from buckwheat seed. *Cereal Chemistry* **63**, 332–335.

Mitsunaga, T., Shimizu, M. and Iwashima, A. (1986b). Occurrence of thiamine-binding proteins in plant seeds. *Journal of Plant Physiology* **124**, 177–180.

Mitsunaga, T., Shimizu, M. and Iwashima, A. (1987). A possible role for thiamin-binding protein in the germination of rice seed. *Journal of Plant Physiology* **130**, 279–284.

Mizote, T. and Nakayama, H. (1989). The *thiM* locus and its relation to phosphorylation of hydroxyethylthiazole in *Escherichia coli*. *Journal of Bacteriology* **171**, 3228–3232.

Mizote, T., Truda, M., Smith, D. D., Nakayama, H. and Nakazawa, T. (1999). Cloning and characterization of the *thiD/J* gene of *Escherichia coli* encoding a thiamin-synthesizing bifunctional enzyme, hydroxymethylpyrimidine kinase/phosphomethylpyrimidine kinase. *Microbiology* **145**, 495–501.

Molin, W. T. and Fites, R. C. (1980). Isolation and characterization of thiamin hosphotransferase from *Glycine max* seedlings. *Plant Physiology* **66**, 308–312.

Molin, W. T., Wilkerson, C. G. and Fites, R. C. (1980). Thiamin phosphorylation by thiamin pyrophosphotransferase during seed germination. *Plant Physiology* **66**, 313–315.

Money, B. P., Miernyk, J. A. and Randall, D. D. (2002). The complex fate of alpha-ketoacids. *Annual Review of Plant Biology* **53**, 357–375.

Mozafar, A. and Oertli, J. J. (1992). Uptake and transport of thiamin (vitamin-B1) by barley and soybean. *Journal of Plant Physiology* **139**, 436–442.

Mozafar, A. and Oertli, J. J. (1993). Thiamin (vitamin-B(1))—Translocation and metabolism by soybean seedling. *Journal of Plant Physiology* **142**, 438–445.

Mücke, U., König, S. and Hübner, G. (1995). Purification and characterisation of pyruvate decarboxylase from pea seeds (*Pisum sativum* cv. Miko). *Biological Chemistry Hoppe-Seyler* **376**, 111–117.

Muhitch, M. J., Shaner, D. L. and Stidham, M. A. (1987). Imidazolinones and acetohydroxyacid synthase from higher plants: Properties of the enzyme from maize suspension culture cells and evidence for the binding of imazapyr to acetohydroxyacid synthase *in vivo*. *Plant Physiology* **83**, 451–456.

Müller, C., Schwender, J., Zeidler, J. and Lichtenthaler, H. K. (2000). Properties and inhibition of the first two enzymes of the non-mevalonate pathway of isoprenoid biosynthesis. *Biochemical Society Transactions* **28**, 792–793.

Navarro, D., Zwingmann, C., Hazell, A. S. and Butterworth, R. F. (2005). Brain lactate synthesis in thiamine deficiency: A re-evaluation using 1H-13C nuclear magnetic resonance spectroscopy. *Journal of Neuroscience Research* **79**, 33–41.

Nemeria, N., Korotchkina, L., McLeish, M. J., Kenyon, G. L., Patel, M. S. and Jordan, F. (2007). Elucidation of the chemistry of enzyme-bound thiamin diphosphate prior to substrate binding: Defining internal equilibria among tautomeric and ionization states. *Biochemistry* **46**, 10739–10744.

Nghiêm, H. O., Bettendorff, L. and Changeux, J. P. (2000). Specific phosphorylation of Torpedo 43K rapsyn by endogenous kinase(s) with thiamine triphosphate as the phosphate donor. *The FASEB Journal* **14**, 543–554.

Nishimura, M. and Beevers, H. (1979). Subcellular distribution of gluconeogenetic enzymes in germinating castor bean endosperm. *Plant Physiology* **64**, 31–37.

Nishimura, H., Uehara, Y., Sempuku, K. and Iwashima, A. (1984). Purification and properties of thiamin-binding protein from rice bran. *Journal of Nutritional Science and Vitaminology* **30**, 1–10.

Nishino, Y., Itokawa, N., Nishino, K., Piros, J. R. and Cooper, J. (1983). Enzyme system involved in the synthesis of thiamin triphosphate. I. Purification and characterization of protein-bound thiamin diphosphate: ATP phosphoryltransferase. *Journal of Biological Chemistry* **258**, 11871–11878.

Nosaka, K. (2006). Recent progress in understanding thiamin biosynthesis and its genetic regulation in *Saccharomyces cerevisiae*. *Applied Microbiology and Biotechnology* **72**, 30–40.

Nosaka, K., Kaneko, Y., Nishimura, H. and Iwashima, A. (1993). Isolation and characterization of a thiamin pyrophosphokinase gene, *THI80*, from *Saccharomyces cerevisiae*. *The Journal of Biological Chemistry* **268**, 17440–17447.

Nosaka, K., Nishimura, H., Kawasaki, Y., Tsujihara, T. and Iwashima, A. (1994). Isolation and characterization of the *THI6* gene encoding a bifunctional thiamin-phosphate pyrophosphorylase/hydroxyethylthiazole kinase from *Saccharomyces cerevisiae*. *The Journal of Biological Chemistry* **269**, 30510–30516.

Okuzaki, A., Shimizu, T., Kaku, K., Kawai, K. and Toriyama, K. (2007). A novel mutated acetolactate synthase gene conferring specific resistance to pyrimidinyl carboxy herbicides in rice. *Plant Molecular Biology* **64**, 219–224.

Ontiveros-Palacios, N., Smith, A. M., Grundy, F. J., Soberon, M., Henkin, T. M. and Miranda-Ríos, J. (2008). Molecular basis of gene regulation by the THI-box riboswitch. *Molecular Microbiology* **67**, 793–803.

Ott, K. H., Kwagh, J. G., Stockton, G. W., Sidorov, V. and Kakefuda, G. (1996). Rational molecular design and genetic engineering of herbicide resistant crops by structure modeling and site-directed mutagenesis of acetohydroxyacid synthase. *Journal of Molecular Biology* **263**, 359–368.

Papini-Terzi, F. S., Galhardo, R. S., Farias, L. P., Menck, C. F. and Van Sluys, M. A. (2003). Point mutation is responsible for *Arabidopsis tz-201* mutant phenotype affecting thiamin biosynthesis. *Plant & Cell Physiology* **44**, 856–860.

Paterami, I. and Kanellis, A. K. (2010). Stress and developmental responses of terpenoid osynthetic genes in *Cistus creticus* subsp. creticus. *Plant Cell Reports* **29**, 629–641.

Paxton, R., Scislowski, P. W., Davis, E. J. and Harris, R. A. (1986). Role of branched-chain 2-oxo acid dehydrogenase and pyruvate dehydrogenase in 2-oxobutyrate metabolism. *The Biochemical Journal* **234**, 295–303.

Peapus, D. H., Chiu, H. J., Campobasso, N., Reddick, J. J., Begley, T. P. and Ealick, S. E. (2001). Structural characterization of the enzyme-substrate, enzyme-intermediate, and enzyme-product complexes of thiamin phosphate synthase. *Biochemistry* **40**, 10103–10114.

Peñuelas, J. and Munné-Bosch, S. (2005). Isoprenoids: An evolutionary pool for photoprotection. *Trends in Plant Science* **10**, 166–169.

Phillips, M. A., Walter, M. H. and Ralph, S. G. (2007). Functional identification and differential expression of 1-deoxy-D-xylulose5-phosphate synthase in induced terpenoid resin formation of Norway Spruce (*Picea abies*). *Plant Molecular Biology* **65**, 243–257.

Raines, C. A. (2003). The Calvin cycle revisited. *Photosynthesis Research* **75**, 1–10.

Rakel, D. (ed.) (2007). *In* Integrative Medicine, 2nd edn., Saunders Elsevier, Philadelphia, PA.

Ralser, M., Wamelink, M. M., Kowald, A., Gerisch, B., Heeren, G., Struys, E. A., Klipp, E., Jakobs, C., Breitenbach, M., Lehrach, H. and Krobitsch, S. (2007). Dynamic rerouting of the carbohydrate flux is key to counteracting oxidative stress. *Journal of Biology* **6**, 10.

Randall, D. D., Miernyk, J. A., David, N. R., Gemel, J. and Luethy, M. H. (1996). Regulation of leaf mitochondrial pyruvate dehydrogenase complex activity by reversible phosphorylation. *In* Protein Phosphorylation in Plants, (P. R. Shewry, N. G. Halford and R. Hooley, eds.), pp. 87–103. Oxford Press, Clarendon, UK.

Rapala-Kozik, M. and Kozik, A. (1992). Mechanism of ligand-protein interaction in plant seed thiamin-binding proteins. Probing the binding site of protein isolated from buckwheat seeds with a series of thiamin-related compounds. *Biochimica et Biophysica Acta* **1159**, 209–214.

Rapala-Kozik, M. and Kozik, A. (1996). Mechanism of ligand-protein interaction in plant seed thiamine-binding proteins. Preliminary chemical identification of amino acid residues essential for thiamine binding to the buckwheat-seed protein. *Biochimie* **78**, 77–84.

Rapala-Kozik, M. and Kozik, A. (1998). Purification and preliminary characterisation of a thiamine-binding protein from spruce seeds. *Plant Physiology and Biochemistry* **37**, 539–544.

Rapala-Kozik, M., Chernikevich, I. P. and Kozik, A. (1999). Ligand-protein interaction in plant seed thiamine-binding proteins. Binding of various thiamine analogues to the sepharose-immobilized buckwheat-seed protein. *Journal of Protein Chemistry* **18**, 721–728.

Rapala-Kozik, M., Ostrowska, K., Bednarczyk, K., Duliński, R. and Kozik, A. (2003). Polypeptide components of oligomeric legumin-like thiamin-binding protein from buckwheat seeds characterized by partial amino acid sequencing and photoaffinity labeling. *Journal of Protein Chemistry* **22**, 167–175.

Rapala-Kozik, M., Olczak, M., Ostrowska, K., Starosta, A. and Kozik, A. (2007). Molecular characterization of the *thi3* gene involved in thiamine biosynthesis in *Zea mays*: cDNA sequence and enzymatic and structural properties of the recombinant bifunctional protein with 4-amino-5-hydroxymethyl-2-methylpyrimidine (phosphate) kinase and thiamine monophosphate synthase activities. *The Biochemical Journal* **408**, 149–159.

Rapala-Kozik, M., Kowalska, E. and Ostrowska, K. (2008). Modulation of thiamine metabolism in *Zea mays* seedlings under conditions of abiotic stress. *Journal of Experimental Botany* **59**, 4133–4143.

Rapala-Kozik, M., Golda, A. and Kujda, M. (2009). Enzymes that control the thiamine diphosphate pool in plant tissues. Properties of thiamine pyrophosphokinase and thiamine-(di)phosphate phosphatase purified from *Zea mays* seedlings. *Plant Physiology and Biochemistry* **47**, 237–242.

Raschke, M., Burkle, L., Muller, N., Nunes-Nesi, A., Fernie, A. R., Arigoni, D., Amrhein, N. and Fitzpatrick, T. B. (2007). Vitamin B1 biosynthesis in plants requires the essential iron sulfur cluster protein, THIC. *Proceedings of the National Academy of Sciences of the United States of America* **104**, 19637–19642.

Reddick, J. J., Nicewonger, R. and Begley, T. P. (2001). Mechanistic studies on thiamin phosphate synthase: Evidence for a dissociative mechanism. *Biochemistry* **40**, 10095–10102.

Ribeiro, A., Praekelt, U., Akkermans, A. D., Meacock, P. A., van Kammen, A., Bisseling, T. and Pawlowski, K. (1996). Identification of *agthi1*, whose product is involved in biosynthesis of the thiamine precursor thiazole, in actinorhizal nodules of *Alnus glutinosa*. *The Plant Journal* **10**, 361–368.

Ribeiro, D. T., Farias, L. P., de Almeida, J. D., Kashiwabara, P. M., Ribeiro, A. F., Silva-Filho, M. C., Menck, C. F. and Van Sluys, M. A. (2005). Functional characterization of the thi1 promoter region from *Arabidopsis thaliana*. *Journal of Experimental Botany* **56**, 1797–1804.

Ricci, V. and Rindi, G. (1992). Thiamin uptake by rat isolated enterocytes: Relationship between transport and phosphorylation. *Archives Internationales de Physiologie, de Biochimie et de Biophysique* **100**, 275–279.

Rindi, G. and Laforenza, U. (2000). Thiamine intestinal transport and related issues: Recent aspects. *Proceedings of the Society for Experimental Biology and Medicine* **224**, 246–255.

Rivoal, J., Ricard, B. and Pradet, A. (1990). Purification and partial characterization of pyruvate decarboxylase from *Oryza sativa* L. *European Journal of Biochemistry* **194**, 791–797.

Roche, T. E., Baker, J. C., Yan, X., Hiromasa, Y., Gong, X., Peng, T., Dong, J., Turkan, A. and Kasten, S. A. (2001). Distinct regulatory properties of pyruvate dehydrogenase kinase and phosphatase isoforms. *Progress in Nucleic Acid Research and Molecular Biology* **70**, 33–75.

Rodionov, D. A., Vitreschak, A. G., Mironov, A. A. and Gelfand, M. S. (2002). Comparative genomics of thiamin biosynthesis in procaryotes. New genes and regulatory mechanisms. *Journal of Biological Chemistry* **277**, 48949–48959.

Rontein, D., Dieuaide-Noubhani, M., Dufourc, E. J., Raymond, P. and Rolin, D. (2002). The metabolic architecture of plant cells. Stability of central metabolism and flexibility of anabolic pathways during the growth cycle of tomato cells. *Journal of Biological Chemistry* **277**, 43948–43960.

Said, H. M., Balamurug, K., Subramanian, V. S. and Marchant, J. S. (2004). Expression and functional contribution of hTHTR-2 in thiamin absorption in human intestine. *American Journal of Physiology. Gastrointestinal and Liver Physiology* **286**, G491–G498.

Schenk, G., Duggleby, R. G. and Nixon, P. F. (1998). Properties and functions of the thiamin diphosphate dependent enzyme transketolase. *The International Journal of Biochemistry & Cell Biology* **30**, 1297–1318.

Schnarrenberger, C., Flechner, A. and Martin, W. (1995). Enzymatic evidence for a complete oxidative pentose phosphate pathway in chloroplasts and an incomplete pathway in the cytosol of spinach leaves. *Plant Physiology* **108**, 609–614.

Schroeder, J. I. and Nambara, E. (2006). A quick release mechanism for abscisic acid. *Cell* **126**, 1023–1025.

Serganov, A. (2010). Determination of riboswitch structures: Light at the end of the tunnel? *RNA Biology* **7**, 98–103.

Serganov, A., Polonskaia, A., Phan, A. T., Breaker, R. R. and Patel, D. J. (2006). Structural basis for gene regulation by a thiamine pyrophosphate-sensing riboswitch. *Nature* **441**, 1167–1171.

Shaanan, B. and Chipman, D. M. (2009). Reaction mechanisms of thiamin diphosphate enzymes: New insights into the role of a conserved glutamate residue. *The FEBS Journal* **276**, 2447–2453.

Shimamoto, K. and Nelson, O. E. (1981). Movement of ^{14}C-compounds from maternal tissue into maize seeds grown in vitro. *Plant Physiology* **67**, 429–432.

Shimizu, M., Mitsunaga, T., Inaba, K., Yoshida, T. and Iwashima, A. (1990). Accumulation of thiamine and thiamine-binding protein during development of rice seed. *Journal of Plant Physiology* **137**, 123–124.

Shimizu, M., Inaba, K., Yoshida, T., Toda, T., Iwashima, A. and Mitsunaga, T. (1995). Purification and properties of thiamine-binding proteins from sesame seed. *Physiologia Plantarum* **93**, 93–98.

Shimizu, T., Nakayama, I., Nagayama, K., Miyazawa, T. and Nezu, Y. (2002). ALS inhibitors. *In* Herbicide Classes in Development, (P. Boeger, K. Wakabayashi and K. Hirai, eds.), pp. 1–41. Springer-Verlag, Berlin.

Smith, A. M., Fuchs, R. T., Grundy, F. J. and Henkin, T. M. (2010). Riboswitch RNAs: Regulation of gene expression by direct monitoring of a physiological signal. *RNA Biology* **7**, 104–110.

Song, Q. and Singleton, C. K. (2002). Mitochondria from cultured cells derived from normal and thiamine-responsive megaloblastic anemia individuals efficiently import thiamine diphosphate. *BMC Biochemistry* **3**, 8.

Southan, M. D. and Copeland, L. (1996). Physical and kinetic properties of acetohydroxyacid synthase from wheat leaves. *Physiologia Plantarum* **98**, 824–832.

Sprenger, G. A., Schorken, U., Wiegert, T., Grolle, S., De Graaf, A. A., Taylor, S. V., Begley, T. P., Bringer-Meyer, S. and Sahm, H. (1997). Identifcation of a thiamine-dependent synthase in *Escherichia coli* required for the formation of the 1-deoxy-D-xylulose 5-phosphate precursor to isoprenoids, thiamine and pyridoxol. *Proceedings of the National Academy of Sciences of the United States of America* **94**, 12857–12862.

Stepuro, A. I., Piletskaya, T. P. and Stepuro, I. I. (2005). Role of thiamine thiol form in nitric oxide metabolism. *Biochemistry* **70**, 339–349.

Sudarsan, N., Barrick, J. E. and Breaker, R. R. (2003). Metabolite-binding RNA domains are present in the genes of eukaryotes. *RNA* **9**, 644–647.

Sweetlove, L. J., Heazlewood, J. L., Herald, V., Holtzapffel, R., Day, D. A., Leaver, C. J. and Millar, A. H. (2002). The impact of oxidative stress on Arabidopsis mitochondria. *The Plant Journal* **32**, 891–904.

Tadege, M., Dupuis, I. I. and Kuhlemeier, C. (1999). Ethanolic fermentation: New functions for an old pathway. *Trends in Plant Science* **4**, 320–325.

Tallaksen, C. M., Sande, A., Bøhmer, T., Bell, H. and Karlsen, J. (1993). Kinetics of thiamin and thiamin phosphate esters in human blood, plasma and urine after 50 mg intravenously or orally. *European Journal of Clinical Pharmacology* **44**, 73–78.

Tan, S., Evans, R. R., Dahmer, M. L., Singh, B. K. and Shaner, D. L. (2005). Imidazolinone-tolerant crops: History, current status and future. *Pest Management Science* **61**, 246–257.

Taylor, N. L., Day, D. A. and Millar, A. H. (2004a). Targets of stress-induced oxidative damage in plant mitochondria and their impact on cell carbon/nitrogen metabolism. *Journal of Experimental Botany* **55**, 1–10.

Taylor, N. L., Heazlewood, J. L., Day, D. A. and Millar, A. H. (2004b). Lipoic acid-dependent oxidative catabolism of alpha-keto acids in mitochondria provides evidence for branched-chain amino acid catabolism in Arabidopsis. *Plant Physiology* **134**, 838–848.

Teige, M., Kopriva, S., Bauwe, H. and Süss, K. H. (1995). Chloroplast pentose-5-phosphate 3-epimerase from potato: Cloning, cDNA sequence, and tissue-specific enzyme accumulation. *FEBS Letter* **377**, 349–352.

Teige, M., Melzer, M. and Süss, K. H. (1998). Purification, properties and in situ localization of the amphibolic enzymes D-ribulose 5-phosphate 3-epimerase and transketolase from spinach chloroplasts. *European Journal of Biochemistry* **252**, 237–244.

Thelen, J. J., Miernyk, J. A. and Randall, D. D. (1998). Partial purification and characterization of the maize mitochondrial pyruvate dehydrogenase complex. *Plant Physiology* **116**, 1443–1450.

Thompson, P., Bowsher, C. G. and Tobin, A. K. (1998). Heterogeneity of mitochondrial protein biogenesis during primary leaf development in barley. *Plant Physiology* **118**, 1089–1099.

Thore, S., Leibundgut, M. and Ban, N. (2006). Structure of the eukaryotic thiamine pyrophosphate riboswitch with its regulatory ligand. *Science* **312**, 1208–1211.

Thornalley, P. J. (2005). The potential role of thiamine (vitamin B1) in diabetic complications. *Current Diabetes Reviews* **1**, 287–298.

Tittmann, K., Golbik, R., Uhlemann, K., Khailova, L., Schneider, G., Patel, M., Jordan, F., Chipman, D. M., Duggleby, R. G. and Hübner, G. (2003). NMR analysis of covalent intermediates in thiamin diphosphate enzymes. *Biochemistry* **42**, 7885–7891.

Tovar-Méndez, A., Miernyk, J. A. and Randall, D. D. (2003). Regulation of pyruvate dehydrogenase complex activity in plant cells. *European Journal of Biochemistry* **270**, 1043–1049.

Trucco, F., Hager, A. G. and Tranel, P. J. (2006). Acetolactate synthase mutation conferring imidazolinone-specific herbicide resistance in Amaranthus hybridus. *Journal of Plant Physiology* **163**, 475–479.

Tunc-Ozdemir, M., Miller, G., Song, L., Kim, J., Sodek, A., Koussevitzky, S., Misra, A. N., Mittler, R. and Shintani, D. (2009). Thiamin confers enhanced tolerance to oxidativestress in Arabidopsis. *Plant Physiology* **151**, 421–432.

Valderrama, R., Corpas, F. J., Carreras, A., Gómez-Rodríguez, M. V., Chaki, M., Pedrajas, J. R., Fernández-Ocaña, A., Del Río, L. A. and Barroso, J. B. (2006). A dehydrogenase-mediated recycling of NADPH is a key antioxidant system against salt-induced oxidative stress in olive plants. *Plant, Cell & Environment* **29**, 1449–1459.

Vallabhaneni, R. and Wurtzel, E. T. (2010). From epoxycarotenoids to ABA: The role of ABA 8′-hydroxylases in drought-stressed maize roots. *Archives of Biochemistry and Biophysics* **504**, 112–117.

Voskoboyev, A. I. and Ostrovsky, Y. M. (1982). Thiamin pyrophosphokinase: Structure, properties, and role in thiamin metabolism. *Annals of the New York Academy of Sciences* **378**, 161–176.

Wachter, A. (2010). Riboswitch-mediated control of gene expression in eukaryotes. *RNA Biology* **7**, 67–76.

Wachter, A., Tunc-Ozdemir, M., Grove, B. C., Green, P. J., Shintani, D. K. and Breaker, R. R. (2007). Riboswitch control of gene expression in plants by splicing and alternative 3′ end processing of mRNAs. *The Plant Cell* **19**, 3437–3450.

Wang, D. and Hazell, A. S. (2010). Microglial activation is a major contributor to neurologic dysfunction in thiamine deficiency. *Biochemical and Biophysical Research Communications* **402**, 123–128.

Wang, G., Ding, X., Yuan, M., Qiu, D., Li, X., Xu, C. and Wang, S. (2006). Dual function of rice *OsDR8* gene in disease resistance and thiamine accumulation. *Plant Molecular Biology* **60**, 437–449.

Webb, E. and Downs, D. (1997). Characterization of *thiL*, encoding thiamin-monophosphate kinase, in *Salmonella typhimurium. The Journal of Biological Chemistry* **272**, 15702–15707.

WHO (1999). Thiamine deficiency and its prevention and control in major emergencies. Geneva, World Health Organization (WHO/NDH/99.13).

Widmann, M., Radloff, R. and Pleiss, J. (2010). The thiamine diphosphate dependent enzyme engineering database: A tool for the systematic analysis of sequence and structure relations. *BMC Biochemistry* **11**, 9.

Williams, R. R. and Cline, J. K. (1936). Synthesis of vitamin B_1. *Journal of the American Chemical Society* **58**, 1504–1505.

Williams, M. and Randall, D. D. (1979). Pyruvate dehydrogenase complex from chloroplasts *of Pisum sativum* L. *Plant Physiology* **64**, 1099–1103.

Winkler, W., Nahvi, A. and Breaker, R. R. (2002). Thiamine derivatives bind messenger RNAs directly to regulate bacterial gene expression. *Nature* **419**, 952–956.

Wolak, N., Kujda, M., Banas, A., Kozik, A. and Rapala-Kozik, M. (2010). Involvement of thiamine (vitamin B1) in the response of *Arabidopsis thaliana* seedlings to the environmental stress. *Acta Biochimica Polonica* **57**, 112.

Wong, C. E., Li, Y., Labbe, A., Guevara, D., Nuin, P., Whitty, B., Diaz, C., Golding, G. B., Gray, G. R., Weretilnyk, E. A., Griffith, M. and Moffatt, B. A. (2006). Transcriptional profiling implicates novel interactions between abiotic stress and hormonal responses in Thellungiella, a close relative of Arabidopsis. *Plant Physiology* **140**, 1437–1450.

Woodward, J. B., Abeydeera, N. D., Paul, D., Phillips, K., Rapala-Kozik, M., Freeling, M., Begley, T. P., Ealick, S. E., McSteen, P. and Scanlon, M. J. (2010). A maize thiamine auxotroph is defective in shoot meristem maintenance. *The Plant Cell* **22**, 3305–3317.

Wynn, R. M., Davie, J. R., Meng, M. and Chuang, D. T. (1996). In alpha -Ketoacid Dehydrogenase Complexes (M. S. Patel, T. E. Roche and R. A. Harris, eds.), pp. 101–117. Birkhauser Verlag, Basel.

Yeaman, S. J. (1989). The 2-oxo acid dehydrogenase complexes: Recent advances. *The Biochemical Journal* **257**, 625–632.

Yusa, T. (1961). Thiamine triphosphate in yeasts and some plant materials. *Plant & Cell Physiology* **2**, 471–474.

Zabalza, A., González, E. M., Arrese-Igor, C. and Royuela, M. (2005). Fermentative metabolism is induced by inhibiting different enzymes of the branched-chain amino acid biosynthesis pathway in pea plants. *Journal of Agricultural and Food Chemistry* **53**, 7486–9743.

Zeidler, J., Sayer, B. G. and Spenser, I. D. (2003). Biosynthesis of vitamin B1 in yeast. Derivation of the pyrimidine unit from pyridoxine and histidine. Intermediacy of urocanic acid. *Journal of the American Chemical Society* **125**, 13094–13105.

Zhang, S. X., Weilersbacher, G. S., Henderson, S. W., Corso, T., Olney, J. W. and Langlais, P. J. (1995). Excitotoxic cytopathology, progression, and reversibility of thiamine deficiency-induced diencephalic lesions. *Journal of Neuropathology and Experimental Neurology* **54**, 255–267.

Zhang, Y., Taylor, S. V., Chiu, H. J. and Begley, T. P. (1997). Characterization of the *Bacillus subtilis* thiC operon involved in thiamine biosynthesis. *Journal of Bacteriology* **179**, 3030–3035.

Zhang, S., Zhou, L., Nemeria, N., Yan, Y., Zhang, Z., Zou, Y. and Jordan, F. (2005). Evidence for dramatic acceleration of a C-H bond ionization rate in thiamin diphosphate enzymes by the protein environment. *Biochemistry* **44**, 2237–2243.

Zhang, M., Li, K., Zhang, C., Gai, J. and Yu, D. (2009). Identification and characterization of class 1 DXS gene encoding 1-deoxy-D-xylulose-5-phosphate synthase, the first committed enzyme of the MEP pathway from soybean. *Molecular Biology Reports* **36**, 879–887.

Zhao, R., Gao, F., Wang, Y., Diaz, G. A., Gelb, B. D. and Goldman, I. D. (2001). Impact of the reduced folate carrier on the accumulation of active thiamin metabolites in murine leukemia cells. *The Journal of Biological Chemistry* **276**, 1114–11148.

Zhao, R., Gao, F. and Goldman, I. D. (2002). Reduced folate carrier transports thiamine monophosphate: An alternative route for thiamine delivery into mammalian cells. *American Journal of Physiology. Cell Physiology* **282**, C1512–C1517.

Zhu, J. K. (2002). Salt and drought stress signal transduction in plants. *Annual Review of Plant Biology* **53**, 247–273.

Biosynthesis of Vitamin B_2 and Flavocoenzymes in Plants

MARKUS FISCHER*,†,[1] AND ADELBERT BACHER*,†

*Institute of Food Chemistry, University of Hamburg, Hamburg, Germany
†Ikosatec GmbH, Garching, Germany

ABSTRACT

Riboflavin (vitamin B_2) derivatives serve as cofactors for a very wide variety of redox enzymes but are now also known to participate in the catalysis of certain non-redox reactions and as cofactors of blue-light photoreceptors. In parallel with the unique

[1]Corresponding author: E-mail: markus.fischer@chemie.uni-hamburg.de

Advances in Botanical Research, Vol. 58 0065-2296/11 $35.00

DOI: 10.1016/B978-0-12-386479-6.00003-2

features of its chemical reactivity, the vitamin is biosynthesised from one molecule of GTP and two molecules of ribulose phosphate by a mechanistically unique series of enzyme-catalysed reactions. Although the work on its biosynthesis has predominantly involved microorganisms, a reasonably detailed picture is now also emerging for plants. A central topic of this review is the emerging role of riboflavin biosynthesis enzymes in connection with plant's iron acquisition and pathogen resistance.

I. INTRODUCTION

Vitamin B_2 (riboflavin) is the obligatory precursor of flavocoenzymes, which are involved in a wide variety of redox reactions as well as certain reactions without net exchange of redox equivalents. Flavocoenzymes have been estimated to serve as cofactors for up to 2% of all enzymes and are essential in all organisms.

Riboflavin, the committed precursor of flavocoenzymes, is biosynthesised in plants and in numerous, although by no means all, microorganisms. Animals, as well as microorganisms that are not endowed with the capacity to obtain riboflavin by biosynthesis, are absolutely dependent on extrinsic sources. For animals, plants are the most important source of the vitamin. The potential contribution to the vitamin supply by the gastrointestinal flora, notably in mammals, is incompletely understood.

Riboflavin is widely distributed as an essential constituent of all living cells. The most relevant dietary sources are milk and milk products, meat, eggs, fish and green leafy vegetables (Souci *et al.*, 2008) (Table I). The absorption of animal-derived riboflavin is better than riboflavin from vegetables.

TABLE I
Vitamin B_2 Content of Foods (Souci et al.*, 2008)*

Food	Vitamin B_2 (mg/100 g)
Pork liver	3.2
Beef liver	3.0
Chicken breast	0.9
Wheat germ	0.7
Camembert/Parmesan	0.6
Sardine	0.4
White mushrooms	0.4
Egg	0.3
Spinach	0.23
Pork	0.2
Beef	0.2
Milk/Yoghurt	0.18
Avocado	0.1
Maracuja	0.1

The daily requirement for an adult human is reported to be in the range of around 1.3 mg (Yates *et al.*, 1998). Recommendations for a daily riboflavin intake are to be increased during pregnancy and lactation (to 1.4 and 1.6 mg) (Yates *et al.*, 1998) (Table II). The symptoms of human riboflavin deficiency are poorly characterised, as riboflavin is widely distributed in human nutrients and any deficiency tends to be overshadowed by deficiencies for other essential food components. Riboflavin deficiency may be difficult to recognise because it is normally accompanied by other vitamin deficiencies and is most commonly found among people whose diets are inadequate. The first deficiency symptoms are sores at the corners of the mouth and a sore throat. These may be followed by a magenta coloration of the tongue (*glossitis*), red and raw lips and skin sores.

Urinary excretion of less than 30 mg of riboflavin/g creatinine is associated with clinical signs of riboflavin deficiency. The degree of stimulation of the activity of erythrocyte glutathione reductase activity by the addition of FAD is used to estimate the sufficiency of riboflavin in humans. The use of oral

TABLE II
Riboflavin: Dietary Reference Intake (Yates et al., *1998)*

Age	mg/day
Infants	
0–6 months	0.3[a]
7–12 months	0.4[a]
Children	
1–3 years	0.5
4–8 years	0.6
Males	
9–13 years	0.9
14–18 years	1.3
18+ years	1.3
Females	
9–13 years	0.9
14–18 years	1.0
18+ years	1.1
Pregnancy	
19–30 years	1.4
31–50 years	1.4
Lactation	
≤ 18 years	1.6
19–30 years	1.6
31–50 years	1.6

[a]Values are Adequate Intakes (AI), others are Recommended Daily Allowance (RDA). (Adapted from www.vitaminherbuniversity.com).

contraceptives has been reported to be conducive to reduced riboflavin saturation.

Despite its presence as riboflavin and/or its coenzyme forms in most typical human food components (Table I), vitamin B_2 is produced on an estimated global scale of about 3000 metric tons per year by technically advanced fermentation processes using *Bacillus subtilis*, *Ashbya gossypii* and *Candida famata* strains, which have replaced the chemical synthesis (Stahmann *et al.*, 2000). Commercial riboflavin is predominantly used as feed supplement for animal husbandry and as food colourant. Only minor amounts of the global production become part of vitamin supplements for direct human consumption.

Although plants are the ultimate source of most of the riboflavin that is consumed by animals, the biosynthesis of the vitamin has been predominantly studied in microorganisms (yeasts and bacteria). This may have been due to the relative ease of experimentation with microorganisms as compared to plants. Moreover, the prevailing opinion seems to have been that, with the exception of photosynthesis and directly related aspects, much of primary metabolism in plants is basically a boilerplate variant of microbial and mammalian metabolism. In any case, apart from some minor studies in the 1960s, most work on riboflavin biosynthesis in plants is of relatively recent origin.

Work on the biosynthesis of riboflavin (Fig. 1) in microorganisms started with the seminal observation that riboflavin production by the naturally flavinogenic ascomycete, *Eremothecium ashbyi*, could be increased by the addition of purines to the culture medium (MacLaren, 1952). Subsequent work with radioisotopes by several research groups established that all the nitrogen atoms and the carbon atoms of the vitamin's pyrimidine ring are entirely derived from a purine precursor (for review, see Plaut *et al.*, 1974). Another pioneering discovery was the isolation of a green fluorescent compound (6,7-dimethyl-8-ribityllumazine, **10**, Fig. 1) from the culture fluid of *E. ashbyii* (Masuda, 1957a,b), which was established as the direct biosynthetic precursor of riboflavin around 1960 (Katagiri *et al.*, 1958a,b,c,d,e; Maley and Plaut, 1959; Plaut, 1960, 1963).

For an extended period, the preferred objects of research on riboflavin biosynthesis were naturally occurring flavinogenic microorganisms and, later on, flavinogenic mutants of non-flavinogenic organisms. The reason for that was twofold. (i) Flavinogenic strains were expected to provide favourable conditions for analysis because they were correctly expected to have enhanced levels of flavin biosynthetic enzymes. (ii) Microbial flavinogenesis was of practical interest as a potential technology for the production of

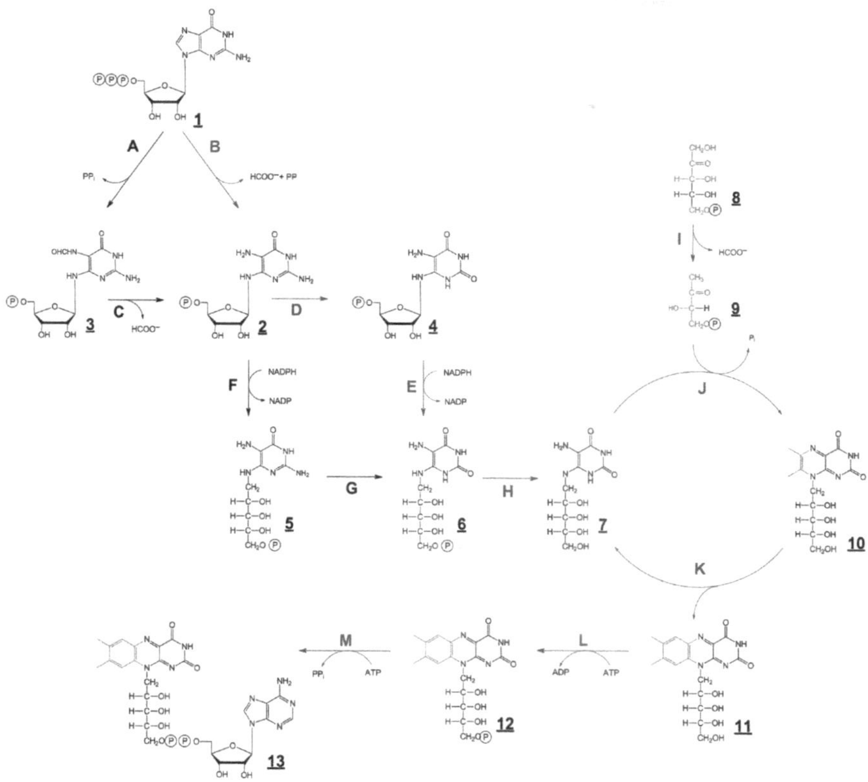

Fig. 1. Biosynthesis of riboflavin and flavocoenzymes. (A) GTP cyclohydrolase III; (B) GTP cyclohydrolase II; (C) 2-amino-5-formylamino-6-ribosylamino-4(3*H*)-pyrimidinone 5′-phosphate hydrolase; (D) 2,5-diamino-6-ribosylamino-4-(3*H*)-pyrimidinone 5′-phosphate deaminase; (E) 5-amino-6-ribosylamino-2,4(1*H*,3*H*)-pyrimidinedione 5′-phosphate reductase; (F) 2,5-diamino-6-ribosylamino-4(3*H*)-pyrimidinone 5′-phosphate reductase; (G) 2,5-diamino-6-ribitylamino-4(3*H*)-pyrimidinedione 5′-phosphate deaminase; (H) hypothetical phosphatase; (I) 3,4-dihydroxy-2-butanone 4-phosphate synthase; (J) 6,7-dimethyl-8-ribityllumazine synthase; (K) riboflavin synthase; (L) riboflavin kinase; (M) FAD synthetase; **1**, GTP; **2**, 2,5-diamino-6-ribosylamino-4(3*H*)-pyrimidinone 5′-phosphate; **3**, 2-amino-5-formylamino-6-ribosylamino-4(3*H*)-pyrimidinone 5′-phosphate; **4**, 5-amino-6-ribosylamino-2,4(1*H*,3*H*)-pyrimidinedione 5′-phosphate; **5**, 2,5-diamino-6-ribitylamino-4(3*H*)-pyrimidinedione 5′-phosphate; **6**, 5-amino-6-ribitylamino-2,4(1*H*,3*H*)-pyrimidinedione 5′-phosphate; **7**, 5-amino-6-ribitylamino-2,4(1*H*,3*H*)-pyrimidinedione; **8**, ribulose 5-phosphate; **9**, 3,4-dihydroxy-2-butanone 4-phosphate; **10**, 6,7-dimethyl-8-ribityllumazine; **11**, riboflavin; **12**, FMN; **13**, FAD. Green arrows mark the plant pathway; red, fate of the 4-carbon precursor **9** derived from ribulose 5-phosphate. (For interpretation of the references to colour in this figure legend, the reader is referred to the Web version of this chapter.)

the vitamin; in fact, fermentation has by now completely superseded the chemical synthesis as a method for the technical production of the vitamin.

Ever since the early studies, there has been an uninterrupted stream of additional data on flavin biosynthesis and function. The inclusion of Archaea and plants began in the 1990s. The early work on riboflavin biosynthesis has been reviewed repeatedly, and readers are directed to these articles (Bacher, 1991; Bacher *et al.*, 1993, 2000, 2001; Brown and Neims, 1982; Brown and Reynolds, 1963; Brown and Williamson, 1987; Fischer and Bacher, 2005, 2006, 2008, 2011; Plaut *et al.*, 1974; Young, 1986).

There are significant differences in the riboflavin pathways of different taxonomic groups. The early steps constitute a moderately complex maze as shown in Fig. 1. Invariably, all pathway variants start with GTP. In eubacteria, fungi and plants, GTP is converted into 2,5-diamino-6-ribosylamino-4-(3*H*)-pyrimidinone monophosphate by GTP cyclohydrolase II, which catalyses the release of C-8 of the purine moiety by cleavage of two different carbon nitrogen bonds, resulting in the formation of formate as well as the cleavage of a phosphoanhydride bond resulting in the formation of inorganic pyrophosphate (Foor and Brown, 1975). Archaea, however, use two enzymes to achieve the same end result, with the formamide **3** as an intermediate (Graham *et al.*, 2002; Grochowski *et al.*, 2009; Spoonamore *et al.*, 2006).

The reaction product of GTP cyclohydrolase II can be converted to 5-amino-6-ribitylamino-2,4(1*H*,3*H*)-pyrimidinedione 5′-phosphate (**6**) via two different reaction sequences. In bacteria and in plants, the pyrimidine ring of 2 undergoes deamination resulting in **4**, and the subsequent reduction of the ribosyl side chain affords 5-amino-6-ribitylamino-2,4(1*H*,3*H*)-pyrimidinedione 5′-phosphate (**6**) (Burrows and Brown, 1978). In yeasts and Archaea, however, the ribosyl residue of 2 is first converted into a ribityl moiety by a reductase affording 2,5-diamino-6-ribitylamino-4(3*H*)-pyrimidinone 5′-phosphate (**5**) (Bacher and Lingens, 1970; Hollander and Brown, 1979). The subsequent deamination of the pyrimidine ring of **5** affords **6**. Notably, the yeast and fungal pathways converge at the level of the 5′-phosphate **6**. However, that compound is not used by the next known enzyme in the pathway, lumazine synthase, which accepts only the dephosphorylated form of **6** as substrate. It follows that **6** must be dephosphorylated in order to enter the final stretch of the biosynthetic pathway, but it is still unknown how the required dephosphorylation occurs.

The final two reactions involve the assembly of the xylene ring of the vitamin from two identical 4-carbon molecules. Work around 1990 established that ribulose 5-phosphate (**8**) is transformed into 3,4-dihydroxy-2-butanone 4-phosphate (**9**) by a mechanistically complex rearrangement resulting in the extrusion of C-4 of the substrate as formate (Richter *et al.*, 1992; Volk *et al.*, 1990; Volk and Bacher, 1988, 1990, 1991) (notably, GTP cyclohydrolase II also affords formate as a second product).

The condensation of the pyrimidine derivative **7** with the carbohydrate **9**, catalysed by lumazine synthase, was shown in the 1990s to afford the green fluorescent pteridine derivative, 6,7-dimethyl-8-ribityllumazine (**10**) (Kis and Bacher, 1995; Volk and Bacher, 1988), several decades after the compound had been established to serve as the sole substrate for the terminal intermediate in the pathway, riboflavin synthase. The latter enzyme catalyses the transfer of a 4-carbon unit between two identical substrate molecules resulting in the formation of equivalent amounts of riboflavin and the pyrimidine **7**. That second product of riboflavin synthase is also the substrate of lumazine synthase and can be recycled; hence, on average, all molecules generated by the early stage of the biosynthetic pathway must be processed twice by lumazine synthase and by riboflavin synthase.

In summary, the biosynthesis of one molecule of riboflavin requires one molecule of GTP and two molecules of ribulose phosphate. No cofactors are required except divalent metal ions and NADPH. The reactions catalysed by GTP cyclohydrolase II, pyrimidine deaminase, 3,4-dihydroxy-2-butanone 4-phosphate synthase, lumazine synthase and riboflavin synthase are all exergonic, and the reverse reaction defies experimental detection. Clearly, the vitamin arises by a unique sequence of chemical reactions. Despite their extraordinary complexity that will be described below in more detail, some of these reactions can proceed at appreciable rates in the absence of enzyme catalysis. That said, the riboflavin pathway is extraordinary in many respects.

Around the turn of the century, when many aspects of flavin biosynthesis had been worked out with microorganisms, the newly available and rapidly growing body of available DNA sequences from a variety of plants provided a basis for rapid identification of candidate genes for riboflavin biosynthesis in plants by way of database mining. The role of candidate genes could then be verified by heterologous expression. In that way, most riboflavin biosynthesis genes and proteins were recovered during a short period. As shown in more detail below, plants, green algae and cyanobacteria all use the bacterial pathway, as opposed to the yeast/archaeal pathways shown in Fig. 1. Specifically, however, the plant mRNAs invariably specify N-terminal extensions which enable the targeting of the enzymes to chloroplasts.

All organisms need to convert riboflavin into the coenzyme forms, FMN (**12**) and FAD (**13**). Remarkably, in species endowed with *de novo* riboflavin biosynthesis, a phosphate residue is removed from the 5′-position of the ribityl side chain of the precursor **6** and needs to be replaced after the formation of the alloxazine ring.

In Fig. 1, the reactions catalysed by the enzymes of Gram-positive bacteria and plants are shown in colour for ease of orientation. The following section is predominantly focused on plant aspects. However, it should be noted that in-depth studies of the respective plant enzymes are few in number. Notably, an X-ray structure has only been reported for lumazine synthase of plant origin, and detailed kinetic and mechanistic studies of the plant enzymes are mostly elusive. However, based on homology aspects, realistic conjectures are possible on the basis of the more extensive knowledge of the bacterial orthologues.

II. A SINGLE PLANT ENZYME, RibAB GENERATES BOTH COMMITTED PRECURSORS FOR THE RIBOFLAVIN PATHWAY

The first committed substrates of the convergent riboflavin synthase are produced by GTP cyclohydrolase II and 3,4-dihydroxy-2-butanone 4-phosphate synthase. The cognate genes *ribA* (Richter *et al.*, 1993) and *ribB* (Richter *et al.*, 1992) of *Escherichia coli* were both cloned by marker rescue in the early 1990s. With that information in hand, it became obvious that the *ribA* gene at the 5′-end of the rib operon of *B. subtilis* (Mironov *et al.*, 1989) specifies a bifunctional enzyme with an N-terminal 3,4-dihydroxy-2-butanone 4-phosphate synthase domain and a C-terminal GTP cyclohydrolase II domain (it should be noted that, for historical reasons, the monofunctional GTP cyclohydrolase II gene of *E. coli* and the bifunctional *B. subtilis* gene are both designated *ribA*).

A cDNA segment of *Arabidopsis thaliana* was cloned in 1998 by marker rescue of a *ribA*⁻ mutant of *E. coli.* Its sequence was similar to the *ribA* gene of *E. coli* (Kobayashi *et al.*, 1998), but no attempt was made to characterise the translation product. Later work showed the reported sequence to be a fragment of an open reading frame specifying a bifunctional 3,4-dihydroxybutanone 4- phosphate synthase/GTP cyclohydrolase II (Herz *et al.*, 2000). The sequence of the two catalytic domains was similar to the bifunctional RibA protein of *B. subtilis.* A homologous gene was also cloned from tomato. The 3,4-dihydroxybutanone 4-phosphate synthase domains of the plant genes are preceded by N-terminal segment of about 120 amino acid residues which fulfil the criteria for chloroplast targeting sequences. The recombinant full-length protein expressed in *E. coli* and a pseudomature form were both enzymatically active.

The *ribA* gene of *A. thaliana* comprises seven exons. The gene fusion has been proposed to guarantee a balanced substrate input from the convergent pathways (Moore, 2004). However, detailed kinetics studies or informations on substrate concentrations inside the chloroplast compartment, which would enable simulations, are not available.

Whereas the bifunctional RibA proteins that are characteristic for plants and Gram-positive bacteria have not been studied yet in closer detail, important insights on the structure of their catalytic domains and their reaction mechanisms can be gleaned from sequence comparison and by extrapolation from the detailed study of the monofunctional RibA and RibB proteins of bacteria and fungi which have been studied in considerable detail.

NMR studies with ^{13}C-labelled substrates showed that the 3,4-dihydroxy-2-butanone 4-phosphate synthase of *E. coli* extracts C-4 of the substrate, ribulose phosphate (**8**), as formate by way of a strictly intramolecular rearrangement (Fischer *et al.*, 2002; Volk and Bacher, 1991) (Fig. 2). Based on these findings, it has been proposed that the reaction is initiated by the formation of the endiol structure **14**. Following the formation of the 3-ketulose **15** and the elimination of water, a Lobry de Bryn van Ekenstein isomerisation is believed to afford the branched aldose intermediate **18** which can subsequently eliminate formate. The reaction is terminated by isomerisation of the resulting endiol **19**. Reprotonation is stereospecific; only the L enantiomer of the product is formed. Mg^{2+} is required for catalysis.

Crystal structures (Fig. 3) have been reported for the (monofunctional) 3,4-dihydroxy-2-butanone 4-phosphate synthases from *E. coli* (Liao *et al.*, 2001a), *Salmonella typhimurium* (Kumar *et al.*, 2010), *Candida albicans* (Echt *et al.*, 2004), *Methanococcus jannaschii* (Steinbacher *et al.*, 2004) and from the

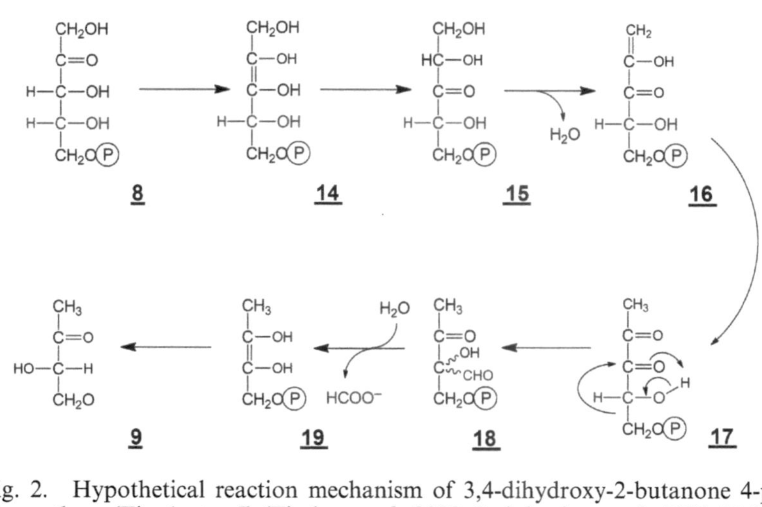

Fig. 2. Hypothetical reaction mechanism of 3,4-dihydroxy-2-butanone 4-phosphate synthase (Fig. 1, step **I**) (Fischer *et al.*, 2002; Steinbacher *et al.*, 2003; Volk and Bacher, 1990).

Fig. 3. Overlay of single subunits of 3,4-dihydroxy-2-butanone 4-phosphate synthases of *C. albicans* (green), *E. coli* (blue), *M. jannaschii* (red) and *M. grisea* (olive green). Only the substrate ribulose 5-phosphate (**8**) of the *C. albicans* complex is shown (yellow) (Echt *et al.*, 2004; Liao *et al.*, 2001a, 2002; Steinbacher *et al.*, 2003, 2004). (For interpretation of the references to colour in this figure legend, the reader is referred to the Web version of this chapter.)

rice-pathogenic fungus, *Magnaporthe grisea* (Liao *et al.*, 2002; Steinbacher *et al.*, 2003). The solution structure of the *E. coli* enzyme has also been determined by NMR (Kelly *et al.*, 2001). The structure of the 3,4-dihydroxy-2-butanone 4-phosphate synthase domain of the bifunctional RibA protein of *Mycobacterium tuberculosis* has also been reported (Singh *et al.*, 2011). More than 20 structures have been reported, some of them with very high resolution in the 1 Å range. The monofunctional proteins are homodimers with subunits comprising about 200 amino acid residues (depending on pH, the recombinant *M. tuberculosis* domain can be monomeric or dimeric, but the quaternary structure of the naturally occurring, bifunctional full-length protein has not been reported).

All available three-dimensional structures of 3,4-dihydroxy-2-butanone 4-phosphate synthases from different organisms show similar $\alpha\beta$ folds with complex connectivity (Fig. 3). The dimers interact predominantly by hydrophobic interactions. The two topologically equivalent active sites are located at the subunit interface, and residues from both subunits are involved in each active site. A segment with the strictly conserved polar residues (arginine 25, glutamates 26 and 28 and aspartates 21 and 30 (*E. coli* numbering)), which are all essential for catalysis, is located in a flexible loop that becomes ordered upon binding of substrate and divalent metal ions.

In 3,4-dihydroxy-2-butanone 4-phosphate synthase from *M. jannaschii*, three of the four hydroxy groups of ribulose 5-phosphate are coordinated by the metal ions. Based on crystallographic refinement, the metals were assigned as zinc and calcium, which were present in the crystallisation buffer (but neither of these metal ions supports catalysis) (Steinbacher *et al.*, 2004).

The reaction mechanism of GTP cyclohydrolase II is characterised by extraordinary complexity (Fig. 4). Specifically, the enzyme catalyses the hydrolytic cleavage of no less than three different bonds: two in the imidazole ring of the purine moiety and one in the triphosphate motif. The available data suggest that the first reaction step involves the formation of a covalent adduct (**20**) between the protein and a GMP moiety under release of inorganic pyrophosphate (Ritz *et al.*, 2001). The X-ray structure of the *E. coli* enzyme (Ren *et al.*, 2005) suggests specifically that a phosphoamide bond is formed between Asp128 and the α-phosphate of the substrate, GTP. The formation of the phosphoamide derivative is followed by the hydrolytic cleavage of the bonds connecting C-8 and N-9 of the purine moiety, which results in the formation of the formamide-type intermediate **21**. In a subsequent step, the formamide motif is hydrolysed under formation of formate. A Zn^{2+} ion is essential for both hydrolysis steps, which are believed to be further supported by Tyr 105 serving as a proton donor (Kaiser *et al.*, 2002). Finally, the phosphoamide bond is cleaved hydrolytically. Very well in line with this scenario is the observation that a fraction of substrate GMP (about 10% in case of the *E. coli* enzyme) is converted to GMP (**22**) by the enzyme, rather than to the riboflavin precursor **2** (Ritz *et al.*, 2001). This implicates that, occasionally, the phosphoamide bond can be hydrolysed even before the opening of the imidazole ring. Also in line with the hypothesis is the observation that 2-amino-5-formamido-6-ribosylamino-4(3*H*)-pyrimidinone 5′-pyrophosphate (**21**) can serve as a non-natural substrate for GTP cyclohydrolase II (Schramek *et al.*, 2001). Most surprisingly, the initial formation of the phosphoamide bond between protein and a GMP moiety at the beginning of the reaction is the rate-determining step (Schramek *et al.*, 2001).

The crystal structure of GTP cyclohydrolase II from *E. coli* has been determined at near-atomic resolution in complex with a non-hydrolysable GTP analogue (Fig. 5) (Ren *et al.*, 2005).

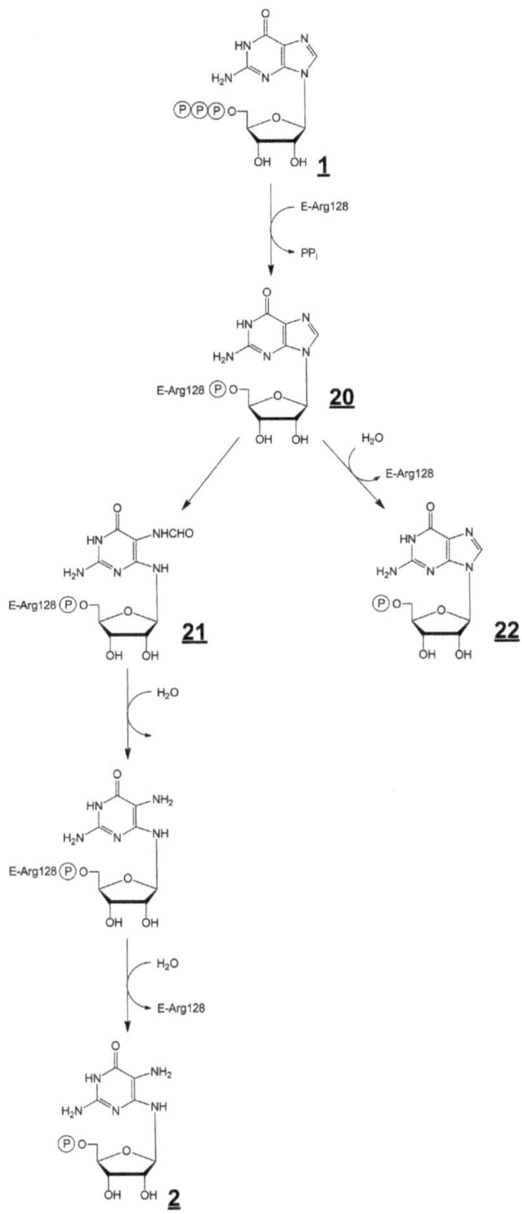

Fig. 4. Hypothetical mechanism for release of formate by GTP cyclohydrolase II (Fig. 1, step **I**). Arg128 refers to GTP cyclohydrolase II from *E. coli*. (Kaiser *et al.*, 2007; Ren *et al.*, 2005).

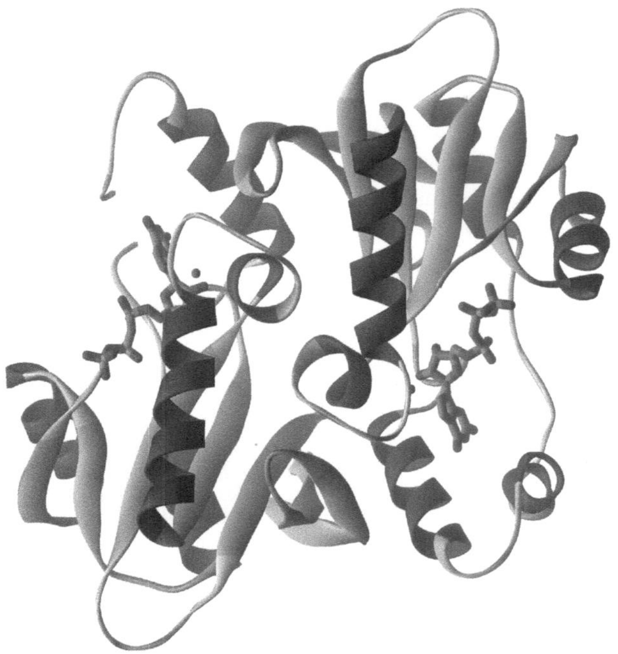

Fig. 5. Structure of GTP cyclohydrolase II from *E. coli* (Ren *et al.*, 2005).

III. DEAMINASE/REDUCTASE

The transformation of the GTP cyclohydrolase II product **2** into lumazine synthase substrate **8** requires the hydrolytic removal of the amino group in position 2 of the pyrimidine ring, the reduction the ribosyl side chain affording a ribityl side chain, and dephosphorylation. In yeasts and Archaea, side chain reduction precedes deamination and the pathway proceeds via 2,5-diamino-6-ribitylamino-4(3*H*)-pyrimidinone 5′-phosphate (**5**) as intermediate (Bacher and Lingens, 1970; Hollander and Brown, 1979). In eubacteria, however, deamination precedes side chain reduction and the reaction sequence proceeds via 5-amino-6-ribosylamino-4(3*H*)-pyrimidinone 5′-phosphate (**4**) as intermediate (Burrows and Brown, 1978). Numerous eubacteria have genes that specify fusion proteins with an N-terminal deaminase domain and a C-terminal reductase domain (Richter *et al.*, 1997).

Recent evidence indicates that plants and blue-green algae use the bacterial pathway via 5-amino-6-ribosylamino-4(3*H*)-pyrimidinone 5′-phosphate (**4**) (Fischer *et al.*, 2004a) (Fig. 1). Specifically, a pyrimidine deaminase from *A. thaliana* (GenBank Accession Code: At4g20960) has been expressed in pseudomature form and is described in more detail below.

The *A. thaliana* open reading frame At4g20960 specifies a putative terminal plastid-targeting sequence followed by a segment with relatively close similarity to eubacterial pyrimidine deaminase, and a C-terminal domain of about 170 amino acid residues whose function is currently unknown (Fig. 6).

Marker rescue experiments using a *ribD* mutant of *E. coli* showed that the plant gene product serves as deaminase but has no reductase activity. The putative deaminase domain (residues 64–250), when expressed in a recombinant *E. coli* strain, catalysed the deamination of **2** at a similar rate as the homologous RibD-protein of *E. coli.* In analogy to bacterial homologues, the plant deaminase domain binds a Zn^{2+} ion which is believed to be essential for catalysis.

The structure of the plant deaminase can be anticipated in significant detail on basis of the X-ray structures of the bifunctional deaminase/reductase orthologues of *E. coli* (RibD-protein) (Stenmark *et al.*, 2007) and of *B. subtilis* (RibG protein) (Chen *et al.*, 2006, 2009) (Fig. 7). The deaminases are members of the pyrimidine deaminase superfamily.

Structures of cytidine deaminases of *E. coli* and the yeast *Saccharomyces cerevisiae* have been determined to very high resolution (< 1.2 Å) and can also serve as models for the deaminase domain from *A. thaliana*. The essential Zn^{2+} ion of the cytosine deaminases and paralogs is bound by two cysteines and one histidine. The highly conserved zinc coordination site is indicated by *asterisks* in Fig. 6.

The eubacterial and archaeal reductases, although catalysing different reactions, are all structurally related to dihydrofolate reductase (Chatwell *et al.*, 2006; Chen *et al.*, 2006; Römisch-Margl *et al.*, 2008). An *in vivo* study on the mechanism of 2,5-diamino-6-ribosylamino-4(3*H*)-pyrimidine 5′-phosphate reductase using the ascomycete *A. gossypii* showed that the incoming hydrogen atom ends up in the *pro-S* position at C-1′ of the ribityl side chain of **5** (Keller *et al.*, 1988) (Fig. 8). Thus, a hypothetical reaction mechanism via an Amadori-type intermediate could be ruled out. Instead, it was proposed that the reaction proceeds via the Schiff base intermediate **23**. This is well in line with a recent crystallographic study on RibG-protein from *B. subtilis* (Chen *et al.*, 2009).

The bifunctional deaminase/reductase of *E. coli* has been shown to transfer the C4-*pro-R* hydrogen of NADPH to C-1 of **4** (Magalhaes *et al.*, 2008); the same stereochemistry is supported for the pyrimidine reductase of *M. jannaschii* (using **2** as substrate) by crystallographic evidence (Chatwell *et al.*, 2006).

Putative genes of *A. thaliana* (Accession No. AX546684) and *Oryza sativa* (Accession No. Q6K6I4) have been suggested to specify pyrimidine reductase on basis of sequence similarity Chatwell *et al.*, 2006 but direct experimental confirmation is as yet not available (Fig. 9).

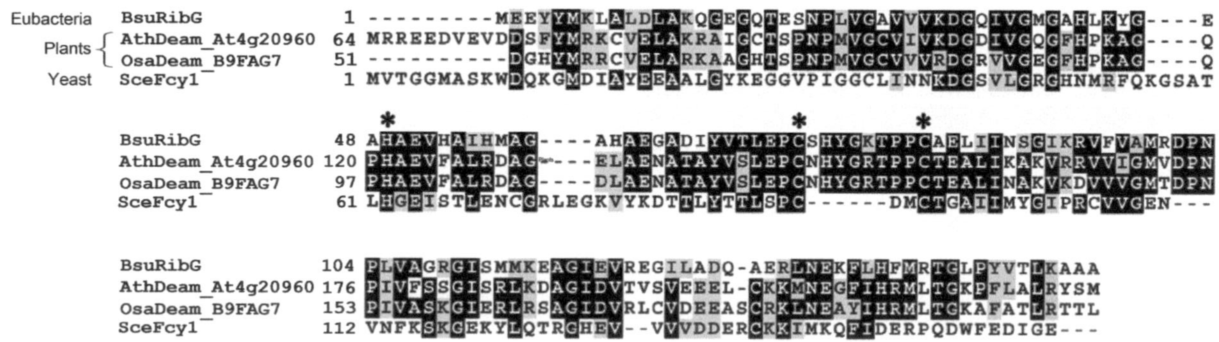

Fig. 6. Sequence alignment of 2,5-diamino-6-ribosylamino-4(3*H*)-pyrimidinone 5′-phosphate deaminase domains from eubacteria and plants with cytosine deaminase from *S. cerevisiae*. AthDeam_At4g20960, *A. thaliana* (At4g20960); *BsuRibG*, *B. subtilis* ribG (P17618); *OsaDEam_B9FAG7*, *Oryza sativa* (*B9FAG7*); and *SceFcy1* (cytosine deaminase of *S. cerevisiae*) (Q12178). The highly conserved zinc coordination site of yeast cytosine deaminase is indicated by *asterisks* (Fischer *et al*., 2004a).

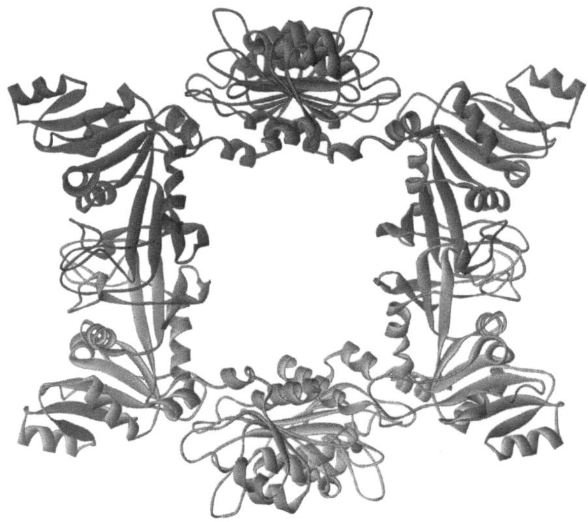

Fig. 7. Structure of RibG-protein from *B. subtilis* (Chen *et al.*, 2006).

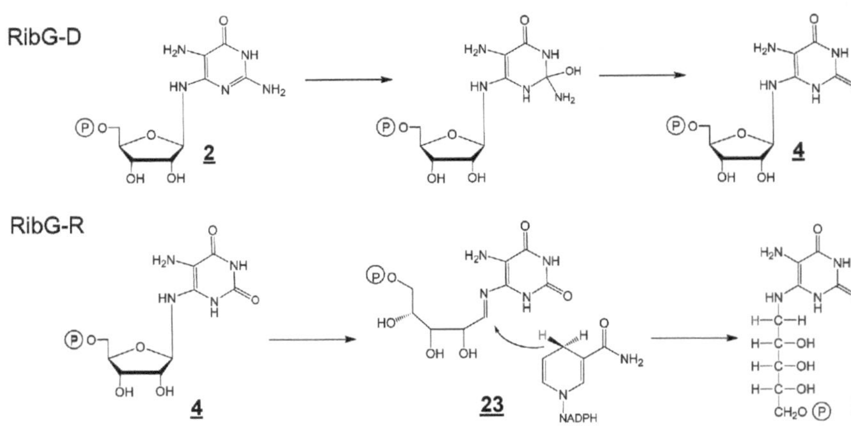

Fig. 8. The proposed deamination and reduction mechanisms catalysed by the deaminase and reductase domains of eubacterial RibG, respectively (Chen *et al.*, 2009).

IV. LUMAZINE SYNTHASE

In the field of riboflavin biosynthesis research, lumazine synthase occupies a commanding position. At least 38 X-ray structures from a variety of pathogenic and non-pathogenic microorganisms have been reported, and the bibliography comprises at least 14 major protein structure papers. Moreover,

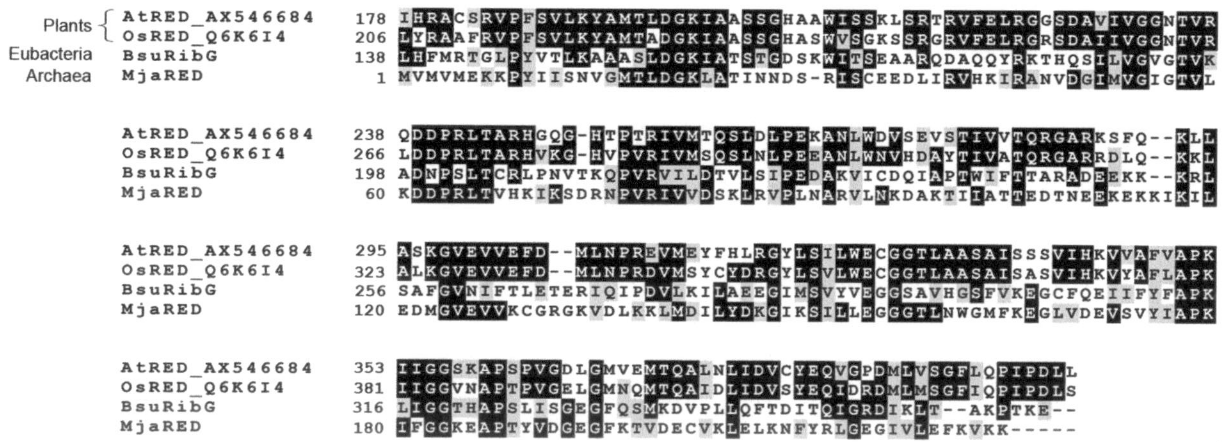

Fig. 9. Sequence alignment of putative 2,5-diamino-6-ribosylamino-4(3*H*)-pyrimidinone 5′-phosphate reductase domains from Archaea, plants and eubacteria. *MjaRED*, *M. jannaschii* (Accession No. Q58085); *Ath_AX546684*, *Arabidopsis thaliana* (Accession No. Ax546684); *Osa_Q6K6I4*, *Oryza sativa* (Accession No. Q6K6I4); *BsuRibG*, *B. subtilis* (Accession No. P17618). Conserved residues are shown in black, homologous in grey (Chatwell *et al.*, 2006).

lumazine synthase is the only riboflavin biosynthesis enzyme with the structure of a plant orthologue published (Jordan *et al.*, 1999).

The lumazine synthase nomenclature is unfortunately confusing for historical reasons. The protein subunit catalysing the condensation of **7** with **9** had been discovered in the 1970s as a component of a large protein complex from *B. subtilis*, which had been designated as "heavy riboflavin synthase" (Bacher *et al.*, 1980; Bacher and Mailänder, 1978). Whereas it was clear that the riboflavin synthase activity was exclusively due to a 25-kDa peptide designated α subunit, the enzymatic role of the 15-kDa peptide (designated "β subunit") was assigned only much later as lumazine synthase (Neuberger and Bacher, 1986). The 1-MDa enzyme complex from *B. subtilis* and other Bacillaceae is now known to be a bifunctional lumazine synthase/riboflavin synthase.

On the surface, the reaction now known to be catalysed by lumazine synthase looks simple enough (Fig. 10). The diaminopyrimidine derivative **7** condenses with the carbohydrate **9** to form one equivalent each of the lumazine derivative **10** and inorganic phosphate as well as two equivalents of water. In line with that, the reaction can proceed uncatalysed in dilute aqueous solutions of the reactants at room temperature at neutral pH (Kis *et al.*, 2001). So fast is the spontaneous process that blanks without enzyme are really mandatory in order to monitor the relatively modest rate acceleration brought about by the enzyme.

On closer inspection, the reaction mechanism of lumazine synthase is anything but simple. The well-known regiochemistry of the reaction (C-2

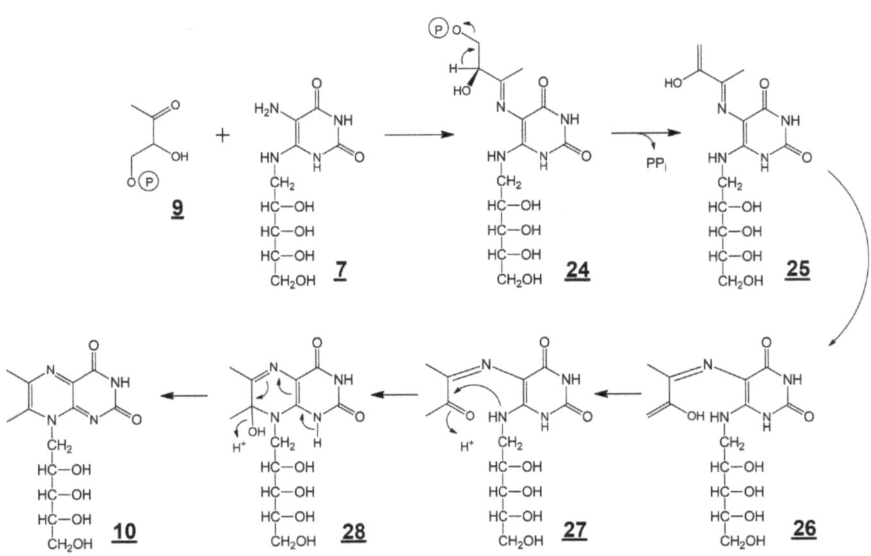

Fig. 10. Hypothetical reaction mechanism of lumazine synthase (Kis *et al.*, 1995).

of **9** becomes directly linked to N-5 of **7**) suggests that the formation of a Schiff base is the first reaction step (Keller *et al.*, 1988; Kis *et al.*, 1995; Nielsen *et al.*, 1986). Stopped flow experiments have identified a major transient with absorption maxima at 455 and 408, and that suggests one or several intermediates with extended conjugated double bond systems, possibly **25** and/or **26** (Haase *et al.*, 2003; Schramek *et al.*, 2003). The elimination of inorganic phosphate could then generate a new carbonyl group that could now react with the position 6 ribitylamino group of the pyrimidine **27**. Ultimately, the elimination of water could generate the final product, either as the neutral molecule **10** or as the exomethylene anion of **33** (cf. Fig. 14) which could be converted to the neutral form by protonation. As already mentioned, the reaction can proceed without enzyme catalysis, but the uncatalysed reaction is only partially regiospecific and is believed to involve partitioning via two different reaction trajectories (Kis *et al.*, 2001).

Studies on the bifunctional lumazine synthase/riboflavin synthase of *B. subtilis* were initiated in the early 1970s. The 1-MDa protein was shown to consist of 3 α subunits, which accounted for its riboflavin synthase activity and of 60 β subunits, which were only later assigned as lumazine synthase. The 60 β subunits were shown to form a spherical shell with icosahedral 532 symmetry, which had some similarity to small virus capsids (Bacher *et al.*, 1986; Ladenstein *et al.*, 1986, 1988) (Fig. 11). The three α subunits were shown to be enclosed in the central core. Following the induced dissociation of the large, hetero-oligomeric particles, it was possible to reassemble the β subunits as hollow, icosahedral capsids of 60 monomers. The assembly of much larger hollow capsids comprising more than 100 monomers was also possible under *in vitro* conditions.

The icosahedral, 60-meric lumazine synthase capsid is best described as a dodecamer of pentamers, although free pentamers have not been observed (Fig. 11). The pentameric modules of the icosahedral/dodecahedral capsids have channels running along their c_5 axes. The central part of the pentamer module can be described as a superhelical assembly of the $\alpha 3$ helices of the five subunits which constitute the pentamer. The subunits are characterised by a flavodoxin-like $\alpha\beta$ fold. The active sites, which will be described in closer detail below, are all located at the interfaces of adjacent subunits in the pentamer modules. Notably, each 60-mer has 60 topologically equivalent active sites.

Over the past three decades, the X-ray structures of lumazine synthases have been reported for a range of widely different species including fungi, eubacteria, Archaea and one plant (spinach) (Braden *et al.*, 2000; Gerhardt *et al.*, 2002a; Klinke *et al.*, 2005, 2007; Koch *et al.*, 2004; Meining *et al.*, 2000; Morgunova *et al.*, 2005, 2006, 2007, 2011; Persson *et al.*, 1999; Zylberman

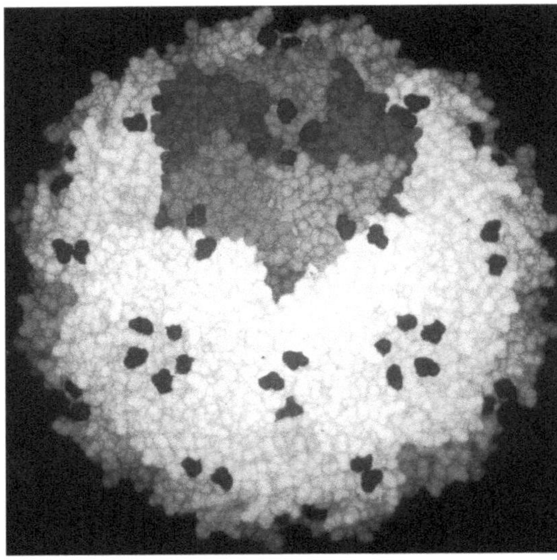

Fig. 11. Space filling model of the lumazine synthase capsid from *B. subtilis.* Subunits in one pentamer are shown in colour. Selected amino acid residues are also shown in colour to indicate areas where the fivefold (red), threefold (violet) and twofold (red) symmetry axes penetrate the molecular surface. (For interpretation of the references to colour in this figure legend, the reader is referred to the Web version of this chapter.)

et al., 2004, 2006) (Fig. 12). The microbial species include several pathogens (e.g. *Bacillus anthracis*, *Brucella abortus*, *M. tuberculosis*, *S. typhimurium*, *C. albicans*). The X-ray structure of lumazine synthase from the important rice pathogen, *M. grisea*, has also been reported (Persson *et al.*, 1999). Despite the large molecular size (15 nm), the X-ray structure of the enzyme from the hyperthermophilic eubacterium *Aquifex aeolicus* could be determined to near-atomic resolution (1.4 Å) (Zhang *et al.*, 2001).

The preferred coverage of human- and plant-pathogenic microorganisms is due to the fact that the enzymes of the riboflavin biosynthesis pathway are under evaluation as potential targets for the design of inhibitors with anti-infective or fungicidal properties. The unifying tenet of these different quests is the fact that the riboflavin biosynthetic enzymes are essential in plants and in many animal and plant pathogens but are absent in humans and animals. Hence, inhibitors used in medicine or plant crop protection (herbicides, fungicides) would be exempt from target-related human toxicity (although that would of course not rule out off-target toxicity).

The extensive comparative studies have revealed the existence of four different quaternary structures for lumazine synthase, namely (i) c_5-pentamers (yeasts and fungi, *B. abortus*, *M. tuberculosis*) (Braden *et al.*, 2000;

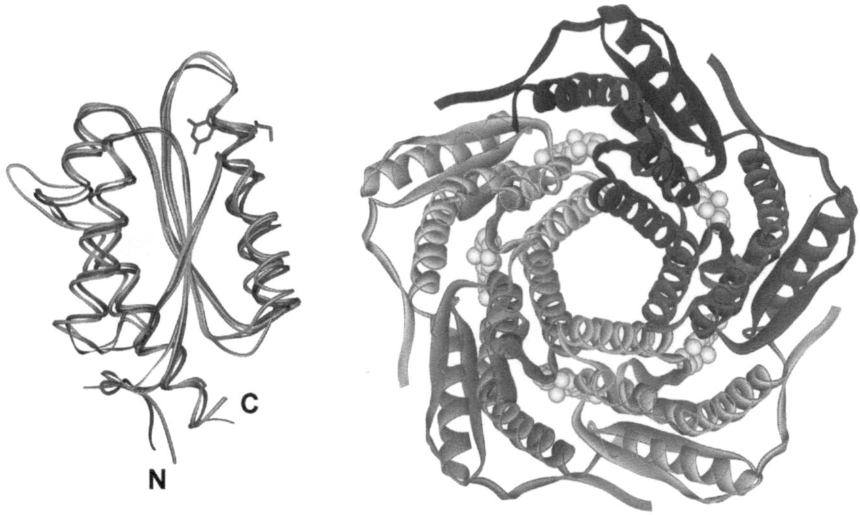

Fig. 12. (Left) Structural superposition of monomers of the pentameric lumazine synthase from *Schizosaccharomyces pombe* (yellow) and the icosahedral lumazine synthases from *Aquifex aeolicus* (green) and spinach (red). A bound substrate analogue inhibitor (5-nitro-6-(D-ribitylamino)-2,4(1*H*,3*H*)-pyrimidinedione, **29**) to mark the substrate binding site is shown (based on the spinach structure, shown in yellow) (Gerhardt *et al.*, 2002a, Persson *et al.*, 1999, Zhang *et al.*, 2001). (Right) Pentameric assembly of *spinach* lumazine synthase viewed along the fivefold non-crystallographic symmetry axis as seen from the inner capsid wall. The active sites are all located at interfaces of adjacent monomers (Persson *et al.*, 1999; Zhang *et al.*, 2003). (For interpretation of the references to colour in this figure legend, the reader is referred to the Web version of this chapter.)

Gerhardt *et al.*, 2002a; Meining *et al.*, 2000; Morgunova *et al.*, 2005); (ii) d_5-symmetric decamers (*B. abortus*) (Zylberman *et al.*, 2004); (iii) hollow icosahedral capsids (certain eubacteria) (Kumar *et al.*, 2011); and (iv) hetero-oligomers consisting of α_3 cores inside β_{60} capsids (Bacillaceae) (Bacher *et al.*, 1986). Recombinant lumazine synthase of spinach is a hollow icosahedral capsid, but it is unknown whether $\alpha_3\beta_{60}$ complexes can be assembled inside plant cells (Persson *et al.*, 1999).

Sequence comparison has been explored as a method to predict the association state of different lumazine synthase orthologues (Fornasari *et al.*, 2004). However, in practice, the quaternary structure outcome of sequence modification has been incompletely predictable. The assembly process of the icosahedral capsids appears to be very complex; notably, the hypothetical pentamer building blocks of the icosahedral ensembles are conjectural, on basis of the symmetry properties and intersubunit interfacing in the capsid, and have not been directly observed as an isolated molecular species.

Pseudomaturated lumazine synthase from spinach has been crystallised as a recombinant, homo-oligomer consisting of subunits with 156 amino acid residues (Persson *et al.*, 1999). The mRNA predicts a plastid-targeting sequence comprising 66 amino acid residues. The structure was solved at a resolution of 3.1 Å. The subunit fold and the general topology of the 532-symmetric capsid closely resemble those of the icosahedral lumazine synthases of bacterial origin. Notably, the N-terminus of each subunit participates as a fifth strand in the central four-stranded β sheet of the respective, adjacent subunit in the pentamer ensemble.

The active site of lumazine synthase has been subject to intense studies employing substrate, intermediate and product analogues; cocrystal structures implying at least 17 different analogues have been reported. The spinach enzyme has been cocrystallised with the substrate analogue, 5-nitro-6-ribitylamino-2,4(1*H*,3*H*)-pyrimidinedione (**29**, Fig. 13); this relatively strong inhibitor (Otto and Bacher, 1981) differs from the natural substrate **7** only by the replacement of the position-5 amino group by a nitro group. However, in contrast to the strongly oxygen-sensitive substrate **7**, the inhibitor **29** is a stable compound. The pyrimidine ring of the inhibitor is located in a hydrophobic pocket involving the residues Phe22, Ala56, Val81 and Val92 and forms an essentially coplanar π complex with the side chain of Phe22 that is believed to play a major role for binding. The carbonyls of the inhibitor interact with the side chain of Asn23 and the backbone NH group of Ala56. The ribityl side chain is involved in hydrogen bonds with a variety of main- and side-chain elements.

Analogues emulating the structural features of the putative intermediate **24** believed to result from the reaction of the pyrimidine substrate **7** with the carbonyl group of the second substrate **9** have played an important role in order to gain an understanding of the substrate and intermediate topology.

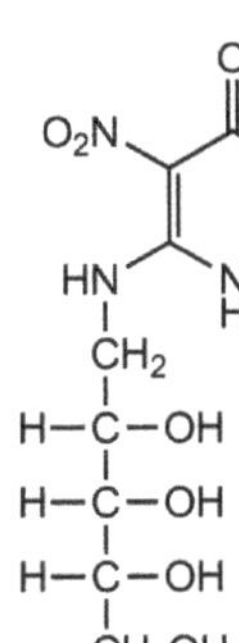

29

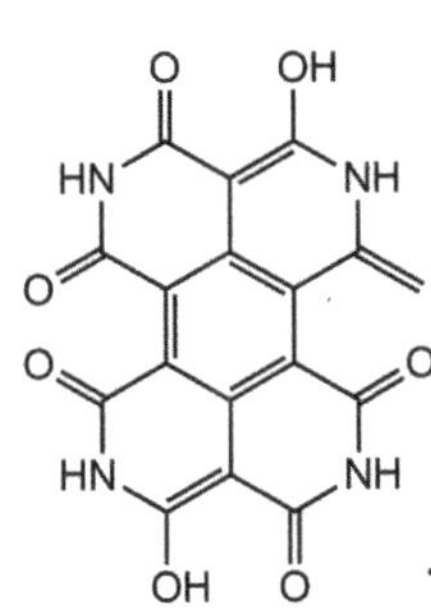

30

Fig. 13. Inhibitors for lumazine synthase (Zhang *et al.*, 2007a).

Specifically, derivatives of ribitylaminopyrimidines, 9-ribitylpurines and 8-ribityllumazines with aliphatic side chains carrying terminal phosphate or phosphonate residues were investigated by X-ray crystallography of cocrystals with various lumazine synthases of microbial or yeast origin. These data leave no doubt that the hypothetical Schiff base intermediate **24** of the lumazine synthase trajectory is bound in an extended conformation, with the negatively charged phosphate residue located in the close proximity of Arg127 and Lys131 and Lys135 residues (*B. subtilis*) (Fischer *et al.*, 2003a; Meining *et al.*, 2000). These results confirmed a hypothesis that had been proposed much earlier on basis of structural work with lumazine synthase from *B. subtilis* that had located an inorganic phosphate residue in that position (Ritsert *et al.*, 1995). It is also noteworthy in this connection that inorganic phosphate is a potent stabiliser of the icosahedral capsid.

The finding that the reaction intermediate **24** must be bound in an extended conformation adds yet another complex twist to the complex reaction mechanism. If indeed the Schiff base **24** is generated with the proposed, extended conformation, a *cis–trans* isomerisation of the imine motif becomes necessary in order to obtain an acceptable reactant topology for the formation of the pyrazine ring of the product.

The extensive work, by several research groups, on substrate and intermediate analogues has afforded potent lumazine synthase inhibitors with K_i values in the nanomolar range (Cushman *et al.*, 1986, 1990, 1991, 1992, 1997, 1998, 1999a,b,c, 2001, 2002a,b, 2004, 2005; Talukdar *et al.*, 2007, 2009; Zhang *et al.*, 2007a,b, 2008; Zhao *et al.*, 2009). Regretfully, none of these compounds caused inhibition of microbial growth, most probably due to difficulties of entry into bacterial or yeast cells.

Screening of compound libraries has also been employed in the search for lumazine synthase inhibitors (Chen *et al.*, 2005). That work has resulted in the discovery of the pentacyclic inhibitor **30** (Zhang *et al.*, 2007a) (Fig. 13).

It is as yet unknown whether plant lumazine synthases can encapsidate riboflavin synthase, in parallel to the $\alpha_3\beta_{60}$ lumazine synthase/riboflavin synthase complexes of bacteria (however, complex formation has been ruled out experimentally for the icosahedral lumazine synthases of certain Gram-negative bacteria) (Bacher *et al.*, 1986). The bifunctional enzyme complex of Bacillaceae does, however, pose numerous unanswered structural and mechanistic problems. Thus, as the homotrimeric α_3 riboflavin synthase is devoid of trimeric symmetry, it is unknown how it could be accommodated inside the 532-symmetric cavity inside the icosahedral capsid. Moreover, at least for the *B. subtilis* enzyme, it remains enigmatic how riboflavin (being the product of the riboflavin synthase moiety in the core of the enzyme complex) can penetrate the capsid wall in order to be unloaded into the bulk solvent compartment. Interestingly, a modelling study has shown that this problem would be non-existent in the case of the spinach enzyme where the channels along the fivefold axis are wide enough to allow the passage of riboflavin into and out of the core of the capsid (Persson *et al.*, 1999).

The detailed structure elucidation of lumazine synthase served as the basis for an extensive mutagenesis study conducted with the *B. subtilis* enzyme (Fischer *et al.*, 2003a). Every residue of the first amino acid shell around the spacious reactive site cavity was subjected to conservative as well as less conservative replacement. Whereas the removal of positively charged residues from the site accommodating the phosphate moiety of the substrate **9** and the early intermediate **24** resulted in a significant decline of enzyme activity, the attempt to nail down any residues that might be involved in covalent catalysis failed. Although the reaction mechanism clearly has a requirement for several proton exchange steps, the search for residues that might facilitate these reactions by acting as acids and/or bases also turned up essentially empty. Based on these data, it must be assumed that lumazine synthase exerts its catalytic effect mainly by establishing a favourable topological relation between the reactants. In other words, the catalytic power of the enzyme appears predominantly based on modulating the activation entropy (Fischer *et al.*, 2003a).

It is in order at this point to remind the readers that the rate acceleration by lumazine synthase is rather modest (Kis *et al.*, 1995). The reaction can run at easily detectable rates without any enzyme catalysis in neutral aqueous solution with concentrations of **7** and **9** in the low millimolar range. In fact, it is mandatory to run blanks without enzyme in order to measure the velocity of the enzyme-catalysed reaction with sufficient accuracy. It is tempting to speculate that there has been significant selective pressure that has locked the evolution of lumazine synthase at the level of a fairly limited rate enhancement.

V. RIBOFLAVIN SYNTHASE

Riboflavin synthase catalyses one of the most puzzling and mechanistically complex reaction trajectories (Fischer and Bacher, 2011). Without any requirement for cofactors, the enzyme engineers the transfer of a 4-carbon unit between two identical substrate molecules. To fully appreciate that catalytic achievement, it must be noted that the 4-carbon segment that must be carved out of the donor substrate is linked to the diaminopyrimidine moiety by two carbon nitrogen bonds which are both part of a heteroaromatic ring system with a significant level of resonance stabilisation. Last but not least, the

4-carbon segment that is cut out of the lumazine donor has been incorporated into the heterocyclic moiety precisely by the preceding reaction step which is catalysed by lumazine synthase; notably, the formation of the lumazine is exergonic and is not detectably reversible. One of the products of riboflavin synthase, the pyrimidine **7**, is structurally identical with the substrate of lumazine synthase; in other words, whereas the lumazine synthase reaction is not *per se* reversible (under regeneration of both substrates, namely **7** and **9**) at a measurable rate, the pyrimidine substrate **7** of lumazine synthase can be regenerated by riboflavin synthase in a reaction that goes to completion; this quasi-reversal of the lumazine formation is enabled, with regard to its thermodynamic aspects, by the resonance stabilisation of the newly formed xylene ring of riboflavin synthase's other product, namely riboflavin. The quantitative conversion of the pyrimidine **7** into riboflavin requires two equivalents of the carbohydrate **9**. That, in conjunction with the unique reaction topology summarised in Fig. 1, has the consequence that, on average, every pyrimidine molecule must be handled twice by both lumazine synthase and riboflavin synthase (note that the emphasis here is on "on average"; stochastically, certain pyrimidine molecules do end up as riboflavin after being handled exactly once by each of the enzymes).

The discovery of riboflavin synthase followed rapidly on the serendipitous discovery of its substrate as a green fluorescent substance (originally named G compound to connotate the colour of its fluorescence) of the naturally flavinogenic ascomycete *E. ashbyii* (Kuwada *et al.*, 1958; Maley and Plaut, 1959; Masuda, 1957b). The structure was obviously similar to riboflavin, and an enzyme activity converting G compound to riboflavin was discovered in cell extracts of *A. gossypii* and also of *E. coli* around 1960. It then came as a major surprise that the enzyme required nothing besides the lumazine (no second substrate and no cofactors) for its conversion into the vitamin (Plaut, 1960, 1963). Classical studies conducted by the group of Plaut then unravelled the 4-carbon transfer described in the previous paragraph.

Also in the 1960s, riboflavin activity was partially purified from spinach (Mitsuda *et al.*, 1965, 1970). However, that early work on riboflavin biosynthesis remained without significant follow-up for more than two decades. The cloning and expression of the genes for lumazine synthase and riboflavin synthase from plants (spinach, tobacco and *A. thaliana*) were reported in patents around 2000 (Bacher and Eberhardt, 2001; Viitanen *et al.*, 2000, 2002). The same patents also revealed the cloning of the genes for both enzymes from the rice pathogen, *M. grisea*. This work was focused on the discovery of potential agents with herbicide or fungicide activity for use in crop protection. Only a single research paper has been published so far on the cloning and expression of riboflavin synthase genes of plant origin (*A. thaliana*) (Fischer *et al.*, 2005).

In contrast to the limited work on plant riboflavin synthase, the enzyme of *E. coli* has been studied in close detail, and the sequence similarity between riboflavin synthases from plants and bacteria suggests that the bacterial data can be extrapolated to the plant enzyme to a significant extent. The basic details of the riboflavin synthase reaction were established in the 1960 (Harvey and Plaut, 1966; Plaut and Harvey, 1971; Plaut *et al.*, 1970a). The early work showed that all carbon atoms of the xylene ring of riboflavin are exclusively derived from the substrate **9**. Specifically, it was shown that the carbon atoms 6α, 6, 7 and 7α of one substrate molecule are sacrificed and serve as building blocks for the extension of the chromophore system of a second substrate molecule that results in the formation of riboflavin (Beach and Plaut, 1970a; Paterson and Wood, 1969, 1972; Sedlmaier *et al.*, 1987). The donor substrate is converted to the pyrimidine **7** in this process. The data strongly suggested that the exchange of the 4-carbon unit between the two substrate molecules was a concerted, intermolecular process that did not proceed via a free 4-carbon unit.

If the enzymatic transformation of a single substrate into a mixture of riboflavin and **7** was puzzling, it was even more surprising that the reaction could proceed without any enzyme catalysis, and under relatively mild conditions (Beach and Plaut, 1970b; Paterson and Wood, 1969; Plaut *et al.*, 1970b). Specifically, riboflavin was formed in high yield when a neutral aqueous solution of the lumazine derivative **10** was refluxed under an inert atmosphere. Later, it was shown that the uncatalysed reaction could also proceed under acidic conditions (Beach and Plaut, 1969).

Based on these observations, initial hypotheses for the reaction mechanisms were proposed by the research groups of Plaut and Wood. These hypotheses set out from the exceptional CH acidity of 8-substituted 6,7-dimethyllumazines. In case of 6,7,8-trimethyllumazine, the abstraction of a proton yields the structurally unusual exomethylene anion **33** (Fig. 14). Deprotonation affords a complex mixture comprising small amounts of the exomethylene anion **33** in equilibrium with at least four molecular species (**34**–**37**) arising by nucleophilic attack of the position 7 carbon atom of **33** by one of the hydroxy groups of the ribityl side chain, whereas the model compound 6,7,8-trimethyllumazine (**31**) has a pK of 9.8 and forms a single anion species (**32**) with a 7-exomethylene group (Beach and Plaut, 1970b, 1971; Bown *et al.*, 1986; Pfleiderer *et al.*, 1971). The riboflavin synthase substrate **10** is even more acidic, with a pK of 8.4 (Pfleiderer and Hutzenlaub, 1973), but the deprotonation affords a complex mixture that is dominated by a series of tricyclic molecular species **34**–**37**, whereas the

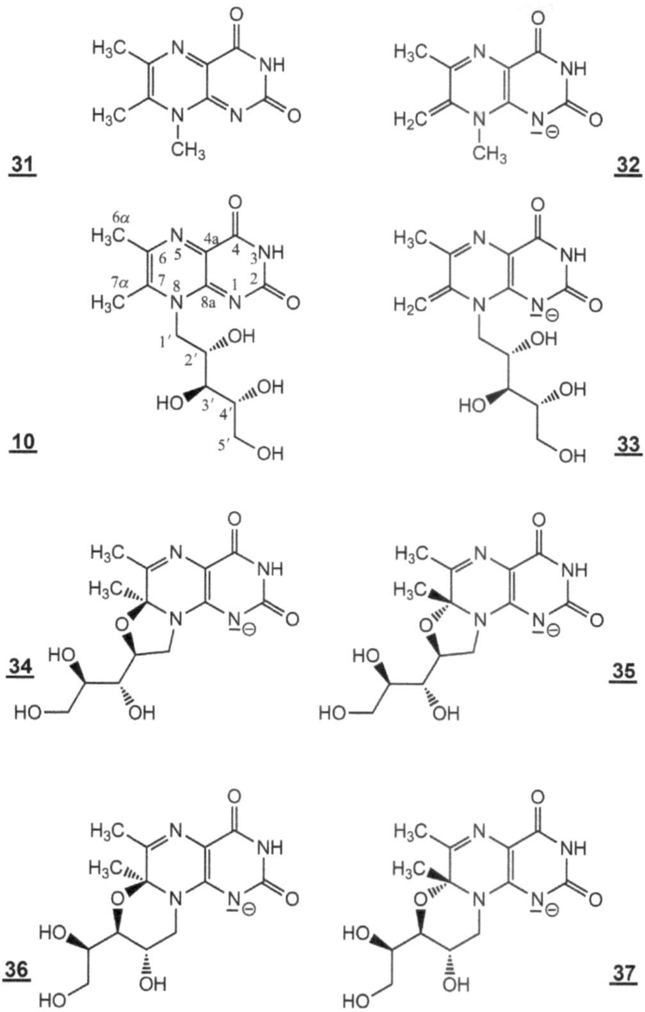

Fig. 14. Structures of lumazine derivatives. **10**, 6,7-dimethyl-8-ribityllumazine; **33**, 7-exomethylene anion of **10**; **34–37**, tricyclic adduct anions of **10**; **31**, 6,7,8-trimethyllumazine; **32**, 6,7,8-trimethyllumazine anion.

exomethylene anion is only present in small amounts. The tricyclic species are believed to arise by nucleophilic attacks of the position 2′- or 3′-hydroxy groups of the ribityl side chain at the C-7 carbon of the exomethylene anion **33**, lumazine derivatives devoid of hydroxy groups in the position 8 substituent can only form the exomethylene anion species (Fig. 15).

Up to date, all mechanistic hypotheses for riboflavin formation use the unique exomethylene anion as a starting point. The initial mechanistic

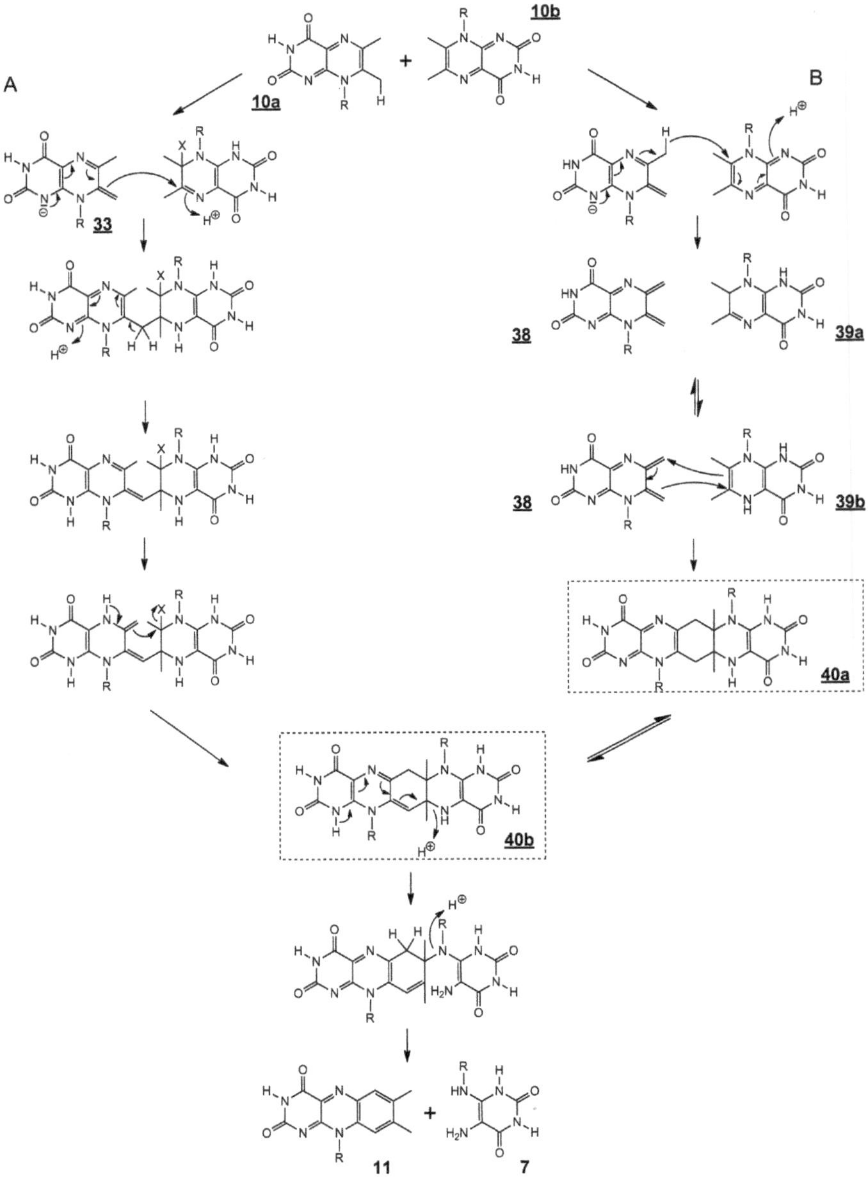

Fig. 15. Hypothetical reaction mechanisms of riboflavin synthase. (A) Mechanism proposed by Plaut, Wood and co-workers (Beach and Plaut, 1970b; Paterson and Wood, 1972; Plaut and Beach, 1976; Plaut *et al.*, 1970b) and modified after discovery of the pentacylic reaction intermediate **40** (Illarionov *et al.*, 2001a); (B) hydride transfer/cycloaddition mechanism (Kim *et al.*, 2010). **10a**, donor lumazine; **10b**, acceptor lumazine; R, ribityl chain.

suggestions of the Plaut and Wood group assumed that the exomethylene anion could execute a nucleophilic attack on a second substrate molecule (Beach and Plaut, 1970b; Plaut and Beach, 1976; Plaut *et al.*, 1970b).

Around 1970, studies with deuterium-labelled substrate established that both the enzyme-catalysed and -uncatalysed reactions are regiospecific in the sense that the incoming and the pre-existing 4-carbon unit of the acceptor substrate are linked in a head-to-tail fashion (Beach and Plaut, 1970a; Fischer *et al.*, 2004b; Paterson and Wood, 1969, 1972; Sedlmaier *et al.*, 1987). That unexpected result necessitated a modification of the mechanism. Later on, Beach and Plaut suggested a mechanism that involves tricyclic intermediates (Rowan and Wood, 1968); that mechanism could be ruled out by later work and is not discussed in detail in this review. Readers are directed to earlier reviews for a detailed discussion of these early mechanistic concepts (Plaut, 1971; Plaut and Beach, 1976; Plaut *et al.*, 1974).

Not surprisingly, as flavocoenzymes are trace metabolites, the enzymes for their biosynthesis have low expression levels. Whereas riboflavin synthase had been purified 4000-fold from yeast (Harvey and Plaut, 1966; Plaut *et al.*, 1970b) and 700-fold from spinach (Mitsuda *et al.*, 1970), the accessibility of reasonably pure protein for biophysical studies was limited. In the 1990s, the accessibility of pure protein in significant amount was improved by recombinant homologous or heterologous expression, and detailed structural and biophysical studies became possible. A major breakthrough was the determination of X-ray structures of riboflavin synthase from *E. coli* (Liao *et al.*, 2001b) and *Schizosaccharomyces pombe* (Gerhardt *et al.*, 2002a) (to be discussed in detail below). Moreover, sufficient amounts of protein were now available for single turnover studies which resulted in the photometric discovery of a transient which could be isolated by preparative rapid quench experiments (Illarionov *et al.*, 2001a). Although transient species was only obtained in substoichiometric amounts (based on the amount of enzyme), its structure could be determined by NMR spectroscopy (the structure determination depended critically on the use of ^{13}C-labelled substrate in order to enhance the sensitivity of the ^{13}C NMR measurements) (Illarionov *et al.*, 2001a, 2003).

The structure of the transient species is shown in Fig. 15; the pentacyclic (**40**) molecule is a covalent dimer of the substrate. Notably, the reaction generates two novel chiral centres. The covalent adduct can serve as a substrate for *E. coli* riboflavin synthase and fulfils the criteria for a kinetically competent intermediate (i.e. it is consumed more rapidly than the substrate **10**) (Illarionov *et al.*, 2001a). The treatment of the adduct with riboflavin synthase was shown to afford three products, namely riboflavin, the pyrimidine **7** and 6,7-dimethyl-8-ribityllumazine (**10**) (upon longer incubation, the latter is consumed under formation of **7** and **11**) (Fig. 16).

The cleavage of the pentacyclic intermediate affording riboflavin and the pyrimidine **7** is easily explained. The tetrahydropyrazine ring of **40b** can be opened by a vinylogous β-elimination, and a second elimination step affords riboflavin and the pyrimidine **7**. However, the formation of the pentacyclic intermediate **40** from the lumazine derivative **10** is anything but easy to explain. An attempt to reconcile the novel information with the Plaut/Wood mechanism is shown in Fig. 15A. However, whereas that hypothesis does not seem to violate established chemical facts, the path from the substrate to the intermediate is long and tortuous.

Figure 15B shows a recent hypothesis that represents a break with established ideas in so far as it introduces a cryptic redox process into the formation of the pentacyclic intermediate (Kim *et al.*, 2010). That hypothesis, which requires further study, will be best discussed after a description of the structures of the bacterial/fungal/plant riboflavin synthase and the independently evolved pentameric riboflavin synthase from Archaea.

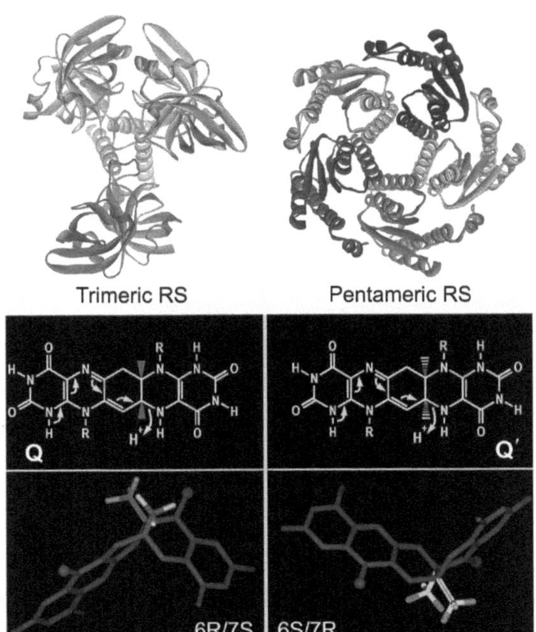

Fig. 16. Stereochemistry of 6,7-dimethyl-8-ribityllumazine conversion into riboflavin catalysed by trimeric eubacterial and pentameric archaeal riboflavin synthase. Q and Q′, pentacyclic reaction intermediates; R, ribityl.

The amino acid sequence of riboflavin synthase from *E. coli* revealed two segments with marked sequence similarity and a C-terminal segment that was devoid of similarity to the other two segments (Schott *et al.*, 1990). This suggested that each subunit could form two domains which would interact to form a pseudo-c_2-symmetric ensemble that would be naturally conducive to the experimentally determined regiospecificity of the riboflavin synthase reaction. The N-terminal domain could be expressed as a stable, recombinant protein which retained the capacity to bind riboflavin with high affinity (Eberhardt *et al.*, 2001), whose three-dimensional structure was determined by X-ray crystallography (Meining *et al.*, 2003) and NMR spectroscopy (Gerhardt *et al.*, 2002b; Liao *et al.*, 2001b; Truffault *et al.*, 2001).

Riboflavin synthase of *B. subtilis* had been shown to be a homotrimer by hydrodynamic studies (Fig. 16). A breakthrough came with the determination of the crystal structure of riboflavin synthase from *E. coli* (Liao *et al.*, 2001b). That structure confirmed that each subunit folds into two domains with closely similar structure. Each domain consists of a six-stranded β-barrel. The two domains of each subunit are related by pseudo-c_2 symmetry. Moreover, the N-terminal domain of one (and only one) subunit forms a dimer contact with the C-terminal domain of one adjacent subunit, and that ensemble is also characterised by pseudo-c_2 symmetry. Last but not least, the homotrimer is kept together by a short triple-helix motif formed by the helical C-termini. Notably, however, the subunits of the homotrimer are not related by trigonal symmetry (the fact that, nevertheless, the Patterson analysis of the refraction data had appeared to indicate some threefold symmetry relations does not in conflict with that, (cf. Liao *et al.*, 2001b)).

Additional information was gleaned from the X-ray structure of riboflavin synthase from the yeast, *S. pombe* (Gerhardt *et al.*, 2002b). Whereas that protein is a trimer in solution, the crystal structure shows only monomers. That is probably due to the breakdown of hydrophobic contacts between the C-terminal helices, which may have been caused by the methylpentanediol (MPD) that had been used as precipitant. Most important, however, each monomeric subunit of the *S. pombe* crystal form carried two molecules of a substrate analogue, with one respective ligand molecule bound to each β-barrel domain. At this point, it was logical to try to model the experimentally observed ligand of the yeast protein into the trimeric structure of the *E. coli* protein (which had been crystallised without ligand) (Gerhardt *et al.*, 2002b).

The model studies leave no doubt that the active site is located between the "special pair" comprising the N-terminal domain of one subunit and the C-terminal domain of the adjacent subunit. Two ligand molecules are bound at the domain interface, each located in a surface depression of one of the adjacent subunits. Due to the pseudo-c_2 symmetric arrangement of the domain pair, the two bound ligands are also related by pseudo-twofold symmetry. This is perfectly in line with the known regiochemistry of the catalytic reaction and had already been predicted on basis of sequence arguments.

Although only just one homotrimer structure is as yet available, one may speculate that the protein has dynamic properties that could allow different domains, at different times, to come close in order to form an active site, although at any one time only a single active site is believed to exist. The "kissing" active site domains could come apart to allow the unloading of products and the loading of new substrate. ^{19}F NMR studies are well in line with the hypothesis that all six domains can bind a given ligand at the same time, but that the ligands loaded to different domains are not in topologically equivalent environments (Fischer *et al.*, 2003b; Scheuring *et al.*, 1994, 1996).

The model study allowed to assign the absolute stereochemistry of two novel stereocentres of the pentameric reaction intermediate **40** that are generated by the dimerisation of **10** (cf. Fig. 15) (Gerhardt *et al.*, 2002b).

The X-ray structure of *S. pombe* riboflavin synthase and the X-ray and NMR structures of the artificial, homodimeric N-terminal domain (obtained by recombinant expression) provide a detailed description of the environment of the bound substrate at the active site (Eberhardt *et al.*, 2001; Meining *et al.*, 2003; Truffault *et al.*, 2001). At both domains, the lumazine chromophore is involved in hydrogen bonding with the backbone amide motifs and in hydrophobic interactions. The ribityl side chain is involved in hydrogen bond contacts with backbone and side chain elements.

Virtually all highly conserved residues of *E. coli* riboflavin synthase have been probed by mutagenesis (Illarionov *et al.*, 2001b). Surprisingly, the replacement or deletion of some residues of the N-terminal MFTG motif caused dramatic reductions of the catalytic activity, although this segment is located at a considerable distance from both substrate binding sites. Replacement of Ser41 or His 102 also has a massive impact on activity, although the amino acids are relatively far remote from the substrate molecules bound at the active site.

The pentacyclic intermediate **40** can be cleaved by the *E. coli* riboflavin synthase in two different directions affording either one equivalent each of riboflavin and the pyrimidine **7** (forward reaction) or two equivalents of 6,7-dimethyl-8-ribityllumazine (**10**) (reverse reaction). The velocities of the forward and backward reaction are similar for the wild-type enzyme but are differentially affected by mutations (Illarionov *et al.*, 2001b). The novel stereocentres of **40** (marked by asterisks in Fig. 15) are quenched by both fragmentation reactions.

Recent ^{13}C NMR studies have shown that the artificial dimer of the N-terminal domain selectively binds and stabilises the exomethylene anion form **33** of the substrate **10** (Kim *et al.*, 2010). None of the tricyclic molecular species can be bound. In full agreement with the crystallographic data, the hypothetical mechanism implying tricyclic lumazine adduct structures such as **34**–**37** is thereby ruled out. The affinity for the exomethylene anion is modulated by replacement of threonine residues **50** and **67** whose side chains are in direct hydrogen bonding contact with N-5 and C-3′ of the bound substrate.

These findings have recently prompted yet another mechanistic hypothesis which introduces the concept of a hidden redox step in order to explain the enigmatic riboflavin synthase reaction trajectory. More specifically, it has been proposed that an exomethylene anion **33** located at the N-terminal domain inside the active site donates a hydride ion to an electroneutral substrate molecule located at the C-terminal domain's binding site. The hydride donor will thereby be converted to the quasi-quinoid dehydrolumazine structure **38**, and the hydride acceptor is converted to the dihydrolumazine **39a**, which is known to be in equilibrium with the tautomer **39b**. The dehydrolumazine **38** and the dihydrolumazine **39b** could then undergo a $4+2$ cycloaddition which is directly conducive to the pentacyclic enantiomer in the tautomer form **40a**. Following conversion to the tautomer **40b**, the pentacyclic intermediate could then be fragmented under formation of riboflavin and **7**. Alternatively, the tautomer **40a** could react backward under regeneration of 6,7-dimethyl-8-riboflavin, a reaction that has indeed been directly observed (see above). In fact, the two tautomers **40a** and **40b** of the pentacyclic intermediate have different, predetermined breaking points which enable the two different exit pathways in the forward and backward direction, respectively (Fischer and Bacher, 2011; Kim *et al.*, 2010) (Fig. 15).

The proposed hydride transfer has precedent in the Cannizzaro reaction which converts aromatic and heterocyclic aldehydes into equimolar mixtures of the cognate alcohol and carboxylic acid. The Cannizzaro reaction proceeds via a hydroxide adduct that is formed under alkaline conditions and donates a hydride to a free aldehyde molecule. Cannizzaro reactions can proceed at room temperature in aqueous solution, literally in the twinkling of an eye, and with excellent yields. It should also be noted that hydride transfers are one of the most common elementary reactions in biochemistry.

Last but not least, it should be noted that the riboflavin synthase domain has significant similarity to certain flavocoenzyme binding domains. Moreover, the topology of the bound riboflavin synthase substrates is essentially the same as that of the flavin cofactor in the respective cofactor binding domains. Obviously, many flavocoenzymes act as professional hydride transponders. By analogy, a hydride transfer as part of the riboflavin synthase is less strange than it may appear at first sight.

Whereas the reaction catalysed by riboflavin synthase shows massive isotope effects (Plaut *et al.*, 1970a), these are unfortunately not suitable to distinguish in a simple way between hypotheses A and B (Fig. 15), as carbon hydrogen bonds need to be broken in both cases. Thus, further data will be required in order to rule out either hypothesis.

Studies in the 1990s showed that Archaea use riboflavin synthases without similarity to the trimeric enzymes from eubacteria, fungi and plants (Eberhardt *et al.*, 1997). The archaeal enzymes are c_5-symmetric homopentamers whose sequence and structure have significant similarity with lumazine synthase (Fischer *et al.*, 2004b; Ramsperger *et al.*, 2006). The reaction proceeds via a pentameric adduct which differs from the intermediate of the homotrimeric enzymes with regard to the stereochemistry. More specifically, the intermediates of trimeric and pentameric riboflavin synthases are diastereomers as shown in Fig. 16. Hence, the pentacyclic intermediate generated by archaeal riboflavin synthase cannot be processed by trimeric riboflavin synthase and vice versa (for review, see Fischer and Bacher, 2008).

VI. RIBOFLAVIN KINASE AND FAD SYNTHETASE

Despite the involvement of early 5′-phosphorylated precursors (compounds **1** and **6**), the *de novo* biosynthetic pathway affords unphosphorylated riboflavin, as 5′-phosphates are unable to serve as substrates for lumazine synthase and riboflavin synthase and must be dephosphorylated prior to being converted into riboflavin by these enzymes. Riboflavin kinase and FAD synthase are therefore required in all organisms to obtain the coenzyme forms, FMN (riboflavin 5′-phosphate) and FAD (flavin adenine dinucleotide), either from biosynthetic of nutritional riboflavin.

Riboflavin kinase (flavokinase, E.C. 2.7.1.26) converts riboflavin into riboflavin 5′-phosphate (FMN) using ATP as phosphate donor (recent work has shown that riboflavin kinase from Archaea uses CTP instead (Mashhadi *et al.*, 2008)). FAD synthetase (EC 2.7.7.2) converts FMN to flavin adenine dinucleotide by the transfer of an adenylate unit under formation of inorganic pyrophosphate as second product. Research on riboflavin kinase and FAD synthetase spans a period of more than five decades. The early work has been reviewed elsewhere (Bacher, 1991).

A bifunctional flavokinase/FAD synthetase was first demonstrated in *Brevibacterium ammoniagenes* (Manstein and Pai, 1986); the N-terminal domain catalyses the formation of FAD, and the C-terminal domain serves as riboflavin kinase. Similar bifunctional proteins are now known to exist in numerous eubacteria. However, riboflavin kinase and FAD synthetase are separate proteins in animals, fungi, plants and certain eubacteria including *E. coli* and *B. subtilis*.

A bifunctional enzyme whose C-terminal domain resembles the monofunctional riboflavin kinase of yeast has been cloned from *A. thaliana*. The N-terminal domain of the plant protein is a member of the haloacid dehydrogenase superfamily which acts as a phosphatase that accepts FMN as a substrate (Sandoval and Roje, 2005). Genes specifying similar riboflavin kinase/FMN hydrolase fusion proteins have been noted in other plants; gymnosperm but not angiosperm orthologues specify N-terminal extensions that are believed to act as organelle-targeting sequences. It has been suggested that the fusion of the riboflavin kinase and FMN phosphatase domain has occurred at an early time in plant evolution.

The *A. thaliana* genome also comprises two genes with similarity to bacterial riboflavin kinase/FAD synthetase. However, whereas the N-terminal domain is clearly a functional FAD synthetase, sequence arguments suggest that the C-terminal domain is unable to function as riboflavin kinase but may have acquired a different, as yet unknown function. Both *A. thaliana* orthologues comprise N-terminal organelle-targeting segments. Orthologous genes are present in other plant genomes. Sequence arguments suggest that the plant FAD synthetases are of cyanobacterial origin (Sandoval *et al.*, 2008; Yruela *et al.*, 2010). The relatively sparse information concerning the cellular topology of riboflavin kinases and FAD synthetases in plants is reported below (see Section VII).

X-ray structures have been reported for monofunctional riboflavin kinases of human (Karthikeyan *et al.*, 2003) and yeast (*S. pombe*) (Bauer *et al.*, 2003) origin, the monofunctional FAD synthetases of yeast (*S. cerevisiae*) (Leulliot *et al.*, 2010) and the thermophilic archaeon *Thermotoga maritima* (Wang *et al.*, 2005), for the bifunctional riboflavin kinase/FAD synthetase of *Corynebacterium ammoniagenes* (Herguedas *et al.*, 2010), and a CTP-dependent archaeal riboflavin kinase of *M. jannaschii* (Ammelburg *et al.*, 2007). Structures of plant orthologues have not been reported as yet.

VII. CELLULAR TOPOLOGY OF FLAVOCOENZYME BIOSYNTHESIS IN PLANTS

The genes that code for the four currently known plant enzymes involved in the biosynthesis of riboflavin (GTP cyclohydrolase II/3,4-dihydroxy-2-butanone 4-phosphate synthase, pyrimidine reductase, lumazine synthase and riboflavin synthase) specify N-terminal segments that fulfil the criteria for plastid-targeting sequences. By comparison with the catalytic domains, the putative targeting segments show a low degree of conservation.

Chloroplasts have been shown to import full-length lumazine synthase *in vitro* (Jordan *et al.*, 1999). More recently, lumazine synthase has been shown to localise to chloroplasts *in vivo* (Wu *et al.*, 2010). Hence, it appears likely that the biosynthesis of riboflavin proceeds entirely inside chloroplasts.

As already mentioned above, the spatial pattern is more complex for riboflavin kinase and FAD synthase which convert riboflavin into the coenzyme forms. Recent data suggest that the conversion of riboflavin into the coenzyme forms can proceed in the cytoplasm as well as inside mitochondria and plastids (Giancaspero *et al.*, 2009; Sandoval *et al.*, 2008; Torchetti *et al.*, 2010; Yruela *et al.*, 2010). Each compartment appears to individually generate its own flavocoenzymes from riboflavin that is excreted into the cytoplasm by chloroplasts and can be extracted from the cytoplasm by mitochondria.

More specifically, the plant-type FAD synthetases carry putative organelle-targeting sequences. However, putative organelle-targeting motifs precede the riboflavin kinase/FMN hydrolase from gymnosperms but not from angiosperms. The present picture is probably incomplete. Notably, the transporters enabling the exchange of riboflavin (and possibly also FAD) between the different compartments are as yet unknown.

VIII. REGULATION OF RIBOFLAVIN BIOSYNTHESIS

The regulation of riboflavin biosynthesis has been studied in considerable detail in Bacillaceae and other Gram-positive bacteria where all riboflavin biosynthesis genes form part of a single operon. The coding region of the operon is preceded by a regulatory region comprising more than 100 bp that had been identified early on by genetic high-resolution mapping (Mironov *et al.*, 1990). Certain mutations in that region are conducive to riboflavin overproduction.

Transcription of the *rib* operon begins with this regulatory region, and the 5′-end of the resulting mRNA folds into a highly specific three-dimensional motif that binds FMN with a K_d of about 10 nM and with remarkably high selectivity. FMN binding is accompanied by an allosteric transition of this so-called riboswitch (Gelfand *et al.*, 1999).

When the *rib* operon (preceding the coding region) is transcribed in the absence of FMN, the riboswitch structure acts as an antiterminator which allows the transcription of the operon to go to completion. However, if FMN binds to the riboswitch at the 5′-end of the nascent mRNA, transcription of the operon is aborted (Vitreschak *et al.*, 2002, 2004; Winkler *et al.*, 2002). Certain missense mutations in the *ribC* gene specifying riboflavin kinase are conducive to riboflavin overproduction by lowering the intracellular concentration of FMN.

The riboswitch mechanism has been initially discovered by studies on the *rib* operon of *B. subtilis* but is now known to be involved in the control of numerous metabolic pathways in bacteria. A detailed description of this regulatory principle exceeds the limits of this chapter, and readers are directed to excellent reviews of the topic (Mandal and Breaker, 2004; Sutak *et al.*, 2008; Vitreschak *et al.*, 2004).

Studies on the regulation of flavin biosynthesis in plants have been mostly centred around aspects of iron availability as summarised below. At a molecular level, these impressive phenomena are still incompletely understood.

IX. EXCRETION AND ENHANCED FORMATION OF RIBOFLAVIN BY IRON-DEFICIENT ROOTS

Whereas iron is one of the most abundant elements in the earth crust, it is not readily available to plants. Ferric iron (Fe^{3+}) is poorly soluble, especially at relatively high pH, and the situation is further aggravated in soil containing calcium carbonate. Ferrous iron (Fe^{2+}) has substantially higher solubility but is subject to oxidation under aerobic conditions. Hence, a substantial fraction of the global flora, and notably crop plants for human nutrition, grow under conditions of iron deficiency.

Two strategies have evolved in the plant community in order to deal with iron deprivation: (i) the excretion of protons, organic acids and flavins in conjunction with upregulation of Fe(III) chelate reductase (Vorwieger *et al.*, 2007) and (ii) the excretion of chelators (Vorwieger *et al.*, 2007).

In plants using strategy I, iron deficiency is conducive to massive enhancements of riboflavin biosynthesis in root tips at the transcriptomic, proteomic and metabolic level.

Evidence for increased production and for excretion of flavins by the hair roots of certain plants under iron deficiency stress has been building over a period of several decades (Higa *et al.*, 2008, 2010; Lopez-Millan *et al.*, 2000; Susin *et al.*, 1994 and references cited therein). Early reports noted that iron-deficient hair roots developed yellow autofluorescence, whose spectral character was well in line with the optical properties of flavins. Sugar beets were shown to accumulate hitherto unknown 3′- and 5′-sulphates of riboflavin under conditions of iron deficiency (Susin *et al.*, 1993). Their concentration in extracts of iron-deficient roots reached values above 1 mM, well above of the solubility limit of unsubstituted riboflavin.

More recent studies have shown dramatic increases in the root concentrations of riboflavin biosynthetic enzymes under iron deficiency conditions. Thus, *Medicago truncultata* roots grown under conditions of iron deficiency and calcium excess showed 99-fold increase of lumazine synthase and 36-fold increase of GTP cyclohydrolase II/3,4-dihydroxy-2-butanone 4-phosphate synthase; the increase of deaminase and riboflavin synthase was 24- and 6-fold, respectively (Rodriguez-Celma *et al.*, 2011). A massive increase of lumazine synthase and the cognate mRNA was also observed in roots of sugar beets (*Beta vulgaris*) (Rellan-Alvarez *et al.*, 2010).

The modulation of biosynthesis and excretion of flavins by roots of iron-deficient plants must be viewed in the context of a complex network including, but not limited to Fe(III) chelate reductase, a member of the flavochrome superfamily which shuttles electrons across membranes, and a transporter for Fe(II) (Li *et al.*, 2004; Mori, 1999; Robinson *et al.*, 1999; Waters *et al.*, 2002). These membrane proteins have been characterised in some detail but cannot be addressed in detail in the present review. Various hypotheses trying to integrate the modules of this network into a coherent model have been proposed. Thus, it has been argued that the enhanced activity of the Fe(III) chelate reductase causes an increased requirement for flavocoenzymes to serve as cofactors. More direct roles for flavins have also been proposed, such as complexation of iron ions. However, the proposed models fail to describe the interactions in the complex system in significant detail and remain speculative.

In *A. thaliana* seedlings, the mRNA level for two transcription factors (AtbHLH38 (At3g56970) and AtbHLH39 (At3g56980)) showed a massive increase within days after initiation of iron deficiency conditions. The expression of these genes in tobacco was conducive to riboflavin synthesis and excretion in transgenic tobacco plants even under conditions of abundant iron supply (Vorwieger *et al.*, 2007).

Despite the abundance of iron in the earth crust, not only plants but also many millions of humans suffer from insufficient iron supplies. Notably, for populations whose nutrition is predominantly plant based, plants are the main dietary iron source. Thus, iron deficiency of crop plants affects human nutrition by reduced crop size on one hand and poor iron supply on the other hand (Mori, 1999).

Iron supply and riboflavin biosynthesis are also known to be connected in certain yeasts and bacteria, but the research extending over many decades cannot be addressed in this chapter (Almeida *et al.*, 2009; Crossley *et al.*, 2007; Hsu *et al.*, 2011; Sutak *et al.*, 2008).

X. EVOLUTION OF FLAVOCOENZYME BIOSYNTHESIS ENZYMES IN PLANTS

In contrast to the shikimate pathway which appears to have been maintained constant throughout evolution, the riboflavin pathway is a theme with variations. For the first time, this has become evident in the 1970s with the discovery that the pathway proceeds via different, although structurally similar intermediates (compounds **4** and **5**, respectively) in eubacteria and fungi. Work in recent years has revealed the riboflavin pathway of plants as a patchwork that appears to have been assembled from pieces acquired from different taxa.

Several features of the plant pathway suggest a bacterial origin. (i) The first committed intermediates of both branches of the converging riboflavin pathway, that is, **2** and **9**, are both generated by a single fusion protein whose general architecture and sequence shows strong resemblance with bacterial orthologues (Herz *et al.*, 2000). (ii) Pyrimidine deamination precedes the reduction of the ribityl side chain; hence, the plant pathway proceeds via the bacterial intermediate **4** (Fischer *et al.*, 2004a). (iii) In line with that, the sequence of the plant deaminase resembles the bacterial orthologues. Notably, however, the bacterial prototypes are fusion proteins comprising a deaminase and a reductase domain which jointly catalyse two consecutive reactions of the riboflavin pathway. (iv) The plant-type FAD synthetase resembles the bifunctional riboflavin kinase/FAD synthetase that is present in numerous eubacteria; notably, however, the C-terminal domain of the plant protein, despite its general sequence similarity to the bacterial riboflavin kinase domain, may now catalyse a different reaction (if any). All of these features may have been brought to plants via the genome of a cyanobacterial endosymbiont.

Several features of the plant pathway are shared with eubacteria and fungi, but not with Archaea. (i) The formation of the pyrimidine type intermediate **2** from GTP is catalysed by a single enzyme, GTP cyclohydrolase II, in plants, eubacteria and fungi, whereas Archaea require the consecutive action of two enzymes, neither of which has similarity to GTP cyclohydrolase II; notably, however, the product of the first-acting archaeal enzyme, the formamide **3**, has been shown to serve as a (non-released) intermediate of GTP cyclohydrolase II. (ii) Plants, eubacteria and fungi use heterotrimeric riboflavin synthases whose subunits fold into two closely similar domains, whereas Archaea use homopentameric riboflavin synthases which are paralogs of lumazine synthase.

Plants, many eubacteria and Archaea use lumazine synthases consisting of 60 identical subunits; the quasispherical proteins have icosahedral 532 symmetry (see Section IV for details). Fungi and certain eubacteria use pentameric lumazine synthases.

XI. RIBOFLAVIN BIOSYNTHETIC ENZYMES AS POTENTIAL HERBICIDE TARGETS

As flavocoenzymes are essential in all organisms, the enzymes of riboflavin biosynthesis are potential targets for herbicides and fungicides for crop protection. However, they are also potential targets for novel anti-infective drugs. In each case, it is advantageous that animals do not synthesise riboflavin; hence, there would be no risk of target-related toxicity.

Assays for high-throughput screening (HTS) with riboflavin biosynthesis enzymes have been reported (Kaiser *et al.*, 2007). Specifically, a HTS assay for 3,4-dihydroxy-2-butanone 4-phosphate synthase can be monitored using riboflavin synthase and/or lumazine synthase as auxiliary enzyme(s) in order to generate a signal that can be recorded photometrically in the visible range. The reactions catalysed by lumazine synthase and riboflavin synthase can be monitored photometrically or flurometrically, respectively. It is also possible to include riboflavin synthase as auxiliary enzyme in order to screen for inhibitors of lumazine synthase (Kaiser *et al.*, 2007).

XII. RIBOFLAVIN AND PLANT RESISTANCE

Over the past decade, a number of papers from different research groups have reported that riboflavin can enhance the resistance of various plant species including rice against certain viral, bacterial and fungal pathogens. The mechanisms of riboflavin-mediated resistance involved enhanced formation of reactive oxygen species (ROS) and lignin. Specifically, riboflavin treatment was found to prime the expression of lipoxygenase and upregulation of phenylalanine ammonia lyase (Aver'yanov *et al.*, 2000; Dong and Beer, 2000; Taheri and Hofte, 2007; Taheri and Tarighi, 2010; Zhang *et al.*, 2009).

Recent studies have shown that the expression of rice lumazine synthase in transgenic tobacco results in elevated levels of jasmonic acid and ethylene production. Moreover, the expression of defence-related genes and resistance to viral, bacterial and oomycete pathogens were increased (Wu *et al.*, 2010).

Earlier, lumazine synthase had already been implicated in the regulation of jasmonate-mediated defence by genetic studies on the *coronatine insensitive1* (*Coi1*) gene of *A. thaliana* (Xiao *et al.*, 2004). The failure of the recessive *coi1* mutant to react to jasmonate results in a pleiotrophic phenotype involving male sterility as well as sensitivity to insect attack and pathogen infection. COI1 protein interacts with CULLIN1, RBX1 and the Skp1-like proteins ASK1 and ASK2 to form a multi-protein E3 ubiquitin ligase complex designated SCF^{COI1}, which catalyses the ubiquitination of proteins destined for proteasomal degradation. The *cos1* (*coi1 supressor1*) mutation restores the defects of the *coi1* mutation. *cos1 coi1*-double mutants have wild-type-like phenotypes with regard to senescence and defence response. The *cos1* gene was isolated by a map-based cloning strategy and was found to specify lumazine synthase. The *cos1* mutation involves a single nucleotide change that results in the destruction of the splice site at the start of the second exon and is believed to result in the formation of a lumazine synthase with a deletion of six amino acids and with reduced catalytic activity. The authors conclude that the riboflavin biosynthesis pathway is relevant for signal transmission downstream from jasmonate.

Virus-induced silencing of the *ribA* gene specifying the bifunctional GTP cyclohydrolase II/3,4-dihydroxy-2-butanone 4-phosphate synthase of *Nicotiana benthamiana* caused the loss of hypersensitivity response via cell death and the synthesis of NO (nitric oxide) and ROS (reactive oxygen species) (Asai *et al.*, 2010). As expected, silencing of the *ribA* gene also reduced the levels of endogenous riboflavin, FMN and FAD in the plants. RibA-silenced leaves showed high sensitivity for infection by the oomycete *Phytophtora infestans.* The compromised phenotypes could be rescued by treatment with riboflavin, FMN or FAD.

Whereas animals do not biosynthesise riboflavin, it is noteworthy that human riboflavin kinase couples TNF receptor 1 (tumour necrosis factor receptor 1) to NADPH oxidase, thus playing an important role in the generation of ROS that function as defence and signalling molecules related to innate immunity and various cellular responses (Yazdanpanah *et al.*, 2009).

XIII. BIOSYNTHESIS OF 5-DEAZA-7,8-DIDEMETHYL-8-HYDROXY-RIBOFLAVIN IN ALGAE

It has been known for some time that cyanobacteria employ DNA photolyases that use 7,8-dimethyl-8-hydroxy-5-deazaflavin (FO, **46**, Fig. 17) as light-harvesting chromophore (Carell *et al.*, 2001). Recent studies indicate that the deazaflavin also serves as the second chromophore of DNA photolyase of green algae (Eker *et al.*, 1988; Glas *et al.*, 2009; Maul *et al.*, 2008; Müller and Carell, 2009; Petersen and Ronan, 2011), although not of higher plants. This implicates that the deazaflavin, which was initially discovered in Archaea, is considerably more widespread than had been originally assumed (Eirich *et al.*, 1978, 1979). As the biosynthesis of FO is a side branch of the

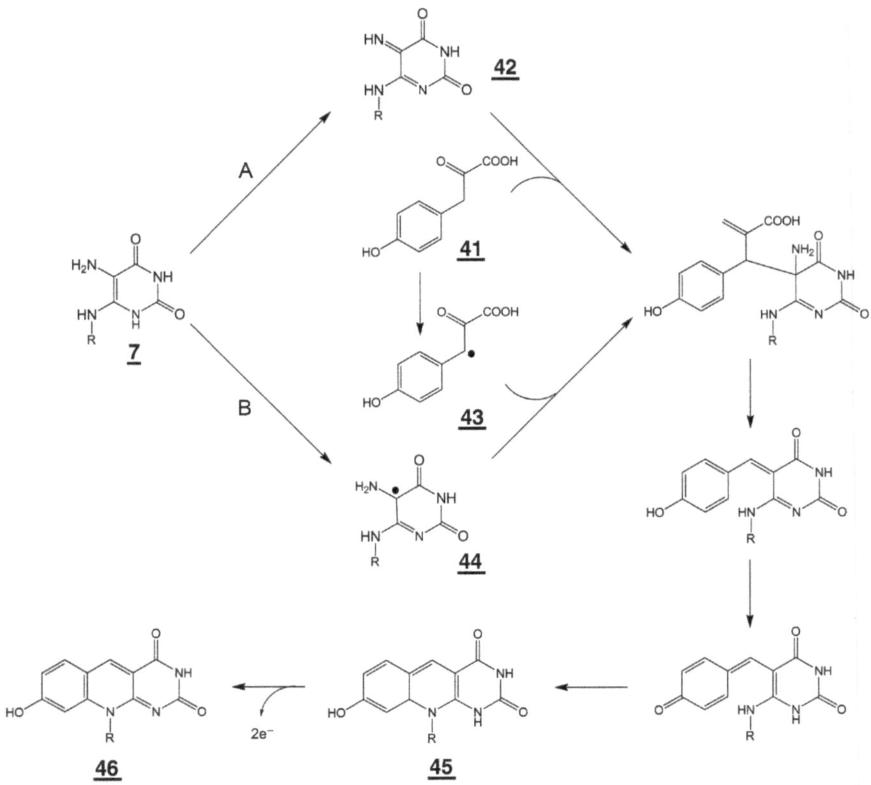

Fig. 17. Hypothetical mechanisms for biosynthesis of 5-deaza-7,8-didemethyl-8-hydroxy-riboflavin (FO, **46**). **7**, 5-Amino-6-ribitylamino-2,4(1*H*,3*H*)-pyrimidinedione; **41**, 4-hydroxyphenylpyruvate. (A) Via the quinoid pyrimidine intermediate **42** (Eisenreich, 1991; Reuke *et al.*, 1992); (B) by free radical recombination of **43** and **44** (Graham *et al.*, 2003).

riboflavin biosynthetic pathway, the state of knowledge on its presence and utilisation in algae and its biosynthesis is summarised below.

DNA photolyases catalyse the light-driven repair of light-induced damage in the DNA of light-exposed cells (for review, see Sancar, 2008). Specifically, two different types of DNA photolyases carry out the repair of cyclobutane pyrimidine dimers and of (6-4) thymidine dimers. DNA photolyases are widespread in nature but are absent in mammals where thymidine dimers are exclusively processed by excision repair.

The reaction catalysed by DNA photolyases starts with the absorption of a photon (for review, see Sancar, 2008). A fully reduced $FADH_2$ chromophore in the optically excited S1 state transfers an electron to the pyrimidine dimer lesion under generation of a radical pair. The free-radical-type intermediate then undergoes fragmentation, and an electron is transferred back to the flavin; thus, there is no net transfer of redox equivalents. All known DNA photolyases use $FADH_2$ as the chromophore for catalysis. Besides that, a second chromophore is used for efficient light harvesting. Energy is then transferred from the antenna chromophore to the catalytic $FADH_2$ centre by Förster transfer.

As light-harvesting chromophore, DNA photolyases can use a tetrahydrofolate derivative (5,10-methenyltetrahydrofolate), a second flavin cofactor (FAD in the fully oxidised state), or factor FO. Higher plants use folate derivatives, and so do many eubacteria.

The field of DNA photolyases has been progressing vigorously, with an average of at least 10 papers per year, and exceeds the scope of this review (cf. Sancar, 2008). However, we will briefly review the evidence for deazaflavin biosynthesis and use in algae and the literature on deazaflavin biosynthesis in Archaea and certain eubacteria.

FO (**46**) was initially identified as the chromophoric moiety of coenzyme F_{420} of methanogenic bacteria, which plays a central role in methanogenesis. FO was also shown to serve as antenna for DNA photolyases of non-methanogenic Archaea and of mycobacteria. Moreover, recent evidence suggests the occurrence of numerous coenzyme F_{420}-dependent enzymes in *M. tuberculosis* (Selengut and Haft, 2010).

Early studies showed a biosynthetic relationship between the deazaflavin cofactor and purine nucleotides (Eisenreich *et al.*, 1991; Jaenchen *et al.*, 1984; Le Van *et al.*, 1985; Schwarzkopf *et al.*, 1990). Moreover, the phenolic ring was shown to be derived from the shikimate pathway. The riboflavin precursor 7 and 4-hydroxyphenylpyruvate (**41**) were subsequently identified as specific biosynthetic precursors (Reuke *et al.*, 1992) (interestingly, *Methanobacterium thermoautotrophicum* was found to convert the radiolabelled **7** into FO (**46**) without noticeable dilution).

More recently, the *cofG* and *cofH* gene products were shown to be sufficient for the formation of FO from **7** and **41** in *M. jannaschii* (Graham *et al.*, 2003). Both peptides comprise CXXXCXXC motifs which are believed to enable the binding of [4Fe–4S] clusters essential for catalysis. Putative orthologues of *cofG* and *cofH* have been found in blue-green algae. The recombinant coexpression of CofG and CofH in *E. coli* enables the formation of small amounts of of FO *in vivo*, and cell extracts catalyse FO formation *in vitro*, albeit at low levels in the range of 100 pmol mg^{-1} min^{-1}.

In *Mycobacterium smegmatis*, the FbiC (EC 2.5.1.77) fusion protein comprising segments with homology to CofG and CofH has been reported to be sufficient for the formation of FO from **7** and **41** (Graham *et al.*, 2003; Guerra-Lopez *et al.*, 2007). Specifically, the N-terminal half of the bifunctional FO synthases resembles CofG, and the C-terminal half is most similar to CofH of *M. jannaschii*.

The *phr1* mutant of the green alga, *Chlamydomonas reinhardtii,* has a severe defect in photoreactivation after ultraviolet irradiation. Recently, it was found that the PHR1 gene is similar to the *fbiC* gene of mycobacteria (Petersen and Ronan, 2011). The gene specifies a putative, N-terminal plastid-targeting sequence which precedes the catalytic domains. A recombinant *E. coli* strain expressing the *phr1* gene (without the plastid-targeting sequence) generated detectable amounts of FO. This is well in line with recent work indicating that the green alga *Ostreococcus tauri* uses FO as antenna for DNA photolyase (Glas *et al.*, 2009).

Whereas the reaction catalysed by FbiC protein has been as yet incompletely characterised, a plausible scenario is shown in Fig. 17 (Eisenreich *et al.*, 1991; Reuke *et al.*, 1992). The primary reaction step, to be discussed in more detail below, could yield a covalent adduct of the pyrimidine **7** and 4-hydroxyphenylpyruvate (**41**), which could subsequently lose ammonia and oxalate by a coordinated elimination reaction. Tautomerisation could prepare the reaction product for a nucleophilic attack of the phenolate moiety (in its quinoid tautomer form, **42**) by the ribityl-substituted amino group, thus affording the tricyclic intermediate **45** which could be converted to FO (**46**) by a two-electron oxidation step; as this reaction is accompanied by aromatisation of the carbocyclic ring, it would be highly exergonic.

An early proposal for the formation of the covalent adduct involved the oxidative conversion of the pyrimidine **7** into a quinoid form (**42**) which could then serve as an electrophile attacking the enolate form of 4-hydroxyphenylpyruvate (**41**). More recently, it has been proposed that the FO synthases (FbiC, Phr1 and CofF/CofG, respectively) could strip one electron each from the pyrimidine **7** and from 4-hydroxyphenylpyruvate (**41**); the two radicals (**43**, **44**) could then combine to form a further intermediate **45**. This proposal is based on the presence of amino acid motifs that are characteristic binding sites for iron sulphur clusters and on reaction accelerations in crude cell extracts upon addition of dithionite, iron ions and S-adenosylmethionine. However, the involvement of iron sulphur clusters and S-adenosylmethionine would *per se* not rule out a quinoid intermediate whose formation they might in fact enable.

XIV. CONCLUSIONS

Whereas plants are the most important producers of flavins for human and animal nutrition, the investigation of the biochemical background of flavin metabolism in plants had a tardive start. Due to the emerging technology of data base mining, however, the knowledge gap could be substantially narrowed during the past decade.

Flavins are essential for all organisms. As enzymes for riboflavin biosynthesis are essential in plants and in many plant pathogens and human pathogens, but absent in humans and in animals, they represent potential targets for novel herbicides, fungicides and anti-infective drugs which should all be exempt from target-related toxicity. Specifically, the silencing of the *ribA* gene of *N. benthamiana* is conducive to a light-sensitive phenotype; thus, inhibitors of riboflavin biosynthesis might have an attractive mode of action as herbicides. Notably, no novel mode of action has been introduced for commercial herbicides since three decades, and the emergence of herbicide resistance in conjunction with the growth of the human population is cause for concern. The knowledge basis for target-oriented search for agents directed against plant enzymes for riboflavin biosynthesis is now available, although there is still a dearth of structural biology work; only a single riboflavin biosynthesis enzyme of plant origin has been studied by X-ray crystallography.

The topology of flavocoenzyme biosynthesis and metabolism in plant cells and the trafficking of proteins and metabolites between different cell compartments are still incompletely known, despite considerable recent progress.

Connections between the regulation of riboflavin biosynthesis, iron acquisition and resistance mechanisms involving the generation of nitrous oxide and ROS are now firmly documented, but their mechanisms are poorly understood. A better understanding of these phenomena at the molecular and cellular level could have important ramifications for plant breeding and biotechnology. Notably, crop plants grow under conditions of iron limitation in large geographic areas; moreover, plants are important providers of iron for human and animal nutrition, and insufficient iron supply is the cause of iron deficiency anaemia affecting 20% of women, 50% of pregnant women and 3% of men.

REFERENCES

Almeida, R. S., Wilson, D. and Hube, B. (2009). *Candida albicans* iron acquisition within the host. *FEMS Yeast Research* **9**, 1000–1012.

Ammelburg, M., Hartmann, M. D., Djuranovic, S., Alva, V., Koretke, K. K., Martin, J., Sauer, G., Truffault, V., Zeth, K., Lupas, A. N. and Coles, M. (2007). A CTP-dependent archaeal riboflavin kinase forms a bridge in the evolution of cradle-loop barrels. *Structure* **15**, 1577–1590.

Asai, S., Mase, K. and Yoshioka, H. (2010). A key enzyme for flavin synthesis is required for nitric oxide and reactive oxygen species production in disease resistance. *The Plant Journal* **62**, 911–924.

Aver'Yanov, A. A., Lapikova, V. P., Nikolaev, O. N. and Stepanov, A. I. (2000). Active oxygen-associated control of rice blast disease by riboflavin and roseoflavin. *Biochemistry (Moscow)* **65**, 1292–1298.

Bacher, A. (1991). Riboflavin kinase and FAD synthetase. *In* Chemistry and Biochemistry of Flavoproteins, (F.Müller, ed.), **1**, pp. 349–370. CRC Press Inc., Boca Raton, FL.

Bacher, A. and Eberhardt, S. (2001). Cloning and characterization of riboflavin synthase from Arabidopsis thaliana and screening for riboflavin synthase-inhibiting herbicides. Application: WO patent application.

Bacher, A. and Lingens, F. (1970). Biosynthesis of riboflavin. Formation of 2,5-diamino-6-hydroxy-4-(1′-D-ribitylamino)pyrimidine in a riboflavin auxotroph. *The Journal of Biological Chemistry* **245**, 4647–4652.

Bacher, A. and Mailänder, B. (1978). Biosynthesis of riboflavin in *Bacillus subtilis*: Function and genetic control of the riboflavin synthase complex. *Journal of Bacteriology* **134**, 476–482.

Bacher, A., Baur, R., Eggers, U., Harders, H. D., Otto, M. K. and Schnepple, H. (1980). Riboflavin synthases of *Bacillus subtilis*. Purification and properties. *The Journal of Biological Chemistry* **255**, 632–637.

Bacher, A., Ludwig, H. C., Schnepple, H. and Ben-Shaul, Y. (1986). Heavy riboflavin synthase from *Bacillus subtilis*. Quaternary structure and reaggregation. *Journal of Molecular Biology* **187**, 75–86.

Bacher, A., Eisenreich, W., Kis, K., Ladenstein, R., Richter, G., Scheuring, J. and Weinkauf, S. (1993). Biosynthesis of flavins. *In* Bioorganic Chemistry Frontiers, (H. Dugas and F. P. Schmidtchen, eds.), pp.147–192. Springer Verlag.

Bacher, A., Eberhardt, S., Fischer, M., Kis, K. and Richter, G. (2000). Biosynthesis of vitamin B_2 (riboflavin). *Annual Review of Nutrition* **20**, 153–167.

Bacher, A., Eberhardt, S., Eisenreich, W., Fischer, M., Herz, S., Illarionov, B., Kis, K. and Richter, G. (2001). Biosynthesis of riboflavin. *Vitamins and Hormones* **61**, 1–49.

Bauer, S., Kemter, K., Bacher, A., Huber, R., Fischer, M. and Steinbacher, S. (2003). Crystal structure of *Schizosaccharomyces pombe* riboflavin kinase reveals a novel ATP and riboflavin-binding fold. *Journal of Molecular Biology* **326**, 1463–1473.

Beach, R. L. and Plaut, G. W. E. (1969). The formation of riboflavin from 6,7-dimethyl-8-ribityllumazine in acid media. *Tetrahedron Letters* **40**, 3489–3492.

Beach, R. L. and Plaut, G. W. (1970a). Stereospecificity of the enzymatic synthesis of the o-xylene ring of riboflavin. *Journal of the American Chemical Society* **92**, 2913–2916.

Beach, R. L. and Plaut, G. W. E. (1970b). Investigations of structures of substituted lumazines by deuterium exchange and nuclear magnetic resonance spectroscopy. *Biochemistry* **9**, 760–770.

Beach, R. L. and Plaut, G. W. E. (1971). Synthesis, properties, and base-catalyzed interactions of 8-substituted 6,7-dimethyllumazines. *The Journal of Organic Chemistry* **36**, 3937–3943.

Bown, D. H., Keller, P. J., Floss, H. G., Sedlmaier, H. and Bacher, A. (1986). Solution structures of 6,7-dimethyl-8-substituted-lumazines. Carbon-13 NMR evidence for intramolecular ether formation. *The Journal of Organic Chemistry* **51**, 2461–2467.

Braden, B. C., Velikovsky, C. A., Cauerhff, A. A., Polikarpov, I. and Goldbaum, F. A. (2000). Divergence in macromolecular assembly: X-ray crystallographic structure analysis of lumazine synthase from *Brucella abortus. Journal of Molecular Biology* **297**, 1031–1036.

Brown, G. M. and Neims, A. (1982). *Advances in Enzymology* **53**, 345.

Brown, G. M. and Reynolds, J. J. (1963). Biogenesis of the water-soluble vitamins. *Annual Review of Biochemistry* **32**, 419–462.

Brown, G. M. and Williamson, J. M. (1987). Biosynthesis of folic acid, riboflavin, thiamine and pantothenic acid. *In Escherichia coli* and *Salmonella thyphimurium*, (F. C. Neidhardt, ed.), **1**, pp.521–538, American Society for Microbiology, Washington, DC.

Burrows, R. B. and Brown, G. M. (1978). Presence of *Escherichia coli* of a deaminase and a reductase involved in biosynthesis of riboflavin. *Journal of Bacteriology* **136**, 657–667.

Carell, T., Burgdorf, L. T., Kundu, L. M. and Cichon, M. (2001). The mechanism of action of DNA photolyases. *Current Opinion in Chemical Biology* **5**, 491–498.

Chatwell, L., Krojer, T., Fidler, A., Römisch, W., Eisenreich, W., Bacher, A., Huber, R. and Fischer, M. (2006). Biosynthesis of riboflavin: Structure and properties of 2,5-diamino-6-ribosylamino-4(3H)-pyrimidinone 5′-phosphate reductase of *Methanocaldococcus jannaschii. Journal of Molecular Biology* **359**, 1334–1351.

Chen, J., Illarionov, B., Bacher, A., Fischer, M., Haase, I., Georg, G., Ye, Q. Z., Ma, Z. and Cushman, M. (2005). A high-throughput screen utilizing the fluorescence of riboflavin for identification of lumazine synthase inhibitors. *Analytical Biochemistry* **338**, 124–130.

Chen, S. C., Chang, Y. C., Lin, C. H. and Liaw, S. H. (2006). Crystal structure of a bifunctional deaminase and reductase from *Bacillus subtilis* involved in riboflavin biosynthesis. *The Journal of Biological Chemistry* **281**, 7605–7613.

Chen, S. C., Lin, Y. H., Yu, H. C. and Liaw, S. H. (2009). Complex structure of *Bacillus subtilis* RibG: The reduction mechanism during riboflavin biosynthesis. *The Journal of Biological Chemistry* **284**, 1725–1731.

Crossley, R. A., Gaskin, D. J., Holmes, K., Mulholland, F., Wells, J. M., Kelly, D. J., van Vliet, A. H. and Walton, N. J. (2007). Riboflavin biosynthesis is associated with assimilatory ferric reduction and iron acquisition by *Campylobacter jejuni. Applied and Environmental Microbiology* **73**, 7819–7825.

Cushman, M., Wong, W. C. and Bacher, A. (1986). Synthesis of 6,7-bis(trifluoromethyl)-8-substituted pteridine-2,4(1*H*,3*H*)-dione (lumazine) hydrates from 4,5-diaminouracil hydrochlorides and perfluorobutane-2,3-dione. Stabilization of the transmolecular covalent hydrates of 8-substituted pteridinediones by trifluoromethyl groups. *Journal of Chemical Society, Perkin Transactions, 1,* 1051–1053.

Cushman, M., Patel, H. H., Patrick, D. A., Bacher, A. and Schott, K. (1990). Synthesis of fluorinated 8-ribityllumazines as fluorine-19 NMR probes and potential inhibitors of the light riboflavin synthase of Bacillus subtilis. *Chem. Biol. Pteridines, 1989 Proc. Int. Symp. Pteridines Folic Acid Deriv.* **9th**, 249–254.

Cushman, M., Patrick, D. A., Bacher, A. and Scheuring, J. (1991). Synthesis of epimeric 6,7-bis(trifluoromethyl)-8-ribityllumazine hydrates. Stereoselective interaction with the light riboflavin synthase of *Bacillus subtilis*. *The Journal of Organic Chemistry* **56**, 4603–4608.

Cushman, M., Patel, H. H., Scheuring, J. and Bacher, A. (1992). Fluorine-19 NMR studies on the mechanism of riboflavin synthase. Synthesis of 6-(trifluoromethyl)-7-oxo-8-(D-ribityl)lumazine and 6-(trifluoromethyl)-7-methyl-8-(D-ribityl)lumazine. *The Journal of Organic Chemistry* **57**, 5630–5643.

Cushman, M., Mavandadi, F., Kugelbrey, K. and Bacher, A. (1997). Design and synthesis of (ribitylamino) uracils bearing fluorosulfonyl, sulfonic acid, and carboxylic acid functionality as inhibitors of lumazine synthase. *The Journal of Organic Chemistry* **62**, 8944–8947.

Cushman, M., Mavandadi, F., Kugelbrey, K. and Bacher, A. (1998). Synthesis of 2,6-dioxo-(1*H*,3*H*)-9-N-ribitylpurine and 2,6-dioxo-(1*H*,3*H*)-8-aza-9-N-ribitylpurine as inhibitors of lumazine synthase and riboflavin synthase. *Bioorganic & Medicinal Chemistry* **6**, 409–415.

Cushman, M., Mavandadi, F., Yang, D., Kugelbrey, K., Kis, K. and Bacher, A. (1999a). Synthesis and biochemical evaluation of bis(6,7-dimethyl-8-D-ribityllumazines) as potential bisubstrate analogue inhibitors of riboflavin synthase. *The Journal of Organic Chemistry* **64**, 4635–4642.

Cushman, M., Mihalic, J. T., Kis, K. and Bacher, A. (1999b). Design and synthesis of 6-(6-D-ribitylamino-2,4-dihydroxypyrimidin-5-yl)-1-hexyl phosphonic acid, a potent inhibitor of lumazine synthase. *Bioorganic & Medicinal Chemistry Letters* **9**, 39–42.

Cushman, M., Mihalic, J. T., Kis, K. and Bacher, A. (1999c). Design, synthesis, and biological evaluation of homologous phosphonic acids and sulfonic acids as inhibitors of lumazine synthase. *The Journal of Organic Chemistry* **64**, 3838–3845.

Cushman, M., Yang, D., Kis, K. and Bacher, A. (2001). Design, synthesis, and evaluation of 9-D-ribityl-1,3,7-trihydro-2,6,8-purinetrione, a potent inhibitor of riboflavin synthase and lumazine synthase. *The Journal of Organic Chemistry* **66**, 8320–8327.

Cushman, M., Yang, D., Gerhardt, S., Huber, R., Fischer, M., Kis, K. and Bacher, A. (2002a). Design, synthesis, and evaluation of 6-carboxyalkyl and 6-phosphonoxyalkyl derivatives of 7-oxo-8-ribitylaminolumazines as inhibitors of riboflavin synthase and lumazine synthase. *The Journal of Organic Chemistry* **67**, 5807–5816.

Cushman, M., Yang, D., Mihalic, J. T., Chen, J., Gerhardt, S., Huber, R., Fischer, M., Kis, K. and Bacher, A. (2002b). Incorporation of an amide into 5-phosphonoalkyl-6-D-ribitylaminopyrimidinedione lumazine synthase inhibitors results in an unexpected reversal of selectivity for riboflavin synthase vs lumazine synthase. *The Journal of Organic Chemistry* **67**, 6871–6877.

Cushman, M., Sambaiah, T., Jin, G., Illarionov, B., Fischer, M. and Bacher, A. (2004). Design, synthesis, and evaluation of 9-D-ribitylamino-1,3,7,9-tetrahydro-2,6,8-purinetriones bearing alkyl phosphate and alpha, alpha-difluorophosphonate substituents as inhibitors of tiboflavin synthase and lumazine synthase. *The Journal of Organic Chemistry* **69**, 601–612.

Cushman, M., Jin, G., Sambaiah, T., Illarionov, B., Fischer, M., Ladenstein, R. and Bacher, A. (2005). Design, synthesis, and biochemical evaluation of 1,5,6,7-tetrahydro-6,7-dioxo-9-D-ribitylaminolumazines bearing alkyl phosphate substituents as inhibitors of lumazine synthase and riboflavin synthase. *The Journal of Organic Chemistry* **70**, 8162–8170.

Dong, H. and Beer, S. V. (2000). Riboflavin induces disease resistance in plants by activating a novel signal transduction pathway. *Phytopathology* **90**, 801–811.

Eberhardt, S., Korn, S., Lottspeich, F. and Bacher, A. (1997). Biosynthesis of riboflavin: An unusual riboflavin synthase of *Methanobacterium thermoautotrophicum*. *Journal of Bacteriology* **179**, 2938–2943.

Eberhardt, S., Zingler, N., Kemter, K., Richter, G., Cushman, M. and Bacher, A. (2001). Domain structure of riboflavin synthase. *European Journal of Biochemistry* **268**, 4315–4323.

Echt, S., Bauer, S., Steinbacher, S., Huber, R., Bacher, A. and Fischer, M. (2004). Potential anti-infective targets in pathogenic yeasts: Structure and properties of 3,4-dihydroxy-2-butanone 4-phosphate synthase of *Candida albicans*. *Journal of Molecular Biology* **341**, 1085–1096.

Eirich, L. D., Vogels, G. D. and Wolfe, R. S. (1978). Proposed structure for coenzyme F420 from Methanobacterium. *Biochemistry* **17**, 4583–4593.

Eirich, L. D., Vogels, G. D. and Wolfe, R. S. (1979). Distribution of coenzyme F_{420} and properties of its hydrolytic fragments. *Journal of Bacteriology* **140**, 20–27.

Eisenreich, W., Schwarzkopf, B. and Bacher, A. (1991). Biosynthesis of nucleotides, flavins, and deazaflavins in *Methanobacterium thermoautotrophicum*. *Journal of Molecular Biology* **266**, 9622–9631.

Eker, A. P. M., Hessels, J. K. C. and van de Velde, J. (1988). Photoreactivating enzyme from the green alga *Scenedesmus acutus*. Evidence for the presence of two different flavin chromophores. *Biochemistry* **27**, 1758–1765.

Fischer, M. and Bacher, A. (2005). Biosynthesis of flavocoenzymes. *Natural Product Reports* **22**, 324–350.

Fischer, M. and Bacher, A. (2006). Biosynthesis of vitamin B_2 in plants. *Physiologia Plantarum* **126**, 304–318.

Fischer, M. and Bacher, A. (2008). Biosynthesis of vitamin B_2: Structure and mechanism of riboflavin synthase. *Archives of Biochemistry and Biophysics* **474**, 252–265.

Fischer, M. and Bacher, A. (2011). Biosynthesis of vitamin B_2: A unique way to assemble a xylene ring. *ChemBioChem* **12**, 670–680.

Fischer, M., Römisch, W., Schiffmann, S., Kelly, M., Oschkinat, H., Steinbacher, S., Huber, R., Eisenreich, W., Richter, G. and Bacher, A. (2002). Biosynthesis of riboflavin in archaea: Studies on the mechanism of 3,4-dihydroxy-2-butanone-4-phosphate synthase of *Methanococcus jannaschii*. *The Journal of Biological Chemistry* **277**, 41410–41416.

Fischer, M., Haase, I., Kis, K., Meining, W., Ladenstein, R., Cushman, M., Schramek, N., Huber, R. and Bacher, A. (2003a). Enzyme catalysis via control of activation entropy: Site-directed mutagenesis of 6,7-dimethyl-8-ribityllumazine synthase. *Journal of Molecular Biology* **326**, 783–793.

Fischer, M., Schott, A. K., Kemter, K., Feicht, R., Richter, G., Illarionov, B., Eisenreich, W., Gerhardt, S., Cushman, M., Steinbacher, S., Huber, R. and Bacher, A. (2003b). Riboflavin synthase of *Schizosaccharomyces pombe*. Protein dynamics revealed by ^{19}F NMR protein perturbation experiments. *BMC Biochemistry* **4**, 18.

Fischer, M., Römisch, W., Saller, S., Illarionov, B., Richter, G., Rohdich, F., Eisenreich, W. and Bacher, A. (2004a). Evolution of vitamin B_2 biosynthesis: Structural and functional similarity between pyrimidine deaminases of eubacterial and plant origin. *The Journal of Biological Chemistry* **279**, 36299–36308.

Fischer, M., Schott, A. K., Römisch, W., Ramsperger, A., Augustin, M., Fidler, A., Bacher, A., Richter, G., Huber, R. and Eisenreich, W. (2004b). Evolution of vitamin B_2 biosynthesis. A novel class of riboflavin synthase in Archaea. *Journal of Molecular Biology* **343**, 267–278.

Fischer, M., Haase, I., Feicht, R., Schramek, N., Kohler, P., Schieberle, P. and Bacher, A. (2005). Evolution of vitamin B_2 biosynthesis: Riboflavin synthase of *Arabidopsis thaliana* and its inhibition by riboflavin. *Biological Chemistry* **386**, 417–428.

Foor, F. and Brown, G. M. (1975). Purification and properties of guanosine triphosphate cyclohydrolase II from *Escherichia coli*. *The Journal of Biological Chemistry* **250**, 3545–3551.

Fornasari, M. S., Laplagne, D. A., Frankel, N., Cauerhff, A. A., Goldbaum, F. A. and Echave, J. (2004). Sequence determinants of quaternary structure in lumazine synthase. *Molecular Biology and Evolution* **21**, 97–107.

Gelfand, M. S., Mironov, A. A., Jomantas, J., Kozlov, Y. I. and Perumov, D. A. (1999). A conserved RNA structure element involved in the regulation of bacterial riboflavin synthesis genes. *Trends in Genetics* **15**, 439–442.

Gerhardt, S., Haase, I., Steinbacher, S., Kaiser, J. T., Cushman, M., Bacher, A., Huber, R. and Fischer, M. (2002a). The structural basis of riboflavin binding to *Schizosaccharomyces pombe* 6,7-dimethyl-8-ribityllumazine synthase. *Journal of Molecular Biology* **318**, 1317–1329.

Gerhardt, S., Schott, A. K., Kairies, N., Cushman, M., Illarionov, B., Eisenreich, W., Bacher, A., Huber, R., Steinbacher, S. and Fischer, M. (2002b). Studies on the reaction mechanism of riboflavin synthase: X-ray crystal structure of a complex with 6-carboxyethyl-7-oxo-8-ribityllumazine. *Structure* **10**, 1371–1381.

Giancaspero, T. A., Locato, V., de Pinto, M. C., de Gara, L. and Barile, M. (2009). The occurrence of riboflavin kinase and FAD synthetase ensures FAD synthesis in tobacco mitochondria and maintenance of cellular redox status. *The FEBS Journal* **276**, 219–231.

Glas, A. F., Maul, M. J., Cryle, M., Barends, T. R., Schneider, S., Kaya, E., Schlichting, I. and Carell, T. (2009). The archaeal cofactor FO is a light-harvesting antenna chromophore in eukaryotes. *Proceedings of the National Academy of Sciences of the United States of America* **106**, 11540–11545.

Graham, D. E., Xu, H. and White, R. H. (2002). A member of a new class of GTP cyclohydrolases produces formylaminopyrimidine nucleotide monophosphates. *Biochemistry* **41**, 15074–15084.

Graham, D. E., Xu, H. and White, R. H. (2003). Identification of the 7,8-didemethyl-8-hydroxy-5-deazariboflavin synthase required for coenzyme F_{420} biosynthesis. *Archives of Microbiology* **180**, 455–464.

Grochowski, L. L., Xu, H. and White, R. H. (2009). An iron(II) dependent formamide hydrolase catalyzes the second step in the archaeal biosynthetic pathway to riboflavin and 7,8-didemethyl-8-hydroxy-5-deazariboflavin. *Biochemistry* **48**, 4181–4188.

Guerra-Lopez, D., Daniels, L. and Rawat, M. (2007). *Mycobacterium smegmatis* mc2 155 fbiC and MSMEG_2392 are involved in triphenylmethane dye decolorization and coenzyme F_{420} biosynthesis. *Microbiology* **153**, 2724–2732.

Haase, I., Fischer, M., Bacher, A. and Schramek, N. (2003). Temperature-dependent presteady state kinetics of lumazine synthase from the hyperthermophilic eubacterium *Aquifex aeolicus. The Journal of Biological Chemistry* **278**, 37909–37915.

Harvey, R. A. and Plaut, G. W. E. (1966). Riboflavin synthase from yeast: Properties of complexes of the enzyme with lumazine derivatives and riboflavin. *The Journal of Biological Chemistry* **241**, 2120–2136.

Herguedas, B., Martinez-Julvez, M., Frago, S., Medina, M. and Hermoso, J. A. (2010). Oligomeric state in the crystal structure of modular FAD synthetase provides insights into its sequential catalysis in prokaryotes. *Journal of Molecular Biology* **400**, 218–230.

Herz, S., Eberhardt, S. and Bacher, A. (2000). Biosynthesis of riboflavin in plants. The ribA gene of *Arabidopsis thaliana* specifies a bifunctional GTP cyclohydrolase II/3,4-dihydroxy-2-butanone 4-phosphate synthase. *Phytochemistry* **53**, 723–731.

Higa, A., Miyamoto, E., UR Rahman, L. and Kitamura, Y. (2008). Root tip-dependent, active riboflavin secretion by *Hyoscyamus albus* hairy roots under iron deficiency. *Plant Physiology and Biochemistry* **46**, 452–460.

Higa, A., Mori, Y. and Kitamura, Y. (2010). Iron deficiency induces changes in riboflavin secretion and the mitochondrial electron transport chain in hairy roots of *Hyoscyamus albus. Journal of Plant Physiology* **167**, 870–878.

Hollander, I. and Brown, G. M. (1979). Biosynthesis of riboflavin: Reductase and deaminase of *Ashbya gossypii. Biochemical and Biophysical Research Communications* **89**, 759–763.

Hsu, P.-C., Yang, C.-Y. and Lan, C.-Y. (2011). *Candida albicans* Hap43 is a repressor induced under low-iron conditions and is essential for iron-responsive transcriptional regulation and virulence. *Eukaryotic Cell* **10**, 207–225.

Illarionov, B., Eisenreich, W. and Bacher, A. (2001a). A pentacyclic reaction intermediate of riboflavin synthase. *Proceedings of the National Academy of Sciences of the United States of America* **98**, 7224–7229.

Illarionov, B., Kemter, K., Eberhardt, S., Richter, G., Cushman, M. and Bacher, A. (2001b). Riboflavin synthase of *Escherichia coli.* Effect of single amino acid substitutions on reaction rate and ligand binding properties. *The Journal of Biological Chemistry* **276**, 11524–11530.

Illarionov, B., Haase, I., Bacher, A., Fischer, M. and Schramek, N. (2003). Presteady state kinetic analysis of riboflavin synthase. *The Journal of Biological Chemistry* **278**, 47700–47706.

Jaenchen, R., Schonheit, P. and Thauer, R. K. (1984). Studies on the biosynthesis of coenzyme F_{420} in methanogenic bacteria. *Archives of Microbiology* **137**, 362–365.

Jordan, D. B., Bacot, K. O., Carlson, T. J., Kessel, M. and Viitanen, P. V. (1999). Plant riboflavin biosynthesis. Cloning, chloroplast localization, expression, purification, and partial characterization of spinach lumazine synthase. *The Journal of Biological Chemistry* **274**, 22114–22121.

Kaiser, J., Schramek, N., Eberhardt, S., Puttmer, S., Schuster, M. and Bacher, A. (2002). Biosynthesis of vitamin B_2. An essential zinc ion at the catalytic site of GTP cyclohydrolase II. *European Journal of Biochemistry* **269**, 5264–5270.

Kaiser, J., Illarionov, B., Rohdich, F., Eisenreich, W., Saller, S., van den Brulle, J., Cushman, M., Bacher, A. and Fischer, M. (2007). A high-throughput screening platform for inhibitors of the riboflavin biosynthesis pathway. *Analytical Biochemistry* **365**, 52–61.

Karthikeyan, S., Zhou, Q., Mseeh, F., Grishin, N. V., Osterman, A. L. and Zhang, H. (2003). Crystal structure of human riboflavin kinase reveals a beta barrel fold and a novel active site arch. *Structure (Cambridge)* **11**, 265–273.

Katagiri, H., Takeda, I. and Imai, K. (1958a). Synthesis of riboflavin by microorganisms. I. The metabolic function of aspartate in the riboflavin synthesis by *Clostridium acetobutylicum*. *The Journal of Vitaminology (Kyoto)* **4**, 156–162.

Katagiri, H., Takeda, I. and Imai, K. (1958b). Synthesis of riboflavin by microorganisms. III. On the role of G compound. *The Journal of Vitaminology (Kyoto)* **4**, 211–216.

Katagiri, H., Takeda, I. and Imai, K. (1958c). Synthesis of riboflavin by microorganisms. IV. The studies of 4-C-donor involved in the enzymatic riboflavin synthesis from 8-N-ribityl-6, 7-dimethyllumazine. *The Journal of Vitaminology (Kyoto)* **4**, 278–284.

Katagiri, H., Takeda, I. and Imai, K. (1958d). Synthesis of riboflavin by microorganisms. V. Properties of enzyme system involved in the enzymatic riboflavin synthesis from 8-N-ribityl-6, 7-dimethyllumazine. *The Journal of Vitaminology (Kyoto)* **4**, 285–298.

Katagiri, H., Yamada, H. and Imai, K. (1958e). On the transglycosidation relating to riboflavin by *Escherichia coli*. II. Formation of riboflavin compounds of oligosaccharides. *The Journal of Vitaminology (Kyoto)* **4**, 126–131.

Keller, P. J., le Van, Q., Kim, S. U., Bown, D. H., Chen, H. C., Kohnle, A., Bacher, A. and Floss, H. G. (1988). Biosynthesis of riboflavin: Mechanism of formation of the ribitylamino linkage. *Biochemistry* **27**, 1117–1120.

Kelly, M. J., Ball, L. J., Krieger, C., Yu, Y., Fischer, M., Schiffmann, S., Schmieder, P., Kuhne, R., Bermel, W., Bacher, A., Richter, G. and Oschkinat, H. (2001). The NMR structure of the 47-kDa dimeric enzyme 3,4-dihydroxy-2-butanone-4-phosphate synthase and ligand binding studies reveal the location of the active site. *Proceedings of the National Academy of Sciences of the United States of America* **98**, 13025–13030.

Kim, R. R., Illarionov, B., Joshi, M., Cushman, M., Lee, C. Y., Eisenreich, W., Fischer, M. and Bacher, A. (2010). Mechanistic insights on riboflavin synthase inspired by selective binding of the 6,7-dimethyl-8-ribityllumazine exomethylene anion. *Journal of the American Chemical Society* **132**, 2983–2990.

Kis, K. and Bacher, A. (1995). Substrate channeling in the lumazine synthase/riboflavin synthase complex of *Bacillus subtilis*. *The Journal of Biological Chemistry* **270**, 16788–16795.

Kis, K., Volk, R. and Bacher, A. (1995). Biosynthesis of riboflavin. Studies on the reaction mechanism of 6,7-dimethyl-8-ribityllumazine synthase. *Biochemistry* **34**, 2883–2892.

Kis, K., Kugelbrey, K. and Bacher, A. (2001). Biosynthesis of riboflavin. The reaction catalyzed by 6,7-dimethyl-8-ribityllumazine synthase can proceed without enzymatic catalysis under physiological conditions. *The Journal of Organic Chemistry* **66**, 2555–2559.

Klinke, S., Zylberman, V., Vega, D. R., Guimaraes, B. G., Braden, B. C. and Goldbaum, F. A. (2005). Crystallographic studies on decameric *Brucella* spp. Lumazine synthase: A novel quaternary arrangement evolved for a new function? *Journal of Molecular Biology* **353**, 124–137.

Klinke, S., Zylberman, V., Bonomi, H. R., Haase, I., Guimaraes, B. G., Braden, B. C., Bacher, A., Fischer, M. and Goldbaum, F. A. (2007). Structural and kinetic properties of lumazine synthase isoenzymes in the order *Rhizobiales*. *Journal of Molecular Biology* **373**, 664–680.

Kobayashi, M., Ohara-Nemoto, Y., Kaneko, M., Hayakawa, H., Sekiguchi, M. and Yamamoto, K. (1998). Potential of *Escherichia coli* GTP cyclohydrolase II for hydrolyzing 8-oxo-dGTP, a mutagenic substrate for DNA synthesis. *The Journal of Biological Chemistry* **273**, 26394–26399.

Koch, M., Breithaupt, C., Gerhardthaase, S., Weber, S., Cushman, M., Huber, R., Bacher, A. and Fischer, M. (2004). Structural basis of charge transfer complex formation by riboflavin bound to 6,7-dimethyl-8-ribityllumazine synthase. *European Journal of Biochemistry* **271**, 3208–3214.

Kumar, P., Singh, M., Gautam, R. and Karthikeyan, S. (2010). Potential antibacterial drug target: Structural characterization of 3,4-dihydroxy-2-butanone-4-phosphate synthase from *Salmonella typhimurium* LT2. *Proteins* **78**, 3292–3303.

Kumar, P., Singh, M. and Karthikeyan, S. (2011). Crystal structure analysis of icosahedral lumazine synthase from *Salmonella typhimurium*, an antibacterial drug target. *Acta Crystallographica, Section D, Biological Crystallography* **67**, 131–139.

Kuwada, S., Masuda, T., Kishi, T. and Asai, M. (1958). On the nomenclature of the fluorescent substances produced in the culture of *Eremothecium ashbyii*. *Chemical & Pharmaceutical Bulletin* **6**, 447.

Ladenstein, R., Meyer, B., Huber, R., Labischinski, H., Bartels, K., Bartunik, H. D., Bachmann, L., Ludwig, H. C. and Bacher, A. (1986). Heavy riboflavin synthase from *Bacillus subtilis*. Particle dimensions, crystal packing and molecular symmetry. *Journal of Molecular Biology* **187**, 87–100.

Ladenstein, R., Schneider, M., Huber, R., Bartunik, H. D., Wilson, K., Schott, K. and Bacher, A. (1988). Heavy riboflavin synthase from *Bacillus subtilis*. Crystal structure analysis of the icosahedral beta 60 capsid at 3.3 Å resolution. *Journal of Molecular Biology* **203**, 1045–1070.

le Van, Q., Schwarzkopf, B., Bacher, A., Keller, P. J., Lee, S. and Floss, H. G. (1985). Biosynthesis of 7,8-didemethyl-8-hydroxy-5-deazariboflavin, the chromophoric moiety of coenzyme F_{420}. *Journal of the American Chemical Society* **107**, 8300–8301.

Leulliot, N., Blondeau, K., Keller, J., Ulryck, N., Quevillon-Cheruel, S. and van Tilbeurgh, H. (2010). Crystal structure of yeast FAD synthetase (Fad1) in complex with FAD. *Journal of Molecular Biology* **398**, 641–646.

Li, L., Cheng, X. and Ling, H. Q. (2004). Isolation and characterization of Fe(III)-chelate reductase gene LeFRO1 in tomato. *Plant Molecular Biology* **54**, 125–136.

Liao, D. I., Calabrese, J. C., Wawrzak, Z., Viitanen, P. V. and Jordan, D. B. (2001a). Crystal structure of 3,4-dihydroxy-2-butanone 4-phosphate synthase of riboflavin biosynthesis. *Structure (Cambridge)* **9**, 11–18.

Liao, D. I., Wawrzak, Z., Calabrese, J. C., Viitanen, P. V. and Jordan, D. B. (2001b). Crystal structure of riboflavin synthase. *Structure (Cambridge)* **9**, 399–408.

Liao, D. I., Zheng, Y. J., Viitanen, P. V. and Jordan, D. B. (2002). Structural definition of the active site and catalytic mechanism of 3,4-dihydroxy-2-butanone-4-phosphate synthase. *Biochemistry* **41**, 1795–1806.

Lopez-Millan, A. F., Morales, F., Andaluz, S., Gogorcena, Y., Abadia, A., De Las Rivas, J. and Abadia, J. (2000). Responses of sugar beet roots to iron deficiency. Changes in carbon assimilation and oxygen use. *Plant Physiology* **124**, 885–898.

Maclaren, J. A. (1952). The effects of certain purines and pyrimidines upon the production of riboflavin by *Eremothecium ashbyii*. *Journal of Bacteriology* **63**, 233–241.

Magalhaes, M. L., Argyrou, A., Cahill, S. M. and Blanchard, J. S. (2008). Kinetic and mechanistic analysis of the *Escherichia coli ribD*-encoded bifunctional deaminase-reductase involved in riboflavin biosynthesis. *Biochemistry* **47**, 6499–6507.

Maley, G. F. and Plaut, G. W. (1959). The isolation, synthesis, and metabolic properties of 6,7-dimethyl-8-ribityllumazine. *The Journal of Biological Chemistry* **243**, 641–647.

Mandal, M. and Breaker, R. R. (2004). Gene regulation by riboswitches. *Nature Reviews. Molecular Cell Biology* **5**, 451–463.

Manstein, D. J. and Pai, E. F. (1986). Purification and characterization of FAD synthetase from *Brevibacterium ammoniagenes. The Journal of Biological Chemistry* **261**, 16169–16173.

Mashhadi, Z., Zhang, H., Xu, H. and White, R. H. (2008). Identification and characterization of an archaeon-specific riboflavin kinase. *Journal of Bacteriology* **190**, 2615–2618.

Masuda, T. (1957a). Application of chromatography. XXXI. Structure of a green fluorescent substance produced by *Eremothecium ashbyii. Pharmaceutical Bulletin* **5**, 28–30.

Masuda, T. (1957b). Application of chromatography. XXXII. Biosynthesis of riboflavin by Eremothecium ashbyii. *Pharmaceutical Bulletin* **5**, 136–141.

Maul, M. J., Barends, T. R., Glas, A. F., Cryle, M. J., Domratcheva, T., Schneider, S., Schlichting, I. and Carell, T. (2008). Crystal structure and mechanism of a DNA (6–4) photolyase. *Angewandte Chemie International Edition in English* **47**, 10076–10080.

Meining, W., Mörtl, S., Fischer, M., Cushman, M., Bacher, A. and Ladenstein, R. (2000). The atomic structure of pentameric lumazine synthase from *Saccharomyces cerevisiae* at 1.85 Å resolution reveals the binding mode of a phosphonate intermediate analogue. *Journal of Molecular Biology* **299**, 181–197.

Meining, W., Eberhardt, S., Bacher, A. and Ladenstein, R. (2003). The structure of the N-terminal domain of riboflavin synthase in complex with riboflavin at 2.6 Å resolution. *Journal of Molecular Biology* **331**, 1053–1063.

Mironov, V. N., Kraev, A. S., Chernov, B. K., Ul'Ianov, A. V. and Golova Iu, B. (1989). Riboflavin biosynthesis genes of *Bacillus subtilis*—Complete primary structure and organization model. *Doklady Akademii Nauk SSSR* **305**, 482–487.

Mironov, V. N., Perumov, D. A., Kraev, A. S., Stepanov, A. I. and Skriabin, K. G. (1990). Unusual structure of the regulatory region of the riboflavin biosynthesis operon in *Bacillus subtilis. Molekuliarnaia Biologiia* **24**, 256–261.

Mitsuda, H., Suzuki, Y., Tadera, K. and Kawai, F. (1965). Biochemical studies on pteridines in plants. I. biogenesis of folic acid in green leaves: Confirmation of enzymatic synthesis of folate compounds by the enzyme system from the spinach. *The Journal of Vitaminology* **11**, 122–138.

Mitsuda, H., Kawai, F., Suzuki, Y. and Yoshimoto, S. (1970). Biogenesis of riboflavin in green leaves. VII. Isolation and characterization of spinach riboflavin synthetase. *The Journal of Vitaminology* **16**, 285–292.

Moore, B. (2004). Bifunctional and moonlighting enzymes: Lighting the way to regulatory control. *Trends in Plant Science* **9**, 221–228.

Morgunova, E., Meining, W., Illarionov, B., Haase, I., Jin, G., Bacher, A., Cushman, M., Fischer, M. and Ladenstein, R. (2005). Crystal structure of lumazine synthase from *Mycobacterium tuberculosis* as a target for rational drug design: Binding mode of a new class of purinetrione inhibitors. *Biochemistry* **44**, 2746–2758.

Morgunova, E., Illarionov, B., Sambaiah, T., Haase, I., Bacher, A., Cushman, M., Fischer, M. and Ladenstein, R. (2006). Structural and thermodynamic insights into the binding mode of five novel inhibitors of lumazine synthase from *Mycobacterium tuberculosis*. *The FEBS Journal* **273**, 4790–4804.

Morgunova, E., Saller, S., Haase, I., Cushman, M., Bacher, A., Fischer, M. and Ladenstein, R. (2007). Lumazine synthase from *Candida albicans* as an antifungal target enzyme: Structural and biochemical basis for drug design. *The Journal of Biological Chemistry* **282**, 17231–17241.

Morgunova, E., Illarionov, B., Saller, S., Popov, A., Sambaiah, T., Bacher, A., Cushman, M., Fischer, M. and Ladenstein, R. (2011). Structural study and thermodynamic characterization of inhibitor binding to lumazine synthase from *Bacillus anthracis*. *Acta Crystallographica, Section D, Biological Crystallography* **66**, 1001–1011.

Mori, S. (1999). Iron acquisition by plants. *Current Opinion in Plant Biology* **2**, 250–253.

Müller, M. and Carell, T. (2009). Structural biology of DNA photolyases and cryptochromes. *Current Opinion in Structural Biology* **19**, 277–285.

Neuberger, G. and Bacher, A. (1986). Biosynthesis of riboflavin. Enzymatic formation of 6,7-dimethyl-8-ribityllumazine by heavy riboflavin synthase from *Bacillus subtilis*. *Biochemical and Biophysical Research Communications* **139**, 1111–1116.

Nielsen, P., Neuberger, G., Fujii, I., Bown, D. H., Keller, P. J., Floss, H. G. and Bacher, A. (1986). Biosynthesis of riboflavin. Enzymatic formation of 6,7-dimethyl-8-ribityllumazine from pentose phosphates. *The Journal of Biological Chemistry* **261**, 3661–3669.

Otto, M. K. and Bacher, A. (1981). Ligand-binding studies on light riboflavin synthase from *Bacillus subtilis*. *European Journal of Biochemistry* **115**, 511–517.

Paterson, T. and Wood, H. C. S. (1969). Deuterium exchange of C7-methyl protons in 6,7-dimethyl-8-D-ribityllumazine, and studies of the mechanism of riboflavin biosynthesis. *Journal of Chemical Society Communications* 290–291.

Paterson, T. and Wood, H. C. (1972). The biosynthesis of pteridines. VI. Studies of the mechanism of riboflavin biosynthesis. *Journal of Chemical Society, Perkin Transactions 1* **8**, 1051–1056.

Persson, K., Schneider, G., Douglas, B. J., Viitanen, P. V. and Sandalova, T. (1999). Crystal structure analysis of a pentameric fungal and icosahedral plant lumazine synthase reveals the structural basis of differences in assembly. *Protein Science* **8**, 2355–2365.

Petersen, J. L. and Ronan, P. J. (2011). Critical role of 7,8-didemethyl-8-hydroxy-5-deazariboflavin for photoreactivation in *Chlamydomonas reinhardtii*. *The Journal of Biological Chemistry* **285**, 32467–32475.

Pfleiderer, W. and Hutzenlaub, W. (1973). Pteridines. LVII. Synthesis and properties of lumazine N-oxides. *Chemische Berichte* **106**, 3149–3174.

Pfleiderer, W., Mengel, R. and Hemmerich, P. (1971). Pteridines. XLIV. Synthesis and structure of N-8-substituted pterins and lumazines. *Chemische Berichte* **104**, 2273–2292.

Plaut, G. W. (1960). Studies on the stoichiometry of the enzymic conversion of 6,7-dimethyl-8-ribityllumazine to riboflavin. *The Journal of Biological Chemistry* **235**, PC41–PC42.

Plaut, G. W. (1963). Studies on the nature of the nature of 6,7-dimethyl-8-ribityllumazine to riboflavin. *The Journal of Biological Chemistry* **238**, 2225–2243.

Plaut, G. W. E. (1971). Metabolism of water-soluble vitamins: The biosynthesis of riboflavin. *In* Comprehensive Biochemistry, (M. Florkin and E. H. Stotz (Hrsg.), eds.), **21**, pp. 11–45. Elsevier, Amsterdam.

Plaut, G. W. and Beach, R. L. (1976). Substrate specificity and stereospecific mode of action of riboflavin synthase. Flavins Flavoproteins. *Flavins and Flavoproteins, Proc. Int. Symp., 5th* 737–746.

Plaut, G. W. E. and Harvey, R. A. (1971). The enzymatic synthesis of riboflavin. *Methods in Enzymology* **18**, 515–538.

Plaut, G. W., Beach, R. L. and Aogaichi, T. (1970a). Studies on the mechanism of elimination of protons from the methyl groups of 6,7-dimethyl-8-ribityllumazine by riboflavin synthetase. *Biochemistry* **9**, 771–785.

Plaut, G. W. E., Beach, R. L. and Aogaichi, T. (1970b). Studies on the mechanisms of elimination of protons from the methyl groups of 6,7-dimethyl-8-ribityllumazine by riboflavin synthetase. *Biochemistry* **9**, 771–785.

Plaut, G. W., Smith, C. M. and Alworth, W. L. (1974). Biosynthesis of water-soluble vitamins. *Annual Review of Biochemistry* **43**, 899–922.

Ramsperger, A., Augustin, M., Schott, A. K., Gerhardt, S., Krojer, T., Eisenreich, W., Illarionov, B., Cushman, M., Bacher, A., Huber, R. and Fischer, M. (2006). Crystal structure of an archaeal pentameric riboflavin synthase in complex with a substrate analog inhibitor: Stereochemical implications. *The Journal of Biological Chemistry* **281**, 1224–1232.

Rellan-Alvarez, R., Andaluz, S., Rodriguez-Celma, J., Wohlgemuth, G., Zocchi, G., Alvarez-Fernandez, A., Fiehn, O., Lopez-Millan, A. F. and Abadia, J. (2010). Changes in the proteomic and metabolic profiles of *Beta vulgaris* root tips in response to iron deficiency and resupply. *BMC Plant Biology* **10**, 120.

Ren, J., Kotaka, M., Lockyer, M., Lamb, H. K., Hawkins, A. R. and Stammers, D. K. (2005). GTP cyclohydrolase II structure and mechanism. *The Journal of Biological Chemistry* **280**, 36912–36919.

Reuke, B., Korn, S., Eisenreich, W. and Bacher, A. (1992). Biosynthetic precursors of deazaflavins. *Journal of Bacteriology* **174**, 4042–4049.

Richter, G., Volk, R., Krieger, C., Lahm, H. W., Röthlisberger, U. and Bacher, A. (1992). Biosynthesis of riboflavin: Cloning, sequencing, and expression of the gene coding for 3,4-dihydroxy-2-butanone 4-phosphate synthase of *Escherichia coli*. *Journal of Bacteriology* **174**, 4050–4056.

Richter, G., Ritz, H., Katzenmeier, G., Volk, R., Kohnle, A., Lottspeich, F., Allendorf, D. and Bacher, A. (1993). Biosynthesis of riboflavin: Cloning, sequencing, mapping, and expression of the gene coding for GTP cyclohydrolase II in *Escherichia coli*. *Journal of Bacteriology* **175**, 4045–4051.

Richter, G., Fischer, M., Krieger, C., Eberhardt, S., Lüttgen, H., Gerstenschlager, I. and Bacher, A. (1997). Biosynthesis of riboflavin: Characterization of the bifunctional deaminase-reductase of *Escherichia coli* and *Bacillus subtilis*. *Journal of Bacteriology* **179**, 2022–2028.

Ritsert, K., Huber, R., Turk, D., Ladenstein, R., Schmidt-Base, K. and Bacher, A. (1995). Studies on the lumazine synthase/riboflavin synthase complex of *Bacillus subtilis*: Crystal structure analysis of reconstituted, icosahedral beta-subunit capsids with bound substrate analogue inhibitor at 2.4 Å resolution. *Journal of Molecular Biology* **253**, 151–167.

Ritz, H., Schramek, N., Bracher, A., Herz, S., Eisenreich, W., Richter, G. and Bacher, A. (2001). Biosynthesis of riboflavin: Studies on the mechanism of GTP cyclohydrolase II. *The Journal of Biological Chemistry* **276**, 22273–22277.

Robinson, N. J., Procter, C. M., Connolly, E. L. and Guerinot, M. L. (1999). A ferric-chelate reductase for iron uptake from soils. *Nature* **397**, 694–697.
Rodriguez-Celma, J., Lattanzio, G., Grusak, M. A., Abadia, A., Abadia, J. and Lopez-Millan, A. F. (2011). Root Responses of *Medicago truncatula* Plants Grown in Two Different Iron Deficiency Conditions: Changes in Root Protein Profile and Riboflavin Biosynthesis. *Journal of Proteome Research* **10**(5), 2590–2601.
Römisch-Margl, W., Eisenreich, W., Haase, I., Bacher, A. and Fischer, M. (2008). 2,5-Diamino-6-ribitylamino-4(3*H*)-pyrimidinone 5′-phosphate synthases of fungi and archaea. *The FEBS Journal* **275**, 4403–4414.
Rowan, T. and Wood, H. C. S. (1968). The biosynthesis of pteridines. Part V. The synthesis of riboflavin from pteridine precursors. *Journal of Chemical Society C* 452–458.
Sancar, A. (2008). Structure and function of photolyase and in vivo enzymology: 50th anniversary. *The Journal of Biological Chemistry* **283**, 32153–32157.
Sandoval, F. J. and Roje, S. (2005). An FMN hydrolase is fused to a riboflavin kinase homolog in plants. *The Journal of Biological Chemistry* **280**, 38337–38345.
Sandoval, F. J., Zhang, Y. and Roje, S. (2008). Flavin nucleotide metabolism in plants: Monofunctional enzymes synthesize FAD in plastids. *The Journal of Biological Chemistry* **283**, 30890–30900.
Scheuring, J., Cushman, M., Oschkinat, H. and Bacher, A. (1994). ^{19}F NMR studies on the mechanism of riboflavin synthase. *Flavins and Flavoproteins 1993, Proc. Int. Symp., 11th* 79–82.
Scheuring, J., Fischer, M., Cushman, M., Lee, J., Bacher, A. and Oschkinat, H. (1996). NMR analysis of site-specific ligand binding in oligomeric proteins. Dynamic studies on the interaction of riboflavin synthase with trifluoromethyl-substituted intermediates. *Biochemistry* **35**, 9637–9646.
Schott, K., Kellermann, J., Lottspeich, F. and Bacher, A. (1990). Riboflavin synthases of *Bacillus subtilis*. Purification and amino acid sequence of the alpha subunit. *The Journal of Biological Chemistry* **265**, 4204–4209.
Schramek, N., Bracher, A. and Bacher, A. (2001). Biosynthesis of riboflavin. Single turnover kinetic analysis of GTP cyclohydrolase II. *The Journal of Biological Chemistry* **276**, 44157–44162.
Schramek, N., Haase, I., Fischer, M. and Bacher, A. (2003). Biosynthesis of riboflavin. Single turnover kinetic analysis of 6,7-dimethyl-8-ribityllumazine synthase. *Journal of the American Chemical Society* **125**, 4460–4466.
Schwarzkopf, B., Reuke, B., Kiener, A. and Bacher, A. (1990). Biosynthesis of coenzyme F_{420} and methanopterin in *Methanobacterium thermoautotrophicum*. Studies with [1′-14C] adenosine. *Archives of Microbiology* **153**, 259–263.
Sedlmaier, H., Muller, F., Keller, P. J. and Bacher, A. (1987). Enzymatic synthesis of riboflavin and FMN specifically labeled with ^{13}C in the xylene ring. *Zeitschrift für Naturforschung, C* **42**, 425–429.
Selengut, J. D. and Haft, D. H. (2010). Unexpected abundance of coenzyme F(420)-dependent enzymes in *Mycobacterium tuberculosis* and other actinobacteria. *Journal of Bacteriology* **192**, 5788–5798.
Singh, M., Kumar, P. and Karthikeyan, S. (2011). Structural basis for pH dependent monomer-dimer transition of 3,4-dihydroxy 2-butanone-4-phosphate synthase domain from *Mycobacterium tuberculosis*. *Journal of Structural Biology* **174**, 374–384.
Souci, S. W., Fachmann, W. and Kraut, H. (2008). Food Composition and Nutrition Tables. MedPharm Scientific Publishers Taylor & Francis, Boca Raton, FL.

Spoonamore, J. E., Dahlgran, A. L., Jacobsen, N. E. and Bandarian, V. (2006). Evolution of new function in the GTP cyclohydrolase II proteins of *Streptomyces coelicolor*. *Biochemistry* **45**, 12144–12155.

Stahmann, K. P., Revuelta, J. L. and Seulberger, H. (2000). Three biotechnical processes using *Ashbya gossypii*, *Candida famata*, or *Bacillus subtilis* compete with chemical riboflavin production. *Applied Microbiology and Biotechnology* **53**, 509–516.

Steinbacher, S., Schiffmann, S., Richter, G., Huber, R., Bacher, A. and Fischer, M. (2003). Structure of 3,4-dihydroxy-2-butanone 4-phosphate synthase from *Methanococcus jannaschii* in complex with divalent metal ions and the substrate ribulose 5-phosphate: Implications for the catalytic mechanism. *The Journal of Biological Chemistry* **278**, 42256–42265.

Steinbacher, S., Schiffmann, S., Bacher, A. and Fischer, M. (2004). Metal sites in 3,4-dihydroxy-2-butanone 4-phosphate synthase from *Methanococcus jannaschii* in complex with the substrate ribulose 5-phosphate. *Acta Crystallographica, Section D, Biological Crystallography* **60**, 1338–1340.

Stenmark, P., Moche, M., Gurmu, D. and Nordlund, P. (2007). The crystal structure of the bifunctional deaminase/reductase RibD of the riboflavin biosynthetic pathway in *Escherichia coli*: Implications for the reductive mechanism. *Journal of Molecular Biology* **373**, 48–64.

Susin, S., Abian, J., Sanchez-Baeza, F., Peleato, M. L., Abadia, A., Gelpi, E. and Abadia, J. (1993). Riboflavin 3′- and 5′-sulfate, two novel flavins accumulating in the roots of iron-deficient sugar beet (*Beta vulgaris*). *The Journal of Biological Chemistry* **268**, 20958–20965.

Susin, S., Abidn, J., Peleato, M. L., Sfinchez-Baeza, F., Abadia, A., Gelpi, E. and Abadia, J. (1994). Flavin excretion from roots of iron-deficient sugar beet (*Beta vulgaris* L.). *Planta* **193**, 514–519.

Sutak, R., Lesuisse, E., Tachezy, J. and Richardson, D. R. (2008). Crusade for iron: Iron uptake in unicellular eukaryotes and its significance for virulence. *Trends in Microbiology* **16**, 261–268.

Taheri, P. and Hofte, M. (2007). Riboflavin-induced resistance against rice sheath blight functions through the potentiation of lignin formation and jasmonic acid signalling pathway. *Communications in Agricultural and Applied Biological Sciences* **72**, 309–313.

Taheri, P. and Tarighi, S. (2010). Riboflavin induces resistance in rice against *Rhizoctonia solani* via jasmonate-mediated priming of phenylpropanoid pathway. *Journal of Plant Physiology* **167**, 201–208.

Talukdar, A., Illarionov, B., Bacher, A., Fischer, M. and Cushman, M. (2007). Synthesis and enzyme inhibitory activity of the s-nucleoside analogue of the ribitylaminopyrimidine substrate of lumazine synthase and product of riboflavin synthase. *The Journal of Organic Chemistry* **72**, 7167–7175.

Talukdar, A., Breen, M., Bacher, A., Illarionov, B., Fischer, M., Georg, G., Ye, Q. Z. and Cushman, M. (2009). Discovery and development of a small molecule library with lumazine synthase inhibitory activity. *The Journal of Organic Chemistry* **74**, 5123–5134.

Torchetti, E. M., Brizio, C., Colella, M., Galluccio, M., Giancaspero, T. A., Indiveri, C., Roberti, M. and Barile, M. (2010). Mitochondrial localization of human FAD synthetase isoform 1. *Mitochondrion* **10**, 263–273.

Truffault, V., Coles, M., Diercks, T., Abelmann, K., Eberhardt, S., Lüttgen, H., Bacher, A. and Kessler, H. (2001). The solution structure of the N-terminal domain of riboflavin synthase. *Journal of Molecular Biology* **309**, 949–960.

Viitanen, P. V., Jordan, D. B. and Bacot, K. O. (2000). Lumazine synthase and riboflavin synthase genes from plants and Magnaporthe grisea. USA patent application.

Viitanen, P. V., Jordan, D. B. and Bacot, K. O. (2002). Riboflavin synthase and lumazine synthase genes and enzymes from plants and fungi. USA patent application.

Vitreschak, A. G., Rodionov, D. A., Mironov, A. A. and Gelfand, M. S. (2002). Regulation of riboflavin biosynthesis and transport genes in bacteria by transcriptional and translational attenuation. *Nucleic Acids Research* **30**, 3141–3151.

Vitreschak, A. G., Rodionov, D. A., Mironov, A. A. and Gelfand, M. S. (2004). Riboswitches: The oldest mechanism for the regulation of gene expression? *Trends in Genetics* **20**, 44–50.

Volk, R. and Bacher, A. (1988). Biosynthesis of riboflavin. The structure of the four-carbon precursor. *Journal of the American Chemical Society* **110**, 3651–3653.

Volk, R. and Bacher, A. (1990). Studies on the 4-carbon precursor in the biosynthesis of riboflavin. Purification and properties of L-3,4-dihydroxy-2-butanone-4-phosphate synthase. *The Journal of Biological Chemistry* **265**, 19479–19485.

Volk, R. and Bacher, A. (1991). Biosynthesis of riboflavin. Studies on the mechanism of L-3,4-dihydroxy-2-butanone 4-phosphate synthase. *The Journal of Biological Chemistry* **266**, 20610–20618.

Volk, R., Koenig, A., Kohnle, A. and Bacher, A. (1990). Biosynthesis of 6,7-dimethyl-8-ribityllumazine. Purification and properties of 3,4-dihydroxy-2-butanone 4-phosphate synthase. *Chem. Biol. Pteridines, 1989 Proc. Int. Symp. Pteridines Folic Acid Deriv.* 344–347.

Vorwieger, A., Gryczka, C., Czihal, A., Douchkov, D., Tiedemann, J., Mock, H. P., Jakoby, M., Weisshaar, B., Saalbach, I. and Baumlein, H. (2007). Iron assimilation and transcription factor controlled synthesis of riboflavin in plants. *Planta* **226**, 147–158.

Wang, W., Kim, R., Yokota, H. and Kim, S. H. (2005). Crystal structure of flavin binding to FAD synthetase of Thermotoga maritima. *Proteins* **58**, 246–248.

Waters, B. M., Blevins, D. G. and Eide, D. J. (2002). Characterization of FRO1, a pea ferric-chelate reductase involved in root iron acquisition. *Plant Physiology* **129**, 85–94.

Winkler, W. C., Cohen-Chalamish, S. and Breaker, R. R. (2002). An mRNA structure that controls gene expression by binding FMN. *Proceedings of the National Academy of Sciences of the United States of America* **99**, 15908–15913.

Wu, T., Guo, A., Zhao, Y., Wang, X., Wang, Y., Zhao, D., Li, X., Ren, H. and Dong, H. (2010). Ectopic expression of the rice lumazine synthase gene contributes to defense responses in transgenic tobacco. *Phytopathology* **100**, 573–581.

Xiao, S., Dai, L., Liu, F., Wang, Z., Peng, W. and Xie, D. (2004). COS1: An Arabidopsis coronatine insensitive1 suppressor essential for regulation of jasmonate-mediated plant defense and senescence. *The Plant Cell* **16**, 1132–1142.

Yates, A. A., Schlicker, S. A. and Suitor, C. W. (1998). Dietary reference intakes: The new basis for recommendations for calcium and related nutrients, B vitamins, and choline. *Journal of the American Dietetic Association* **98**, 699–706.

Yazdanpanah, B., Wiegmann, K., Tchikov, V., Krut, O., Pongratz, C., Schramm, M., Kleinridders, A., Wunderlich, T., Kashkar, H., Utermohlen, O.,

Bruning, J. C. Schutze, S. *et al.* (2009). Riboflavin kinase couples TNF receptor 1 to NADPH oxidase. *Nature* **460**, 1159–1163.

Young, D. W. (1986). The biosynthesis of the vitamins thiamin, riboflavin, and folic acid. *Natural Product Reports* **3**, 395–419.

Yruela, I., Arilla-Luna, S., Medina, M. and Contreras-Moreira, B. (2010). Evolutionary divergence of chloroplast FAD synthetase proteins. *BMC Evolutionary Biology* **10**, 311.

Zhang, X., Meining, W., Fischer, M., Bacher, A. and Ladenstein, R. (2001). X-ray structure analysis and crystallographic refinement of lumazine synthase from the hyperthermophile *Aquifex aeolicus* at 1.6 Å resolution: Determinants of thermostability revealed from structural comparisons. *Journal of Molecular Biology* **306**, 1099–1114.

Zhang, X., Meining, W., Cushman, M., Haase, I., Fischer, M., Bacher, A. and Ladenstein, R. (2003). A structure-based model of the reaction catalyzed by lumazine synthase from *Aquifex aeolicus*. *Journal of Molecular Biology* **328**, 167–182.

Zhang, Y., Illarionov, B., Bacher, A., Fischer, M., Georg, G. I., Ye, Q. Z., Vander Velde, D., Fanwick, P. E., Song, Y. and Cushman, M. (2007a). A novel lumazine synthase inhibitor derived from oxidation of 1,3,6,8-tetrahydroxy-2,7-naphthyridine to a tetraazaperylenehexaone derivative. *The Journal of Organic Chemistry* **72**, 2769–2776.

Zhang, Y., Jin, G., Illarionov, B., Bacher, A., Fischer, M. and Cushman, M. (2007b). A new series of 3-alkyl phosphate derivatives of 4,5,6,7-tetrahydro-1-D-ribityl-1H-pyrazolo[3,4-d]pyrimidinedione as inhibitors of lumazine synthase: Design, synthesis, and evaluation. *The Journal of Organic Chemistry* **72**, 7176–7184.

Zhang, Y., Illarionov, B., Morgunova, E., Jin, G., Bacher, A., Fischer, M., Ladenstein, R. and Cushman, M. (2008). A new series of N-[2,4-dioxo-6-d-ribitylamino-1,2,3,4-tetrahydropyrimidin-5-yl]oxalamic acid derivatives as inhibitors of lumazine synthase and riboflavin synthase: Design, synthesis, biochemical evaluation, crystallography, and mechanistic implications. *The Journal of Organic Chemistry* **73**, 2715–2724.

Zhang, S., Yang, X., Sun, M., Sun, F., Deng, S. and Dong, H. (2009). Riboflavin-induced priming for pathogen defense in *Arabidopsis thaliana*. *Journal of Integrative Plant Biology* **51**, 167–174.

Zhao, Y., Bacher, A., Illarionov, B., Fischer, M., Georg, G., Ye, Q. Z., Fanwick, P. E., Franzblau, S. G., Wan, B. and Cushman, M. (2009). Discovery and development of the covalent hydrates of trifluoromethylated pyrazoles as riboflavin synthase inhibitors with antibiotic activity against Mycobacterium tuberculosis. *The Journal of Organic Chemistry* **74**, 5297–5303.

Zylberman, V., Craig, P. O., Klinke, S., Braden, B. C., Cauerhff, A. and Goldbaum, F. A. (2004). High order quaternary arrangement confers increased structural stability to *Brucella* sp. lumazine synthase. *The Journal of Biological Chemistry* **279**, 8093–8101.

Zylberman, V., Klinke, S., Haase, I., Bacher, A., Fischer, M. and Goldbaum, F. A. (2006). Evolution of vitamin B_2 biosynthesis: 6,7-dimethyl-8-ribityllumazine synthases of *Brucella*. *Journal of Bacteriology* **188**, 6135–6142.

Biosynthesis of NAD and Its Manipulation in Plants☆

GRAHAM NOCTOR,[1] JUTTA HAGER AND SHENGCHUN LI

Institut of Biologie des Plantes, UMR8618 CNRS/Université de Paris sud 11, Bâtiment 630, Université de Paris sud 11, 91405 Orsay CEDEX, France

[1]Corresponding author: E-mail: graham.noctor@u-psud.fr

☆In this chapter, NAD^+ and $NADP^+$ are used to refer to oxidized forms while NADH and NADPH denote reduced forms. Where no distinction is drawn between the oxidized and reduced forms, or where both forms may be concerned, only NAD is used.

Advances in Botanical Research, Vol. 58

0065-2296/11 $35.00
DOI: 10.1016/B978-0-12-386479-6.00002-0

ABSTRACT

NAD and NADP are key brokers in cell redox and energy metabolism, but they also play important roles in signalling pathways. Several of these latter roles involve NAD cleavage, requiring continuous resynthesis of the molecule to maintain cellular pools. Mammals synthesize their cellular NAD from pyridine precursors that are obtained as niacin (vitamin B_3) or produced from tryptophan. In many foods, the major source of niacin can be the pyridine ring of NAD and NADP. Plant NAD contents show considerable plasticity in response to environmental conditions, and recent advances have opened up new perspectives for rational manipulation of tissue NAD contents. In the model plant Arabidopsis, the *de novo* pathway of plant NAD synthesis has been identified, as have proteins involved in the subcellular transport of NAD and in NAD cleavage and recycling. Here, we discuss emerging concepts on the regulation of NAD contents in plants, and we report the use of transformation technology to enrich NAD several-fold in plant tissues through overexpression of the *de novo* synthesis pathway. Microarray analyses reveal that these increases trigger a characteristic transcriptomic signature, notably involving specific genes in defence phytohormone signalling. Thus, manipulation of plant NAD biosynthesis may offer interesting perspectives to (1) dissect redox-related signalling in stress responses, (2) enhance the nutritional value of plants, and (3) modify plant resistance to environmental stress.

I. INTRODUCTION

Life on earth is driven by reactions in which energy is extracted from thermodynamically favourable respiratory electron flow from relatively reducing compounds to more oxidizing ones. Photoautotrophic organisms such as plants continually replenish the biosphere's supply of reducing compounds through the process of photosynthesis. Both respiration and photosynthesis involve membrane-bound electron transfer chains coupled to metabolic reactions in the aqueous phase. Chief among the compounds that link the soluble and membrane-bound components are the pyridine nucleotides, NAD(H) and NADP(H). During aerobic respiration, NADH formed by mitochondrial matrix enzymes is reoxidized by dehydrogenases bound to the inner mitochondrial membrane (Douce and Neuburger, 1989; Rasmusson *et al.*, 2004). In photosynthesis, NADPH generated by ferredoxin-$NADP^+$ reductase (FNR) is reconverted to $NADP^+$ by the photosynthetic isoform of glyceraldehyde-3-phosphate dehydrogenase (GAPDH) in the single reductive reaction of the Calvin–Benson cycle, or reductive pentose phosphate pathway (Harbinson *et al.*, 1990; Scheibe *et al.*, 2005). As well as these essential core processes, both pyridine nucleotides are involved in a host of other reactions. It is estimated that hundreds of enzymes depend on NAD or NADP in a redox-carrying role (Waller *et al.*, 2010 and references therein). As a very general rule, NAD is more closely linked to catabolic reactions whose physiological role is to generate energy while NADP is the preferred cofactor of enzymes involved in biosyntheses and cellular redox homeostasis. However, this distinction is far from absolute. For example, NADPH as well as NADH can be oxidized by the plant mitochondrial electron transport chain (Douce and Neuburger, 1989; Rasmusson *et al.*, 2004), NADH is required for assimilatory nitrate reductase (NR) activity (Kaiser *et al.*, 2000, 2002), and the antioxidative enzyme monodehydroascorbate reductase can use either NADH or NADPH (Hossain and Asada, 1985).

In addition to its cofactor function for oxidoreductases, NAD has been known for some time to be required by other enzymes, such as ADP-ribose transferases (Hayaishi and Ueda, 1977; Henderson, 1983). As these reactions do not involve oxidoreduction, they can be termed "non-redox". From a physiological standpoint, the redox versus non-redox distinction may be somewhat simplistic. Even if the reaction itself does not involve redox exchange, it may still be important in redox metabolism, homeostasis, or signalling. Nevertheless, a key difference is that the non-redox roles involve consumption (cleavage) of the nucleotide, and therefore, resynthesis or new synthesis is required to avoid NAD depletion.

Recent years have seen the appearance of several reviews on NAD synthesis and metabolism in plants (Hashida *et al.*, 2009; Hunt *et al.*, 2004; Katoh and Hashimoto, 2004; Noctor *et al.*, 2006; Roje, 2007) and interest in the direct, targeted manipulation of intracellular NAD concentrations has grown. Among the reasons for this burgeoning interest are (1) the plant biosynthetic pathway has been clarified; (2) it has become evident that NAD contents can be quite plastic in plants; (3) ever-growing attention is being paid to the roles of redox-related factors not only in cellular energy metabolism but also in signalling; and (4) the non-redox reactions mentioned above have continued to receive increasing attention. The aim of the present chapter is to discuss some recent advances in understanding NAD metabolism and turnover within the context of the perspectives for manipulation of these pathways. With this aim in mind, we present previously unpublished data from an ongoing study seeking to engineer increased NAD contents in plant tissues.

II. NAD IN HUMAN NUTRITION AND HEALTH

The fundamental redox exchange function of pyridine nucleotides involves the nicotinamide moiety, which is identical in NAD and NADP (Fig. 1). Thus, the two nucleotides undergo the same redox reactions with $E_m = -320$ mV at pH 7. The difference between the two compounds resides in the phosphate group attached to the 2′ of the ribose ring of the adenosine moiety of NADP (Fig. 1), a group which is absent in NAD, explaining the preference that enzymes may have for one form rather than the other. The only known pathway for the synthesis of NADP is from NAD, a reaction catalysed by NAD kinases (NADKs). Thus, in order to maintain these nucleotide pools or to produce new nucleotides for growth, synthesis of NAD is required. This requires either a dietary source of the pyridine ring or its production from other molecules. Humans can obtain the NAD pyridine precursors either from niacin present in food or from degradation of the indole ring of tryptophan.

A. NIACIN (VITAMIN B_3)

The best known pyridine precursors are nicotinic acid and nicotinamide, collectively known as vitamin B_3 or niacin (or sometimes niacinamide in the case of nicotinamide). Nicotinic acid was identified as the substance able

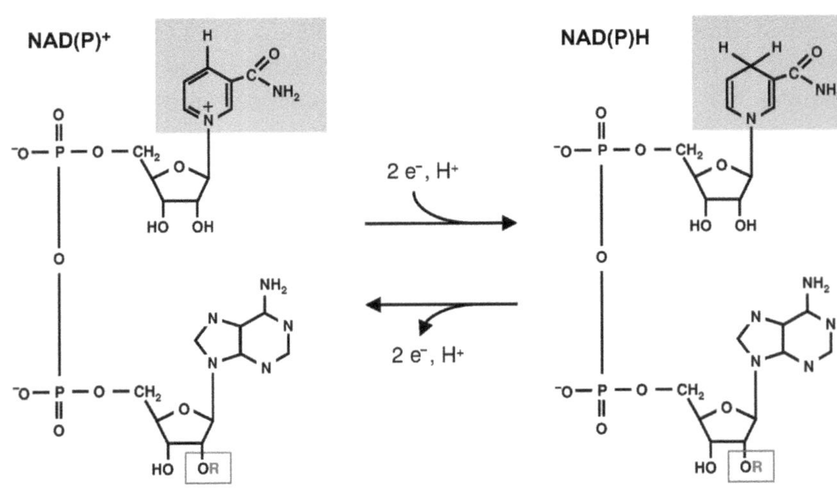

Fig. 1. Basic structures of NAD(P)$^+$ and NAD(P)H. For simplicity, not all C and H atoms are shown. The redox-active pyridine ring is highlighted within the blue square. The difference between NAD and NADP, at C-2 of the AMP ribose ring, is highlighted red. In NAD, R = H. In NADP, R = phosphate group. (For interpretation of the references to colour in this figure legend, the reader is referred to the Web version of this chapter.)

to cure canine black-tongue disease (Elvehjem *et al.*, 1938). This advance led to the recognition that niacin deficiency was responsible for the related human disease, pellagra, outbreaks of which were prevalent in the early part of the twentieth century. Although it is still found in undeveloped regions during certain seasons, pellagra has largely disappeared from the developed world, except in certain conditions such as chronic alcoholism (Bogan and Brenner, 2008).

The U.S. recommended daily amount (RDA) of niacin (nicotinic acid + nicotinamide) is 14–16 mg (Bogan and Brenner, 2008), while the EU recommends between 9 and 18 mg (SCF, 1993). Like many other organisms, however, humans can make the NAD pyridine ring from tryptophan. Thus, provided tryptophan intake is sufficient, niacin may not be required at all in many conditions (SCF, 1993). Tryptophan has been assigned a "niacin equivalent" value of 1/60 (60 mg tryptophan = 1 mg niacin), though this value is necessarily approximate (Horwitt *et al.*, 1981). In many food sources (including milk, meat, and cereals), tryptophan is much more abundant than niacin, but the capacity to convert tryptophan is variable between individuals (Horwitt *et al.*, 1981).

Especially good sources of niacin are tuna, chicken, and liver (Table I). Espresso coffee is rich in niacin, though instant coffee contains much less. Milk and eggs are also niacin-poor (Horwitt *et al.*, 1981). Enrichment with niacin means that one or two modest servings of many breakfast cereal products can be sufficient to meet the above RDAs (Table I; Rose-Sallin *et al.*, 2001; Windahl *et al.*, 1998). Cereal grains naturally contain quite high amounts of niacin, but much of it is complexed, mainly with polysaccharides, leaving only a minor part available for uptake (Lahély *et al.*, 1999). The unavailability of niacin in corn (maize) is thought to explain outbreaks of pellagra in the first decades of the twentieth century in regions where much of the diet was corn-based. This problem may have been conditional on food preparation (Henderson, 1983). Pellagra is uncommon in Central America, even though corn makes up a large part of the calorie intake. However, corn in Central America is traditionally prepared by soaking in limewater (nixtamalization), a procedure that releases much of the bound niacin. An additional factor that could favour the development of pellagra is removal of the embryo (germ) from cereal grains, causing depletion of both niacin and tryptophan, a low-abundance amino acid in corn endosperm (Henderson, 1983). Plant foods that are quite good sources of niacin include potatoes and peas (Table I). Cooking does not generally diminish niacin contents in a wide range of meats and vegetables, and may increase the amount of the vitamin in some cases (Windahl *et al.*, 1998).

TABLE I
Typical Amounts of Niacin in Some Foods

Foodstuff	Weight (g)	Niacin (mg)	% RDA
Fruits and vegetables			
Apple	140	0.13	0.9
Banana	120	0.79	5.3
Tomato (raw)	120	0.73	4.9
Lettuce	100	0.36	2.3
Potato (boiled)	150	2.0	13.0
Peas (boiled)	160	2.4	16.0
Mushrooms (raw)	100	3.5	23.9
Cereals and derived foods			
Rice (white, boiled)	160	2.3	15.3
Wheat flour	120	5.9	39.3
French bread	25	1.2	7.9
Multigrain bread	25	1.1	7.3
Maize (sweetcorn)	80	1.3	8.7
Cornflakes (enriched)	30	7.0	46.7
Fish			
Cod (dry-cooked)	100	1.3	8.6
Haddock	100	4.1	27.5
Tuna (canned)	100	13.3	88.7
Meats			
Chicken (breast, roasted)	100	13.7	91.2
Beef (sirloin steak, grilled)	100	8.2	54.9
Pork (chops, grilled)	100	8.1	54.1
Lamb (roasted)	100	7.0	46.8
Beef liver	100	17.4	115.8
Beverages			
Milk	250 ml	0.2	1.3
Orange juice	250 ml	0.7	4.7
Beer	250 ml	1.0	6.7
Wine (red)	250 ml	0.6	3.9
Tea	250 ml	0	0
Coffee (espresso)	60 ml	3.1	20.7

Source: the USDA National Nutrient Database. A more detailed list can be found at http://www.ars.usda.gov/SP2UserFiles/Place/12354500/Data/SR23/nutrlist/sr23a406.pdf. The recommended daily amount (RDA) is taken as 15 mg. Apart from breakfast cereals, values are for unenriched foodstuffs.

Nicotinic acid, nicotinamide, NAD, and NADP are all recognized niacin vitamers. At least in relatively unprocessed foods, particularly meats, NAD and NADP are considered to provide the largest part of the available niacin (Bogan and Brenner, 2008; Henderson, 1983; Lahély *et al.*, 1999). Although the forms taken up across the gut mucosa for intracellular conversion to NAD are predominantly nicotinic acid and nicotinamide, these are produced from the nucleotides by digestive enzymes in the gut (Henderson, 1983; SCF,

1993). In other foodstuffs, free nicotinic acid or nicotinamide may make a more substantial contribution to the total intake. In most cases, however, identifying the chemical nature of the different vitamers is not a primary aim of niacin assays in foodstuffs (see Section II.C), and so uncertainty remains in this area.

Vitamin B_3 present in supplements can consist of either nicotinic acid or nicotinamide. High doses in the range of 100 mg to several grams per day of nicotinic acid have been used for decades to treat hyperlipidaemias (Bogan and Brenner, 2008; SCF, 1993). However, the mechanisms responsible for the effect of nicotinic acid are not clearly established. The mechanisms may be independent of NAD production, because nicotinamide is much less effective as a treatment, though it is possible that the different efficacies rather reflect preferential use of nicotinic acid for NAD synthesis in some tissues (Bogan and Brenner, 2008). In high doses, nicotinic acid has clear side effects, including painful facial flushing linked to hypotension, though this is often temporary and tolerance develops within a few days (Henderson, 1983; SCF, 1993).

B. BIOSYNTHESIS OF NAD IN HUMANS

Nicotinic acid and nicotinamide acquired from the gut are converted to pyridine mononucleotides (Fig. 2). These are then adenylated to produce the dinucleotide. In the case of NAD synthesis from nicotinic acid, a terminal amination step is required and so the conversion involves three enzymes that constitute the "Preiss–Handler" pathway (Preiss and Handler, 1958). Conversion of nicotinamide to NAD was thought for some years to involve nicotinic acid production by nicotinamidase, and therefore also to require the same three intracellular enzymes. However, it was established in the mid-1970s that conversion of nicotinamide to NAD requires only two enzymes, a phosphoribosyltransferase and an adenylyltransferase (Henderson, 1983). The second of these enzymes shows dual substrate specificity. It either converts nicotinate mononucleotide (NaMN) to nicotinate adenine dinucleotide (NaAD), which is then aminated by NAD synthetase (NADS) using glutamine or ammonia, or produces NAD directly from nicotinamide mononucleotide (NMN; Fig. 2).

Another dietary source of the pyridine ring has been described in foodstuffs, such as milk, which are low in other recognized forms of niacin (Bieganowski and Brenner, 2004). This is nicotinamide riboside. As well as being present in some foodstuffs, it may be a product of NAD(P) digestion via NMN in the gut (Bogan and Brenner, 2008). Whatever its source, it can be either phosphorylated to produce NMN directly (Fig. 2) or, alternatively, cleaved to nicotinamide and ribose with nicotinamide being subsequently

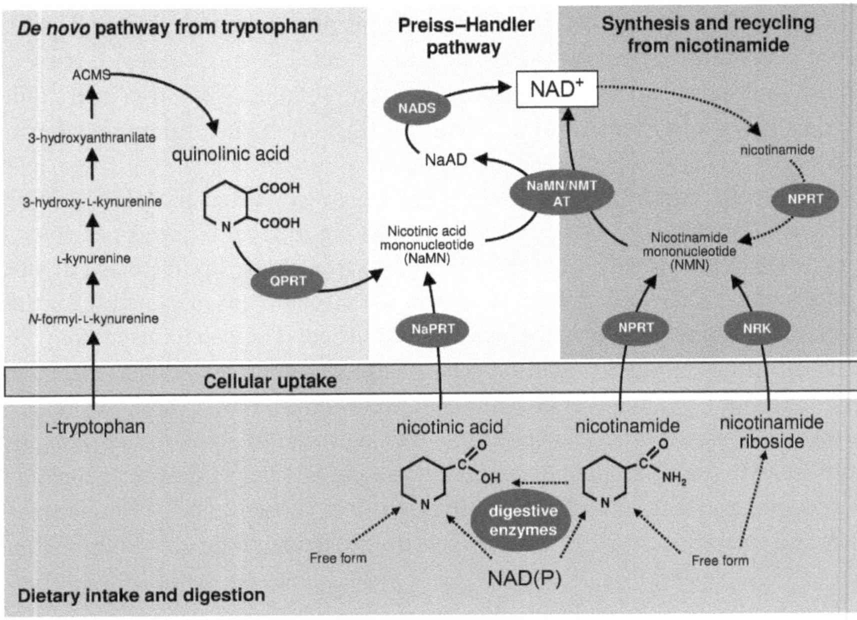

Fig. 2. NAD synthesis pathways in humans. Niacin obtained in the diet as nicotinic acid, nicotinamide, NAD, or NADP can be converted to pyridine nucleotides, which are then converted to NAD. Alternatively, the pyridine group can be synthesized *de novo* by degradation of the indole group of tryptophan. NADS, NAD synthetase; NaMN/NMNAT, nicotinic acid mononucleotide/nicotinamide mononucleotide adenylyltransferase; NaPRT, nicotinic acid phosphoribosyltransferase; NPRT, nicotinamide phosphoribosyltransferase; NRK, nicotinamide riboside kinase; QPRT, quinolinate phosphoribosyltransferase. As well as being obtained in food, nicotinamide is also produced from intracellular NAD by several enzymes (see Fig. 3) and can be recycled into NAD by the two-step pathway shown on the right. (See color plate section at the back of the book.)

converted to NMN through phosphoribosyltransferase activity (Belenky *et al.*, 2009).

Like other animals, humans can synthesize the pyridine ring of NAD^+ through a *de novo* route. In the so-called eukaryotic pathway of *de novo* NAD synthesis, quinolinate is obtained by degradation of the indole ring of tryptophan and then converted to NaMN by quinolinate phosphoribosyltransferase (QPRT). NaMN can then be adenylated and aminated through the final two enzymes of the Preiss–Handler pathway (Fig. 2). Thus, the NAD synthesis pathways from quinolinate and nicotinate converge at NaMN and thereafter involve the same enzymes. While still a subject of ongoing research, human tissues have different capacities for the different pathways of NAD synthesis and turnover. For further details, the reader is referred to the recent review by Bogan and Brenner (2008).

It should be noted that the term "eukaryotic pathway of NAD synthesis" is misleading. First, some prokaryotes can use this pathway (Gazzaniga *et al.*, 2009; Kurnasov *et al.*, 2003). Second, as we discuss later in this chapter, the best characterized pathway in plants is not this sequence but rather the so-called prokaryotic pathway (Katoh *et al.*, 2006).

C. MEASURING NIACIN, NAD, AND RELATED COMPOUNDS

In determining the vitamin B_3 content of foodstuffs, the contribution of the different vitamers is often not considered. Common extraction treatments such as alkaline or acid hydrolysis mean that any NAD, NADP, and nicotinamide present in the sample are converted to nicotinic acid, which is the form that is then measured as niacin (Lahély *et al.*, 1999; Rose-Sallin *et al.*, 2001; Windahl *et al.*, 1998). Acid extraction is considered to yield the total bioavailable niacin in a sample (i.e. that which can be absorbed from the gut) while alkaline hydrolysis can also release bound forms of niacin (Lahély *et al.*, 1999; Windahl *et al.*, 1998).

Widely used methods for niacin determination in foods are microbial bioassay and a chemical colorimetric method based on the reaction of pyridine with cyanogen bromide (Fuller, 1980). More recent developments include high-performance liquid chromatography (HPLC) and capillary electrophoresis (CE) methods. Nicotinic acid can be readily detected by CE or by HPLC with UV absorbance detection (Windahl *et al.*, 1998). A more sensitive method using HPLC-fluorescence measured nicotinic acid in various foodstuffs (Lahély *et al.*, 1999; Rose-Sallin *et al.*, 2001). Although nicotinic acid is not volatile, derivatization allows ready detection by gas chromatography–mass spectrometry (GC–MS). While this method only offers relative quantification, it can be useful for large-scale metabolite profiling of plant tissues (e.g. Hager *et al.*, 2010).

While HPLC techniques exist for separating and quantifying NAD and NADP, both can readily be quantified through well-established and sensitive enzymatic cycling assays. It is a simple matter to distinguish between NAD and NADP through the use of specific enzymes (or by separation on HPLC columns). More problematic is discrimination between the reduced and oxidized forms. Usually, this is achieved by exploiting the differential lability of the two forms in acid and alkaline extraction buffers. High-throughput methods have been developed that can measure the two nucleotides at the same time as other redox compounds (Queval and Noctor, 2007).

A full understanding of NAD(P)-related biological questions is likely to benefit from techniques that are able to report on the status of specific pools in different compartments. The fluorescent properties of reduced pyridine nucleotides have been exploited in the analysis of plant mitochondrial metabolism (Kasimova *et al.*, 2006; Neuburger *et al.*, 1984). However, these measurements do not separate the NADH and NADPH signals from each other, or give unambiguous information on redox states. The limitations of the information generated by the different methods of measuring pyridine nucleotides have been nicely summarized by Hagedorn *et al.* (2007). A recent interesting technique coupled CE separation to an enzymatic recycling method with fluorescence detection to measure NAD^+ and NADH concentrations in a single rat cell (Xie *et al.*, 2009).

As plant biologists have focused on measuring NAD and NADP rather than their metabolites or precursors, and food nutritionists are mainly interested in total niacin, quantitative information on the different compounds involved in NAD synthesis and turnover in plants is relatively scarce. This is likely to change in the coming years, as metabolite profiling techniques continue to become increasingly powerful and widespread. As well as nicotinate, GC–MS can be used to measure quinolinate and other similar compounds (Smythe *et al.*, 2002), though not large, phosphorylated compounds such as nucleotides. A recent report described an LC–MS method able to profile NAD alongside most of its precursors and metabolites in mouse erythrocytes (Yamada *et al.*, 2006). This analysis showed that NaMN and NMN could be detected but that they were at least 10-fold less abundant than NAD itself, whereas nicotinate, nicotinamide, and NaAD were below the threshold of detection (Yamada *et al.*, 2006). In Arabidopsis rosette leaves, CE–MS quantification of NaMN and NaAD revealed that both were about 20-fold less abundant than NAD (Takahashi *et al.*, 2009). Using the same technique in rice leaves, the contents of quinolinate, nicotinate, nicotinamide, NaMN, and NaAD were quantified (Takahara *et al.*, 2010). While nicotinamide, NaMN, and NaAD were present at about 1 nmol g^{-1} leaf fresh weight (20-fold less abundant than NAD and NADP), quinolinate and nicotinate were considerably more abundant, at about one-third of NAD and NADP contents (Takahara *et al.*, 2010). In terms of niacin, the summed amounts of the measured pyridine compounds would represent about 0.8 mg/100 g fresh leaf weight, a fairly similar value to those recorded for edible leaf vegetables such as lettuce (Table I). Based on this value, and discounting any contribution from tryptophan, it would require almost 2 kg of rice leaves to supply the U.S. and European RDAs of niacin, with NAD supplying 44%, free nicotinate 12%, and free nicotinamide 1.5% of the total. However, a modest serving of rice grains satisfies about one-sixth of the recommended daily niacin intake (Table I).

III. NAD IN PLANT BIOLOGY

Given the plethora of NAD-dependent enzymes, it is not possible to cover the redox roles of NAD in this chapter. As a prelude to the subsequent sections on NAD synthesis, the following discussion focuses on reactions that consume NAD, including NADP synthesis. Redox aspects are largely confined to a brief overview of the relationship between NAD and NADP, because this issue is relevant to any attempt to evaluate the functional interaction between the two nucleotides.

A. INTERCONVERSION OF NAD AND NADP

NAD is converted to NADP through ATP-dependent phosphorylation catalysed by NADK. A single gene has been described encoding NADK in humans, whereas three genes are found in Arabidopsis (Waller *et al.*, 2010). Chloroplast NADK activity has long been known to be influenced by light, which may in some cases be reflected in light-dependent changes in chloroplast NAD^+ or $NADP^+$ contents (Heineke *et al.*, 1991; Takahama *et al.*, 1981). Based on effects on NADPH, H_2O_2, and oxidative stress sensitivities in human cell lines in which the cytosolic NADK was knocked down or overexpressed, it has been proposed that this enzyme works together with NADP-linked dehydrogenases in ensuring NADPH production. Perhaps unsurprisingly, however, the dehydrogenases were found to play the major role in producing NADPH during oxidative stress (Pollak *et al.*, 2007).

As well as the two NAD^+-specific kinases found in the chloroplast (NADK2) and cytosol (NADK1), a third enzyme shows preference for NADH and is thus able to produce NADPH directly (Turner *et al.*, 2005). Of the different Arabidopsis knockout mutants, only *nadk2* shows a clear phenotype in optimal growth conditions (Chai *et al.*, 2005; Takahashi *et al.*, 2009). However, this mutation is not lethal, suggesting either that the other NADKs can partly replace NADK2 and that the chloroplast envelope can take up NADP produced by these enzymes, or that other chloroplast enzymes competent in NADP synthesis await discovery. Mutants for *nadk3* lacking the NADH-preferring kinase are hypersensitive to several types of stress (Chai *et al.*, 2006). Interestingly, NADK3, previously considered to be cytosolic, has recently been shown to be targeted to the peroxisome (Waller *et al.*, 2010). A thorough study showed that targeting required a previously undescribed peroxisomal targeting sequence 1 (PTS1) composed of the C-terminal tripeptide SRY and accessory targeting amino acids immediately upstream of SRY (Waller *et al.*, 2010). Some NADKs interact with calmodulin (Delumeau *et al.*, 2000; Harding *et al.*, 1997; Turner *et al.*, 2004) although the *in vivo* significance of some of the interactions has been questioned (Waller *et al.*, 2010).

Overexpression of NADK in Arabidopsis or rice produced marked effects on photosynthesis and carbon and nitrogen metabolism (Takahara *et al.*, 2010; Takahashi *et al.*, 2009). These observations add to others that point to close links between NAD(P) status, and the coordination of plant C and N metabolism (Dutilleul *et al.*, 2005; Hager *et al.*, 2010; Pellny *et al.*, 2008). Despite the effects of altering NADK capacity, neither mutants nor overexpressors have been shown to exhibit the changes in either NAD(H) or NADP(H) that might be predicted, though decreased NADPH was reported in *nadk3* mutants (Chai *et al.*, 2006).

NADP can be converted to NAD by a phosphatase activity. This activity has been studied in plants, notably in seeds (Gallais *et al.*, 2000a,b). A recent study revealed an inverse correlation between NADK and NADP phosphatase activities that was associated with the differing dormancy of Arabidopsis ecotypes (Hunt and Gray, 2009). Enzymes and genes responsible for NADP phosphatase activity remain to be fully characterized.

How do NADKs and NADP-dependent dehydrogenases interact in supplying NADPH? NADK has been implicated in NADPH oxidase function linked to pathogenesis and related responses (Harding *et al.*, 1997), while other studies have underlined the importance of enhanced respiratory flux through enzymes such as glucose-6-phosphate dehydrogenase (G6PDH; Pugin *et al.*, 1997). However, G6PDH is only one of several relatively abundant NADP-linked cytosolic dehydrogenases in plants (Foyer and Noctor, 2009; Hayashi *et al.*, 2005). Recent data suggest that the cytosolic NADP-isocitrate dehydrogenase may also play some role during oxidative stress and pathogenesis responses (Mhamdi *et al.*, 2010a). Despite the attention drawn to the roles of NADK in stress responses, our own studies on plants showing obvious oxidative stress do not point to marked reciprocal changes in NAD and NADP contents in these conditions relative to controls (Mhamdi *et al.*, 2010a,b). Whether NAD–NADP cycling is stimulated under stress conditions remains to be established.

B. REDOX LINKS BETWEEN THE NAD AND NADP POOLS

Metabolite pools in mitochondria, chloroplast, and cytosol link NAD and NADP pools through redox transfer. Plants have been known for many years to contain an irreversible cytosolic NADP-linked GAPDH, alongside the better known reversible cytosolic NAD–GAPDH (Kelly and Gibbs, 1973). The NADP-linked enzyme is called non-phosphorylating, because it does not involve the incorporation of phosphate during glyceraldehyde-

3-phosphate oxidation, and so produces 3-phosphoglycerate (3-PGA) directly without ATP formation by PGA kinase. Several roles for this enzyme have been proposed, including export of NADPH from the chloroplast, production of NADPH for biosyntheses or, more recently, generation of reducing power in oxidative stress responses (Kelly and Gibbs, 1973; Rius *et al.*, 2006). The last function is of particular interest, given that the NAD-dependent GAPDH undergoes numerous stress-related post-translational modifications that tend to inhibit its activity (Holtgrefe *et al.*, 2007). Other major soluble dehydrogenases such as isocitrate and malate dehydrogenases are also found as isoforms that can use NAD^+, $NADP^+$, or both (Lancien *et al.*, 2000; Rasmusson and Møller, 1990).

Direct redox transfer between NAD and NADP pools is possible via transhydrogenase activity through the following reversible reaction:

$$NADH + NADP^+ \rightarrow NAD^+ + NADPH$$

This enzyme, which is found in bacteria and the inner mitochondrial membrane, is usually but not always coupled to transmembrane proton transfer. In the presence of an electron transport chain-dependent proton gradient, the reaction is driven in the direction shown above, and this is thought to contribute to the high NADP reduction state compared to NAD in bacteria (Jackson, 2003). In mitochondria, the enzyme capacity is lower than in bacteria, and the main role of transhydrogenase may be in fine-tuning the TCA cycle (Jackson, 2003). Biochemical studies in plants have reported two independent transhydrogenase activities associated with the mitochondrial inner membrane, one of which was associated with complex I (Bykova *et al.*, 1999).

C. NON-REDOX ROLES OF NAD

The NAD molecule can be considered either as a dinucleotide or as an ADP-ribose group joined to nicotinamide (Fig. 3). Cleavage of the pyrophosphate bond that joins the two mononucleotides or the β-*N*-glycosidic linkage between the ADP-ribose and the nicotinamide group is energetically favourable. Several types of enzyme are able to cleave the glycosidic bond. This liberates the pyridine ring as nicotinamide and produces several different compounds as partner products, including cyclic ADP-ribose, free ADP-ribose, acetylated ADP-ribose, or protein-bound ADP-ribose (Fig. 3). Cyclic ADP-ribose synthase has been implicated in abscisic acid (ABA) signalling in plants (Sánchez *et al.*, 2004). Like cADP-ribose, NaADP, which is produced by deamidation of NADP, is involved in calcium signalling in a variety of organisms, including plants (Guse and Lee, 2008; Navazio *et al.*, 2000).

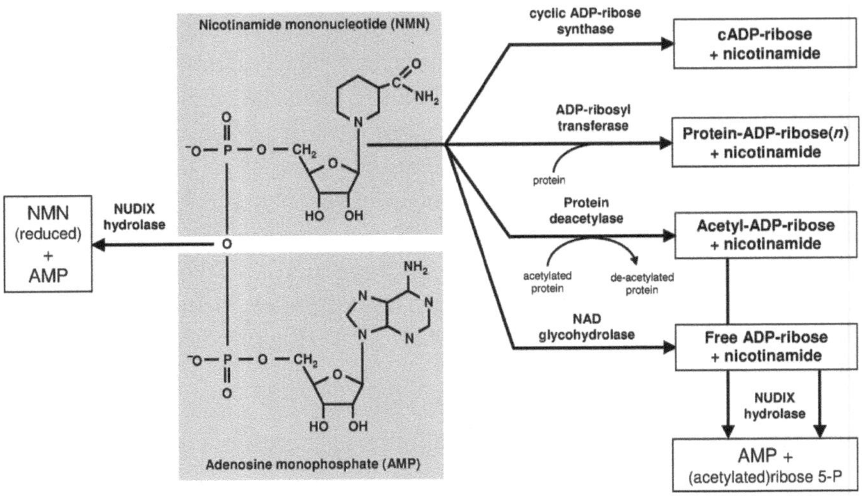

Fig. 3. Enzymes involved in NAD turnover. Different NUDIX hydrolases have different substrate specificities. NUDIX hydrolases so far described that cleave NAD preferentially use NADH rather than NAD^+.

NAD-dependent deacetylases (named SIRTUINS in humans and yeast) have been linked to cell death and oxidative stress responses in plants (Huang *et al.*, 2007). In other organisms, SIRTUINS have been implicated in the regulation of various biological processes, notably ageing (Imai *et al.*, 2000; Lin *et al.*, 2000). Recently, enzymes of NAD recycling and NAD itself have been implicated in the regulation of mammalian circadian clock function (Ramsey *et al.*, 2009).

A major demand on NAD (and potentially, therefore, NAD synthesis) may come from poly(ADP-ribose) polymerase (PARP). This protein transfers ADP-ribose groups from NAD to acceptor proteins, a modification that has been implicated in regulation of gene expression and DNA repair. PARP has been implicated in stress responses in different types of organisms, including plants (Adams-Phillips *et al.*, 2010; De Block *et al.*, 2005; Graziani *et al.*, 2005; Pellny *et al.*, 2009; Vanderauwera *et al.*, 2007; Ying *et al.*, 2003). The RADICALCELLDEATH1 (RCD1) protein has a PARP-type domain and has been implicated in oxidative stress responses in Arabidopsis (Ahlfors *et al.*, 2004).

Other potential NAD-cleaving enzymes that have recently received attention in plants are NUDIX hydrolases. The original member of this family (MutT) was discovered in *Escherichia coli* as a suppressor of mutations linked to oxidized dGTP: all members of the family cleave a *nu*cleoside *di*sphophate bond linked to some other group, *X*, with variable specificity for nucleoside triphosphates, nucleotide sugars, and NADH (Bessman *et al.*,

1996). NUDIX hydrolases that cleave NADH produce the two mononucleotides (with the pyridine ring of NMN in the reduced form) while others can metabolize ADP-ribose (Fig. 3). The physiological functions of these reactions may be to prevent excessive accumulation of NADH, which would inhibit oxidative metabolism, or ADP-ribose, which could affect a number of processes (Bessman *et al.*, 1996; McLennan, 2006). Yeast deficient in the yeast NUDIX hydrolase Ysa1 had decreased reactive oxygen species (ROS) levels and oxidative stress sensitivity linked to increased ADP-ribose. It was suggested that these effects were caused by ADP-ribose inhibition of GAPDH and mitochondrial complex I, thus decreasing ROS production and supporting NADPH generation through the oxidative pentose phosphate pathway (Tong *et al.*, 2009). In Arabidopsis, there are more than 20 genes predicted to encode NUDIX hydrolases (At*NUDT* or At*NUDX*). Analysis of nine of the corresponding recombinant proteins showed that NUDT2, NUDT6, NUDT7, and NUDT10 were able to use ADP-ribose, with the first three also able to use NADH (Ogawa *et al.*, 2005). Knockout mutants for NUDT7 show pleiotropic phenotypes, most notably increased salicylic acid (SA) and pathogen resistance (Bartsch *et al.*, 2006; Ge *et al.*, 2007; Jambunathan and Mahalingam, 2006). Differences in phenotypes have been reported between some of the studies and may be explained by different growth nutrition regimes (Jambunathan *et al.*, 2010). A large part of the total hydrolase activity against ADP-ribose and NADH may be due to NUDT7 (AtNUDX7) activity and modified expression of this enzyme affected *in vivo* contents of both compounds (Ishikawa *et al.*, 2009). Other data point to a role for NUDT7 in metabolizing PARP-derived ADP-ribose during oxidative stress (Ishikawa *et al.*, 2009). In a subsequent study, NUDT6/NUDX6 was implicated in regulation of NADH contents during SA signalling through the redox-regulated NONEXPRESSOROFPATHOGENESISRELATEDGENES1 (NPR1) pathway (Ishikawa *et al.*, 2010). A third member of this family with activity against ADP-ribose, AtNUDX2, has been implicated in oxidative stress tolerance (Ogawa *et al.*, 2009).

The above discussion underlines that the non-redox roles of NAD may be important in stress, of which redox changes are a key part. Emerging evidence from work on other organisms suggests that in some circumstances, NAD^+ concentrations may be limiting for these enzymes (Belenky *et al.*, 2009 and references cited therein). A key issue here could be that nucleotide concentrations determined by direct assay are unlikely to reflect the biochemically relevant "free" concentration, which is probably significantly lower than the measured concentration. This is because the high abundance of redox enzymes that use NAD(H) means that a large fraction of the measurable nucleotide pool within any compartment is bound at any one time to

NAD in plants (Heineke *et al.*, 1991; Igamberdiev and Gardeström, 2003; Rasmusson and Møller, 1990), it may also be important in evaluating the regulation of enzymes that use NAD^+ or NADH as a cleavable substrate.

IV. COMPARTMENTATION AND TRANSPORT OF NAD IN PLANTS

The distribution of NAD between compartments is a key issue in understanding NAD function and regulation of NAD synthesis and turnover. To date, most studies in plants have focused on NAD and NADP themselves, and this is reflected in the discussion below. These nucleotides are generally considered unlikely to move at significant rates between cells, though their precursors could be transported between tissues. The discussion below focuses on the relatively limited information currently available on their distribution and transport within plant cells.

A. INTRACELLULAR DISTRIBUTION

While the fluorescence of the reduced pyridine ring can be used to measure relative changes in NAD(P)H, determination of nucleotide concentrations and redox states (reduced:oxidized ratios) has thus far used techniques that require extraction. A major challenge in such measurements is avoiding alterations in metabolite status during sample preparation, a problem that becomes particularly acute when fractionation of cells is required prior to assay. Artefacts can arise from exchange between compartments during fractionation and/or changes in redox state as cellular integrity and functions are perturbed. Approaches that have been used to circumvent these problems in plants include non-aqueous fractionation and rapid filtration of protoplasts into acid or alkaline buffers. While the former technique is relatively slow, it avoids leakage and inhibits metabolism. The latter is much more rapid as it can arrest metabolism in a fraction of a second, but it requires prior preparation of protoplasts that can be easily ruptured by passage though a nylon mesh with an appropriate pore size (Gardeström, 1987; Stitt *et al.*, 1982).

A summary of some of the data available using these techniques is presented in Table II. NAD and NADP concentrations in the chloroplasts, mitochondria, and cytosol are between 0.1 and 2 mM (Heineke *et al.*, 1991; Igamberdiev and Gardeström, 2003; Szal *et al.*, 2008). These values are in the range of intracellular concentrations measured in other organisms such as yeast (Bogan and Brenner, 2008). Concentrations measured in isolated intact chloroplasts are also similar (Takahama *et al.*, 1981). Based on data obtained for protein. While this issue has been discussed in relation to redox functions of

TABLE II

Pyridine Nucleotide Concentrations (mM) in Three Compartments of Plant Cells in the Light and the Dark (n.m., not measured; n.d., not detected)

Compartment	Condition	NAD^+	NADH	Total NAD	$NADP^+$	NADPH	Total NADP	Reference
Chloroplast (stroma)	Light	0.21	0.05	0.26	n.m.	n.m.	n.m.	*c*
	Light	0.19	n.d.	0.19	0.59	0.29	0.88	*a*
	Dark	0.92	n.d.	0.92	0.51	0.12	0.63	*a*
Mitochondria	Light	0.68	1.76	2.44	n.m.	n.m.	n.m.	*c*
	Light	1.55	0.46	2.01	0.08	0.24	0.32	*b*
	Dark	1.52	0.08	1.60	0.27	0.05	0.32	*b*
Cytosol	Light	0.72	0.09	0.81	n.m.	n.m.	n.m.	*c*
	Light	0.52	0.06	0.58	0.14	0.18	0.32	*b*
	Dark	0.57	0.02	0.59	0.15	0.17	0.32	*b*

[a]Data from Heineke *et al.* (1991); chloroplasts obtained by non-aqueous fractionation of spinach leaves.

[b]Data from Igamberdiev and Gardeström (2003); rapid fractionation of pea protoplasts by filtration. "Light" values are for protoplasts incubated in limiting CO_2.

[c]Data from Szal *et al.* (2008); rapid fractionation of cucumber protoplasts by filtration. The concentrations shown were calculated from values given in nmol mg^{-1} chlorophyll assuming subcellular volumes reported for spinach leaves (Winter *et al.*, 1994).

three compartments in leaves or leaf cell protoplasts, NAD concentrations are highest in the mitochondria and lowest in the chloroplast (Table II). However, given the relative contributions of the different compartments to the intracellular volume (in leaf mesophyll cells, cytosol volume $> 6\times$ total mitochondrial volume: Queval *et al.*, 2011; Winter *et al.*, 1994), the cytosolic pool would make the major contribution to overall leaf mesophyll cell NAD contents.

NADP concentrations are about twice as high in the chloroplast as in the other two compartments (Table II). Interestingly, the mitochondrial NADP redox state is quite strongly affected by light (Igamberdiev and Gardeström, 2003). Mitochondrial NAD pools are more reduced than either cytosolic or chloroplastic NAD pools. However, this is less apparent when photorespiration is not active, for example, in the dark or at high CO_2 (Igamberdiev and Gardeström, 2003). A highly oxidized cytosolic NAD pool is also observed in some animal cells and is probably important to drive the GAPDH reaction in the glycolytic direction (Heineke *et al.*, 1991). A large cytosolic NAD^+ pool explains the redox state of global pools extracted from whole tissues, only 5–20% of which is NADH even in samples taken in the light (see Fig. 4). Light–dark changes in chloroplast NAD^+ may reflect the activity of light-activated stromal NADK (Heineke *et al.*, 1991).

B. SUBCELLULAR TRANSPORT

Reduced and oxidized forms of pyridine nucleotides do not cross most biological membranes at high rates. Instead, shuttle systems allow redox equivalents to be transferred between subcellular compartments through fast exchange of metabolites such as malate and oxaloacetate or triose phosphate and 3-PGA (Day and Wiskich, 1981; Heineke *et al.*, 1991; Journet *et al.*, 1981; Reumann *et al.*, 1994). These systems allow the chloroplast stromal and mitochondrial matrix to be redox-coupled to the cytosol and, potentially therefore, to each other (Foyer *et al.*, 2009; Raghavendra and Padmasree, 2003; Scheibe *et al.*, 2005). In animal and yeast mitochondria, a glycerol-3-phosphate dehydrogenase (G3PDH) located in the inner mitochondrial membrane has long been known to participate in the oxidation of cytosolic redox equivalents. More recently, the enzymes involved in this shuttle have been studied in Arabidopsis (Shen *et al.*, 2003, 2006). As in other organisms, the catalytic site of the mitochondrial G3PDH is oriented towards the outer surface. Since most small molecules traverse the outer mitochondrial membrane quite freely, the enzyme would act to oxidize cytosolic NADH in a similar though more indirect manner than the externally oriented inner membrane NADH dehydrogenase activity (Douce and Neuburger, 1989; Rasmusson *et al.*, 2004). The mitochondrial G3PDH could be particularly important in oxidizing glycerol produced from lipid reserves in germinating seeds and young seedlings (Shen *et al.*, 2003). Observations in knockout mutants for cytosolic G3PDH (which is likely to produce some of the G3P for mitochondrial oxidation) suggest that a shuttle similar to those occurring in other organisms is required for redox homeostasis in Arabidopsis (Shen *et al.*, 2006).

Despite the existence of these high-flux redox exchangers, current knowledge of NAD(P) synthesis implies that lower capacity transporters must operate to allow entry of nucleotides themselves into organelles. Indeed, slow uptake of NAD^+ has been observed in purified mitochondria (Neuburger and Douce, 1983; Neuburger *et al.*, 1985). Recently, the first specific proteins responsible for NAD^+ transport across plastid and mitochondrial membranes have been identified and studied. Two types of proteins that are competent in transmembrane nucleotide transport via exchange mechanisms have been identified. First, nucleotide transporter-type (NTT) carriers are found in plastids, as well as in bacteria that obtain their nucleotides from the extracellular milieu such as a host cell. They catalyse phosphate-neutral reactions involving ATP versus ADP plus phosphate or, in some cases, NAD^+ versus ADP (Haferkamp *et al.*, 2004; Trentmann *et al.*, 2008). However, the three Arabidopsis NTTs do not transport NAD^+ (Palmieri *et al.*, 2009). The second type belongs to the mitochondrial carrier family (MCF). Despite the nomenclature, it is now clear that these transporters are also found in other intracellular membranes such as those bounding the plastids and peroxisomes (Palmieri *et al.*, 2009). Based on homology to an MCF yeast gene encoding the first identified mitochondrial NAD^+ carrier (Todisco *et al.*, 2006), the Arabidopsis proteins *At*NDT1 and *At*NDT2 were shown to be competent in NAD^+ transport and to localize, respectively, to chloroplasts and mitochondria (Palmieri *et al.*, 2009). Their K_M values for NAD^+ were 0.15–0.24 mM, suggesting that they would work effectively at typical intracellular substrate concentrations even if, as discussed above, the free concentrations were somewhat lower than those shown in Table II.

Less information is available on transport of pyridine mononucleotides in plants. Both *At*NDT1 and *At*NDT2 were shown to exchange NAD^+ against NaMN, NMN, or NaAD as well as other nucleotides such as AMP or ADP but not nicotinamide or nicotinate (Palmieri *et al.*, 2009). Indeed, evidence has been presented that NMN and (at lower rates) NaMN can be taken up and converted to NAD by mitochondria isolated from Jerusalem artichoke tubers (Di Martino and Pallotta, 2011). Although yeast contain an NADH-preferring NADK that is localized in the mitochondria (Strand *et al.*, 2003), the Arabidopsis NADHK (NADK3) has been localized to peroxisomes (Waller *et al.*, 2010). Uptake of NAD for the peroxisomal NAD(H) kinase (and other NAD-dependent functions) are still to be described. A mitochondrial (and chloroplast) NADP transporter may await discovery: neither of the two recently described NDTs showed significant exchange of $NADP^+$ or NADPH against NAD^+ (Palmieri *et al.*, 2009). Comparison of NAD turnover rates in nucleated and enucleated mammalian cells suggests that a substantial proportion of breakdown and recycling of NAD may take place in the nucleus (Imai, 2009). Whether this is the case in plants remains unclear. Likewise, the extent to which the nuclear and cytosolic NAD pools are in equilibrium remains to be established.

C. PLASTICITY OF TISSUE NAD(H) CONTENTS

As described above, there are significant differences in both NAD concentrations and redox states between different cellular compartments. However, perhaps because the key roles of pyridine nucleotides imply careful control of

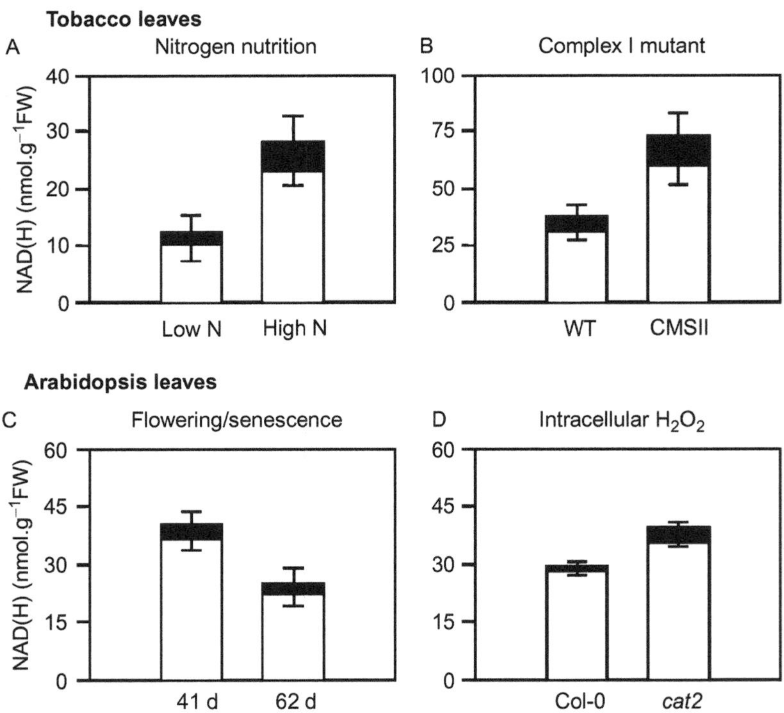

Fig. 4. Response of NAD(H) pools to environmental and developmental factors. (A) Effect of N nutrition on NAD contents. *N. sylvestris* plants grown at low N (0.1 mM NH_4NO_3) or high N (5 mM NH_4NO_3). Data are redrawn from Hager *et al.* (2010). (B) Increases in NAD(H) in tobacco mitochondrial mutants lacking complex I (CMSII; Dutilleul *et al.*, 2005). (C) Changes in leaf NAD(H) during Arabidopsis rosette development. Data are redrawn from Queval and Noctor (2007). (D) NAD levels in oxidative stress conditions. The data show contents in an Arabidopsis catalase-deficient oxidative stress mutant (*cat2*) and the corresponding wild-type plants (Col-0).

their status, there has been little focus on whether these pools are modified by environmental or other factors. Most of the attention has been paid to flexibility or stability of redox states (reduced:oxidized ratios) in response to short-term changes in metabolic conditions (e.g. Igamberdiev and Gardeström, 2003; Kasimova *et al.*, 2006). The recognition of the potential importance of NAD as a cleavable substrate begs the question of the flexibility of tissue contents. How fixed are NAD pools in plants? A better understanding of this issue could be relevant both to plant biology and to human nutrition.

Indications of the flexibility of pyridine nucleotide pools have come from the analysis of mutants lacking dehydrogenases that consume NADH. In an Arabidopsis mutant lacking cytosolic G3PDH, an appreciable increase in the leaf NADH:NAD^+ ratio was accompanied by changes in the antioxidative system. In this case, the overall NAD pool size was the same as in wild-type plants (Shen *et al.*, 2006). The *Nicotiana sylvestris* CMSII mutant, which lacks complex I activity (Gutierres *et al.*, 1997; Pineau *et al.*, 2005), has proved to a very useful genetic system for analyzing *in vivo* responses to perturbation of the mitochondrial electron transport chain. This mutant respires through alternative NADH-oxidizing activities (Sabar *et al.*, 2000) and shows multiple effects on leaf metabolism, including perturbed photosynthesis, enhanced nitrogen assimilation and nitrogen-containing compounds, altered redox homeostasis, and modified responses to the environment (Dutilleul *et al.*, 2003a,b, 2005; Hager *et al.*, 2010; Pellny *et al.*, 2008; Priault *et al.*, 2006). Among the biochemical consequences of the shift from complex I to alternative dehydrogenase activity are increases in leaf contents of both NAD^+ and NADH (Dutilleul *et al.*, 2005; Hager *et al.*, 2010; Szal *et al.*, 2008). Increases in NADH may be caused by the lower substrate affinity of the alternative dehydrogenase compared to complex I (Rasmusson *et al.*, 2004). Strikingly, the NADH:NAD^+ ratio was little or not affected by the CMSII mutation, as both forms were increased to a similar extent (Fig. 4B). Enhanced total NAD in response to a potential over-reduction may reflect the importance of avoiding inhibition of flux through glycolysis and the TCA cycle. However, there was no overall increase in NAD in the G3PDH mutants (Shen *et al.*, 2006). Alongside NAD, adenylates were also substantially increased in the CMSII line (Szal *et al.*, 2008). The flexibility of NAD contents in response to mutations has also been revealed by studies of the *old5* mutant, in which an increase in recycling capacity acts to maintain and even increase the pool when *de novo* synthesis is perturbed (Schippers *et al.*, 2008).

Changes in NAD contents are not confined to mutants but are also affected by development and the environment. In Arabidopsis leaves, NAD contents can change throughout rosette development (Queval and Noctor,

2007), with lower amounts in senescing leaves than in younger ones (Fig. 4C). This may reflect senescence-linked changes in nitrogen nutrition and also, perhaps, oxidative stress (Zimmermann *et al.*, 2006). Tobacco plants grown on abundant N have higher NAD^+ and NADH contents than N-limited plants (Fig. 4A). Since the NAD pyridine and purine rings contain a total of seven N atoms (Fig. 1), this could be part of a two-way regulatory loop between NAD synthesis and N assimilation (Hager *et al.*, 2010). Oxidative stress in a catalase-deficient Arabidopsis mutant also influences NAD contents to some extent (Fig. 4D). Increases in total NAD have also been linked to oxidative stress resulting from loss of NADP-linked dehydrogenase function in yeast (Minard and McAlister-Henn, 2010).

V. PLANT PATHWAYS OF NAD SYNTHESIS AND RECYCLING

Studies on the synthesis and metabolism of pyridine nucleotides and related compounds, such as nicotine, ricinin, and trigonelline, have appeared over the years (Frost *et al.*, 1967; Mann and Byerrum, 1974a,b; Ryrie and Scott, 1969; Sinclair *et al.*, 2000; Wagner *et al.*, 1986; Waller *et al.*, 1966). Annotations of plant genes based on homologies to genes in other organisms have been performed only recently, as genomic information has become available (Hunt *et al.*, 2004; Katoh and Hashimoto, 2004). This advance has been accompanied by reverse genetics studies to analyse the function of the identified genes and detailed labelling studies to identify metabolic pathways and intermediates. Together, this information has opened up new possibilities for understanding NAD synthesis and its regulation and has cleared the way for rationally designed manipulation of this part of plant metabolism.

It has become clear that NAD synthesis and recycling pathways are complex, with considerable variation between taxonomically different organisms. Similar to other nucleotides, the synthesis of NAD is divided into two types of pathways: (1) *de novo* synthesis and (2) recycling routes that salvage the pyridine ring released by NAD cleavage. To date, two *de novo* pathways are well characterized while several recycling pathways have been described.

A. *DE NOVO* PRODUCTION OF NAMN IN BACTERIA AND PLANTS

The *de novo* pathway involves the production of NaMN from the pyridine precursor, quinolinate. This compound is produced from aspartate or from tryptophan. While many eukaryotes and even some prokaryotes have the tryptophan pathway (Kurnasov *et al.*, 2003), evidence that this pathway operates in plants is limited. No predicted gene sequences are found in Arabidopsis, though they have been reported in rice (Katoh and Hashimoto, 2004; Katoh *et al.*, 2006). This observation, together with conflicting results obtained from earlier labelling studies conducted in species such as tobacco and maize (Arditti and Tarr, 1979; Henderson *et al.*, 1959; Tarr and Arditti, 1982), may possibly indicate differences in the *de novo* pathways of NAD synthesis between different plant groups (Katoh *et al.*, 2006).

Both rice and Arabidopsis genomes contain sequences that are homologous to the *E. coli nadA*, *nadB*, and *nadC* genes, encoding quinolinate synthase (QS), L-aspartate oxidase (AO), and QPRT, respectively (Fig. 5). The first enzyme of the pathway, AO, generates an unstable product, iminoaspartate, through an FAD-dependent reaction using either O_2 or fumarate as the electron acceptor, producing either H_2O_2 or succinate as the

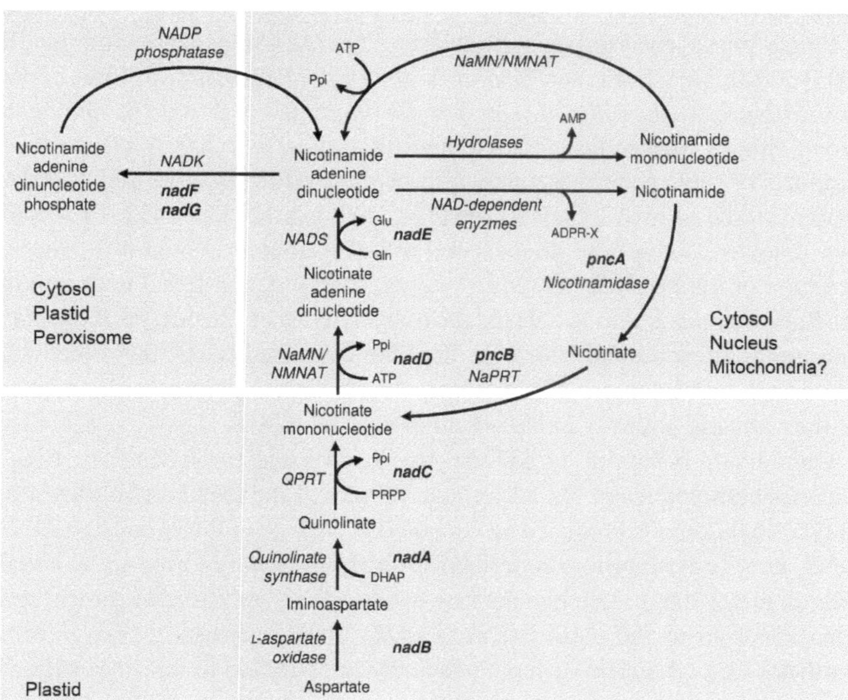

Fig. 5. Pathways of NAD synthesis and recycling in plants. ADPR-X, ADP-ribose or derivative (see Fig. 3); NADK, NAD kinase; NADS, NAD synthetase; NaMN/NMNAT, nicotinate mononucleotide/nicotinamide mononucleotide adenylyltransferase; NaPRT, nicotinate phosphoribosyltransferase; PRPP, 5-phosphoribosyl-1-pyrophosphate; Ppi, pyrophosphate; QPRT, quinolinate phosphoribosyltransferase. The names of the bacterial genes are also given for enzymes involved in NAD(P) synthesis.

reduced by-product (Marinoni *et al.*, 2008). Iminoaspartate is then condensed with DHAP by the second enzyme, QS, to produce quinolinate. The presence of these genes confirms the earlier description of AO activity in plants (Hosokawa *et al.*, 1983).

Studies on *E. coli* NadA established that this protein was the site of O_2 sensitivity. Based on a consensus binding sequence in the polypeptide, it was proposed that this effect is due to oxidative inactivation of an iron–sulphur cluster (Draczynska-Lusiak and Brown, 1992; Gardner and Fridovich, 1991). Recent studies have confirmed the presence of an [4Fe–4S] prosthetic group (Cicchillo *et al.*, 2005; Ollagnier de Choudens *et al.*, 2005). Only three of the iron atoms are coordinated to amino acids in the *E. coli* QS, leaving one iron free to coordinate the reactants during catalysis (Rousset *et al.*, 2008; Saunders *et al.*, 2008). Since this is reminiscent of other O_2-sensitive enzymes like aconitase, it may explain the sensitivity of QS to O_2 (Rousset *et al.*, 2008; Saunders *et al.*, 2008).

Single genes encoding homologues of NadA (At5g14760) and NadB (At5g50210) have been identified in Arabidopsis. Expression of these genes in the respective *E. coli nadA*- and *nadB*- mutants conferred the ability to grow without exogenous nicotinic acid (Katoh *et al.*, 2006). The protein products of these genes were shown to be directed to plastids, and T-DNA mutants to be embryo-lethal (Katoh *et al.*, 2006). Arabidopsis QS consists of two domains, a synthase domain and a SufE3 domain required for incorporation of the Fe–S cluster (Murthy *et al.*, 2007). A non-lethal mutation in the SufE domain is responsible for the early senescence phenotype of the *old5* mutant (Schippers *et al.*, 2008). Rather than decreased NAD, this effect was associated with increased NAD, which was attributed to increased activities of the recycling pathway in *old5* (Schippers *et al.*, 2008).

Quinolinate produced by AO and QS is converted to NaMN by QPRT, which is homologous to the bacterial *nadC* gene. Like the first two enzymes, only a single gene is predicted to encode this enzyme in Arabidopsis and T-DNA knockout mutations described so far have been reported to be lethal (Katoh *et al.*, 2006). Also like the first two enzymes, the encoded protein has been localized to the plastid (Katoh *et al.*, 2006), suggesting that NaMN synthesis through the *de novo* pathway may be restricted to this organelle, at least in some plants.

B. CONVERSION OF NAMN TO NAD

The conversion of NaMN to NAD requires the final two enzymes of the Preiss–Handler pathway (Fig. 2). These enzymes, NaMN adenylyltransferase (NaMNAT) and NADS, are probably located in the cytosol, and so current concepts suggest that the *de novo* pathway requires export of NaMN from the chloroplast. Like the enzymes that produce NaMN via quinolinate, NaMNAT is encoded by a single gene in Arabidopsis (At5g55810) and homozygous mutant plants cannot be recovered (Hashida *et al.*, 2007). While heterozygote mutants showed no obvious phenotype during the sporophyte (diploid) generation, the development of haploid pollen (but not haploid embryo sacs) produced from the mutant parent plants was perturbed (Hashida *et al.*, 2007). This effect involved disruption of pollen tube extension elongation, though pollen grain germination was not affected. This perhaps suggests that NAD synthesis is required to replenish the initial NAD complement of the pollen grain as this is consumed during tube elongation (Hashida *et al.*, 2007). NaMNAT has also been implicated in guard cell responses in Arabidopsis (Hashida *et al.*, 2010a). Like the human enzymes, the Arabidopsis enzyme can use either NMN or NaMN (Hashida *et al.*, 2007).

The final enzyme of NAD synthesis from NaMN, NADS, is predicted also to be encoded by a single gene in Arabidopsis (At1g55090). No detailed studies of mutants have yet appeared to establish whether this gene is indispensable or not. In the pathogenic bacterium *Francisella tularensis*, a novel enzyme, named NMN synthetase, has recently been discovered that can amidate NaMN to NMN (Sorci *et al.*, 2009). In this case, in contrast to the Preiss–Handler pathway (Fig. 2), the amidation step comes before the adenylation step. Unlike NADS, NMN synthetase shows preference for the mononucleotide NaMN over the dinucleotide NaAD while the *F. tularensis* adenylyltransferase prefers NMN to NaMN (Sorci *et al.*, 2009). Given the uncertainties in plants over the terminal reactions of NAD synthesis, further work might reveal whether any species- or tissue-specificity exists in the use of NMN or NaMN to make NAD.

C. RECYCLING PATHWAYS

The well-characterized bacterial recycling pathway depends on nicotinamidase to produce nicotinic acid from nicotinamide formed by the activities shown in Fig. 3 (right). This pathway clearly exists in plants, and nicotinamidase activity has been measured in several plant species (Katahira and Ashihara, 2009; Schippers *et al.*, 2008). Three Arabidopsis genes have been shown to encode proteins with nicotinamidase activity and have been designated *NIC1*, *NIC2*, and *NIC3* (Hunt *et al.*, 2007; Wang and Pichersky, 2007). The major nicotinamidase appears to be NIC1, and insertion mutants show decreased pools of pyridine nucleotides in many tissues, pointing to a role for the recycling pathway in maintaining NAD pools (Wang and Pichersky, 2007). The *NIC2* gene is most strongly expressed in seeds: knockout mutants show normal vegetative growth but germination in the absence of seed stratification was decreased (Hunt *et al.*, 2007). Decreased germination was not linked to changes in ABA but the observed increase in NAD contents could reflect inhibition of PARP by accumulated nicotinamide. However, not only nicotinamide but also NAD^+ or $NADP^+$ (but not NADH or NADPH) was shown to inhibit seed germination (Hunt *et al.*, 2007).

Conversion of nicotinate to NaMN occurs via NaPRT (Fig. 5). Two genes encode this enzyme in Arabidopsis (Hunt *et al.*, 2004). As yet, there is little information on mutants for these genes, so it is not clear whether they are essential or have any redundancy. NaPRT activity can be detected in potato tubers (Katahira and Ashihara, 2009). Mammals lack nicotinamidase (although this enzyme is present in gut bacteria) and directly convert nicotinamide to NMN through a nicotinamide phosphoribosyltransferase (NPRT; Fig. 2). However, there is no evidence for NPRT activity in plants (Katahira and Ashihara, 2009; Wang and Pichersky, 2007; Zheng *et al.*, 2005). Thus, the dual activity of plant NaMN/NMNAT may be required to adenylate NaMN produced by QPRT (*de novo* pathway), NaMN produced by NaPRT (recycling pathway from nicotinamide) and NMN generated by other sources. NMN could be metabolized to nicotinamide prior to recycling to NAD (Ashihara *et al.*, 2005, 2008) or to NaMN via a nicotinamide-independent pathway (Katahira and Ashihara, 2009). In potato tubers, no enzymes were detected that were able to deamidate NMN to NaMN or convert NMN to NPRT directly (Katahira and Ashihara, 2009). While the same study reported clear labelling of NAD(P) from [^{3}H]-quinolinate, the most heavily labelled pool was nicotinate glycoside (NaG). Based on detection of the required enzymes, it was suggested that NaG originates from a two-step degradation of NaMN, and that it may be important as a temporary storage product. In potato leaf discs the major products from [^{3}H]-quinolinate were trigonelline and, at a somewhat lower level, NAD(P) (Katahira and Ashihara, 2009).

Other reactions have been described, and significant inter-species differences in NAD recycling pathways appear to exist. For example, nicotinate riboside kinase activity has been found in both mung bean seedlings and potato tubers, providing an alternative route to NaMN (Katahira and Ashihara, 2009; Matsui and Ashihara, 2008). A recent study reported that isolated Jerusalem artichoke mitochondria were able to convert NMN to NAD, though the enzyme responsible for the activity remains to be characterized (Di Martino and Pallotta, 2011).

D. REGULATION OF NAD SYNTHESIS

Measurements of extractable enzyme activities in rice seedlings, Arabidopsis, and potato suggest that QPRT, NaMNAT, and NADS capacities are quite similar (Hayashi *et al.*, 2005; Katahira and Ashihara, 2009; Schippers *et al.*,

2008). By comparison, a recent study of potato tubers reported that NaPRT activity was 60-fold lower than that of NaMNAT (Katahira and Ashihara, 2009). Except in tissues producing ricinin or nicotine, NaPRT and QPRT activities were within the same order of magnitude, though both were lower than nicotinamidase (Mann and Byerrum, 1974b). The notion that phosphoribosyltransferases are important regulatory steps in the recycling pathway is consistent with information on the mammalian and bacterial pathways, where NPRT or NaPRT activities are considered the rate-limiting steps in NAD synthesis from nicotinamide and nicotinic acid, respectively (Imai, 2009). Phosphoribosyltransferases use phosphoribosylpyrophosphate (PRPP) as the ribose-phosphate donor. Interestingly, overexpression of PRPP synthetase promoted biomass accumulation in both Arabidopsis and tobacco (Koslowsky *et al.*, 2008). Whether this effect is linked to synthesis of NAD or other PRPP-requiring compounds is not yet clear.

Labelling studies are also consistent with nicotinamidase not being a limiting enzyme in NAD recycling. In canola embryos and potato tuber slices, exogenous nicotinamide is rapidly converted to nicotinic acid (Ashihara *et al.*, 2008; Katahira and Ashihara, 2009), consistent with high nicotinamidase activity, which was more than 10-fold higher than NaPRT in the potato tubers (Katahira and Ashihara, 2009). Studies of mutants show that nicotinamidase is necessary for full maintenance of NAD pools (Hunt *et al.*, 2007; Wang and Pichersky, 2007) although whether this enzyme plays an important regulatory role in wild-type plants remains unclear. Increased *NIC1* expression and nicotinamidase activity was part of the response of *old5* mutants to perturbed activity of the *de novo* pathway (Schippers *et al.*, 2008). If allowed to accumulate, nicotinamide itself may play influential roles. It is a competitive inhibitor of NAD-cleaving enzymes in yeast and mammals (Bogan and Brenner, 2008) and has been shown to affect stress responses in plants (Berglund *et al.*, 1993, 1996). Studies in other organisms clearly point to an important role for nicotinamidase (PNC1; Fig. 5). Increased copy number of PNC1 affects ageing in yeast (Anderson *et al.*, 2003) while in the worm, *Caenorhabditis elegans*, mutations in PNC1 affect reproductive development (Vrablik *et al.*, 2009).

NADS activity was increased in rice overexpressing a detoxifying enzyme with NADPH-dihydroflavonol reductase activity (Hayashi *et al.*, 2005) and in the *old5* Arabidopsis mutant (Schippers *et al.*, 2008). In contrast, QPRT was not affected by the *old5* mutation (Schippers *et al.*, 2008). This latter observation might reflect specific stimulation of the recycling pathway in conditions where the supply of quinolinate to QPRT by QS is impaired. NAD synthesis genes have been reported to be strongly expressed in guard cells, and NaMNAT proposed to be a regulator of ABA-induced changes in NAD (Hashida *et al.*, 2010a). A preliminary report suggests that overexpression of the NADS is not sufficient to increase NAD levels (Hashida *et al.*,

2010b), consistent with the notion that enzymes earlier in the pathway are more strongly limiting.

Little information is available on the capacities of the first two enzymes of the *de novo* pathway, although both have been detected in plant extracts (Hosokawa *et al.*, 1983; Schippers *et al.*, 2008). It seems reasonable to assume that their capacities are significantly below those of the subsequent steps. Indeed, the notion that quinolinate production is limiting for *de novo* NAD synthesis is consistent with our recent data (see Section VI). The bacterial NadA and NadB may function together as a complex. While there are several arguments in favour of this notion, such as the instability of the product, iminoaspartate, a physical association has not yet been conclusively demonstrated (Marinoni *et al.*, 2008). Quinolinate production may be light-regulated to some extent. Two vicinal cysteines in NadA (QS) that do not participate in coordinating the Fe–S cluster were found to interact with thioredoxin (TRX), suggesting they can form a disulfide bond (Rousset *et al.*, 2008). An independent study reported that the enzyme activity is stimulated by disulfide bond formation, that the activation/inactivation could be achieved by oxidized and reduced TRX, respectively, and that mutated TRX with a single active site cysteine formed a stable complex with NadA (Saunders and Booker, 2008). While the importance of this potential regulatory mechanism remains to be established in plants, it could have parallels with the oxidative activation of chloroplast enzymes involved in cysteine and glutathione synthesis under conditions of oxidative stress (Bick *et al.*, 2001; Hothorn *et al.*, 2006). Oxidative activation of NAD synthesis may be favoured under conditions in which NAD^+ is being consumed by stress-related reactions.

While nicotinamide is known to inhibit several NAD-cleaving enzymes, NaMN and NAD have been reported to feedback-inhibit bacterial NaPRT (Dulyaninova *et al.*, 1998). In humans, however, NPRT but not NaPRT was inhibited by NAD (Hara *et al.*, 2007). Quinolinate may regulate NADK activity in some bacteria (Garavaglia *et al.*, 2003). In plants, analysis of QPRT purified from castor bean endosperm revealed little or no inhibition by nicotinate or NAD (Mann and Byerrum, 1974a). Now that plant genes have been identified for the *de novo* and recycling pathways, analysis of the purified recombinant enzymes should throw more light on their regulation.

VI. MANIPULATING NAD CONTENTS IN PLANTS

From a physiological point of view, it might be predicted that the *de novo* pathway will be important in young tissues undergoing rapid growth with the recycling pathway more influential in mature tissues, being solicited, in particular, during increased use of NAD, for example, in stress conditions. However, our knowledge of the interplay between the two pathways is still limited. As Fig. 4 shows, leaf NAD contents are affected by environmental, developmental, or genetic factors, though the variation is not much greater than twofold. Our previous analyses established that increases of this order in mitochondrial complex I mutants are associated with marked effects on metabolism (Dutilleul *et al.*, 2005). These plants also show enhanced stress resistance, including resistance to pathogens (Dutilleul *et al.*, 2003b). More recently, a good correlation was observed between leaf NAD and leaf amino acid contents, indicating tight coordination of pyridine nucleotide and plant nitrogen status (Hager *et al.*, 2010). To investigate the contribution of modified NAD status to these effects, we developed a project to enhance this status in a targeted fashion. This section discusses initial results from this study.

A. NAD ENRICHMENT IN BACTERIA AND HUMAN CELLS

E. coli strains carrying additional copies of the *nadA* or *nadB* genes showed a slight increase in quinolinate production, which was much greater when both genes were present in higher copy number (Flachmann *et al.*, 1988). Overexpression of the *E. coli* NaPRT produced a 25-fold increase in extractable enzyme activity, but substantial increases in NAD were dependent on external supply of nicotinate, which enhanced NAD fivefold over control levels (Wubbolts *et al.*, 1990). Overexpression of a deregulated NAPRT in two independent studies also increased NAD levels 1.5- to 2-fold (Berríos-Rivera *et al.*, 2002; Heuser *et al.*, 2007). Increases in NAD of up to sevenfold were achieved by manipulation of multiple genes in *E. coli*, including inactivation of two NUDIX hydrolase genes and overexpression of NaPRT (*pncB*) and NADS (*nadE*; Heuser *et al.*, 2007).

Work on mammalian cells offers a complex picture, possibly due to cell-specific differences in NAD metabolism (Bogan and Brenner, 2008). Most studies have shown that nicotinate is more effective than nicotinamide in enhancing cellular NAD (Hara *et al.*, 2007; Micheli and Sestini, 1997). However, a study on human liver tumour cells reported that NAD was most strongly increased by nicotinamide with quinolinate being ineffective in increasing NAD (Evans *et al.*, 2002).

B. ENRICHING TISSUE NAD IN PLANTS

As a first step to testing the feasibility of manipulating NAD contents in plants, we transformed Arabidopsis plants with bacterial *nadB*, *nadA*, or *nadC* genes encoding the first three enzymes of the *de novo* synthesis pathway, or *pncB*, which produces NaMN in the recycling pathway (Fig. 5; Flachmann *et al.*, 1988; Wubbolts *et al.*, 1990). Heterologous expression of bacterial genes encoding enzymes already present in plants has been used for several years to enhance compounds such as specific amino acids or glutathione (Noctor *et al.*, 1996, 1998; Shaul and Galili, 1992). An advantage of this strategy is that the introduced gene is often not subject to strict control over expression, allowing enzyme capacity to be increased manyfold. Similarly, the introduced protein may be less sensitive to biochemical controls over its activity than the endogenous enzyme. On the other hand, this approach may be limited if the endogenous enzyme has requirements for its activity that the introduced protein is unable to fulfil. One obvious example is a difference in structure that affects interactions with other proteins. Because our aim was to increase NAD in the compartment where it is probably formed in plants, all expression cassettes were constructed for cytosolic expression.

After verification of transgene insertion and selection of homozygotes through classical techniques, independent lines with good expression at the level of transcript abundance were chosen for analysis of NAD contents. Despite apparently strong expression of the transgenes, none of the lines showed marked increases in NAD. However, when leaf discs from the *nadC* overexpressors, with genetically enhanced capacity for QPRT, were given the substrate, quinolinate (Fig. 6A), increases in leaf NAD of up to 10-fold could be observed within 48 h. This system therefore allows inducible, marked increases in NAD contents to be readily achieved. The results of a typical experiment, shown in Fig. 6B, reveal that this dramatic enhancement of NAD was associated with a smaller relative increase in NADP, presumably due to stimulated NADK activity. The reduction states of both pools were much less affected than their overall contents. While the NAD pool remained highly oxidized, with NADH representing about 5% of the leaf pool, NADPH and $NADP^+$ were close to 50% reduced in all genotypes (Fig. 6B). Supplying quinolinate to untransformed control plants had only a slight effect on pyridine nucleotide pools.

These results suggest that the *de novo* path of NAD synthesis in plants is co-limited by QPRT activity and quinolinate because simultaneous increases in both are required to produce substantial increases in NAD. However, subcellular compartmentation could be a key factor. In our *nadC*+ transformants, QPRT capacity is increased in the cytosol, whereas the *de novo* plant pathway produces quinolinate in the plastid. Assuming that the plant QPRT is restricted to the same compartment as the first two enzymes (Katoh *et al.*, 2006), there would be no physiological requirement for quinolinate to cross the chloroplast envelope to allow NAD synthesis. However, our results clearly imply that quinolinate can be taken up from the external milieu into the cytosol.

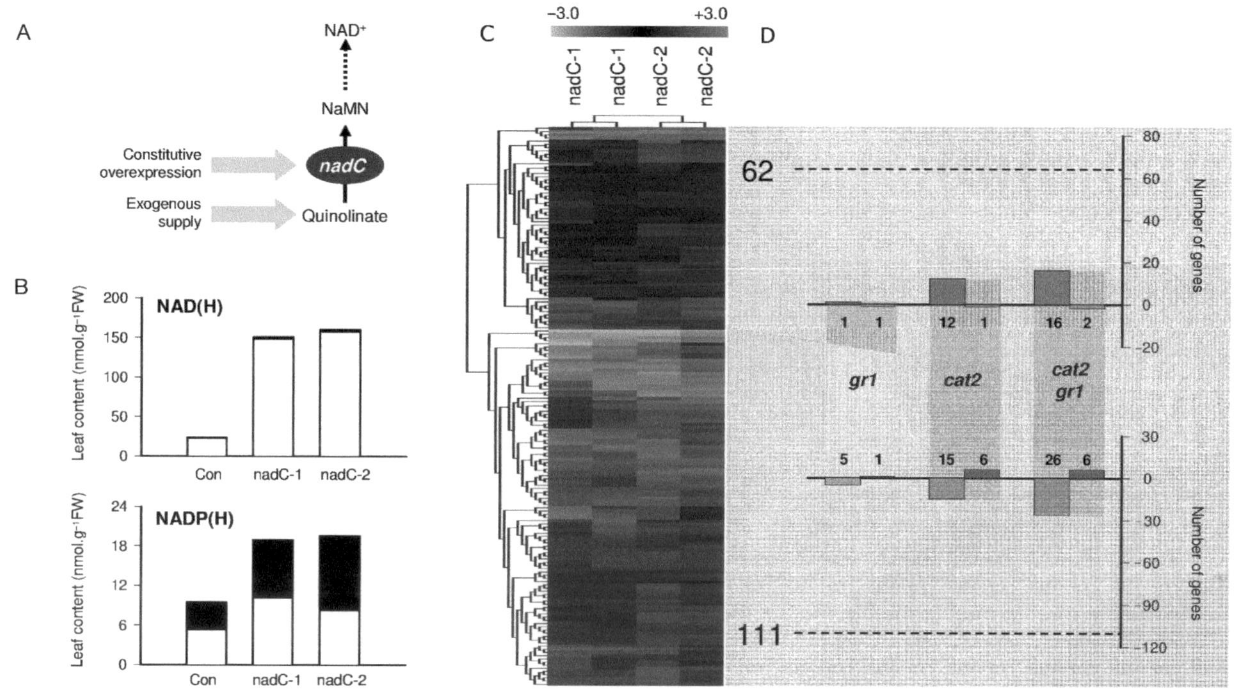

Fig. 6. Changes in gene expression triggered by targeted manipulation of endogenous NAD. Leaf discs from Arabidopsis Col-0 plants transformed with a construct directing the product of the bacterial *nadC* gene to the cytosol (nadC-1, nadC-2) were incubated on buffer containing quinolinate. Samples were harvested after 24-h incubation in the dark. Pyridine nucleotides were quantified as in Queval and Noctor (2007) and parallel samples were used for microarray analysis with the CATMA arrays and approach reported in Mhamdi *et al.* (2010b). Briefly, RNA from two biological replicates (leaf discs of two independent transformed lines supplied with quinolinate) were

The failure of expression of NadB or NadC to increase NAD may be explained either by co-limitation or by a need for functional association between the proteins so that the unstable iminoaspartate can be efficiently channelled from the first enzyme to the second. Substrate limitation in the cytosol is unlikely to be an explanation. The K_M values of the bacterial AO (NadB) for aspartate and QS (NadA) for DHAP have been measured to be 1–1.7 and 0.36 mM, respectively (Marinoni *et al.*, 2008), while the Arabidopsis QS has a K_M for DHAP of 1.1 mM (Murthy *et al.*, 2007). Aspartate is quite an abundant amino acid, while high triose phosphate isomerase activity keeps DHAP at substantial levels in cells. Both metabolites are central to primary plant carbon and nitrogen metabolism. In barley and spinach leaves, their cytosolic concentrations have been estimated to be at least as high or higher than stromal concentrations (Winter *et al.*, 1993, 1994).

Homologous overexpression of the plant AO in the chloroplast could be more successful in producing plants with constitutive increases in NAD, but based on studies in bacteria, it seems unlikely that this alone would suffice to produce the enrichment we report in Fig. 6B. More likely is that such marked constitutive increases would require simultaneous overexpression of the first three enzymes of the *de novo* pathway. We are currently exploring this possibility. It will be interesting to see whether sustainable increases are possible or whether degrading pathways would be up-regulated to

---

dye-swapped twice with control RNA samples from Arabidopsis transformed with an empty vector. Only genes that showed significant same-direction changes in all four analyses are shown. (A) Simple scheme showing the experimental strategy. (B) Leaf NAD(H) and NADP(H) contents in the empty vector controls (Con) and lines expressing nadC supplied with quinolinate (nadC-1 and nadC-2). White blocks, NAD^+ or $NADP^+$. Black blocks, NADH or NADPH. Similar results were obtained in other independent experiments. (C) Overview of genes that were significantly different between nadC+ lines supplied with quinolinate and the controls. Only genes that were significant in all four analyses are shown. The colour bar at top indicates expression fold change relative to controls. A full set of the data can be consulted at http://urgv.evry.inra.fr/cgi-bin/projects/CATdb/consult_project.pl?project_id=161. (D) Comparison of the responses of the significant genes in (C) with their expression in Arabidopsis mutants with targeted manipulation of glutathione or H_2O_2. Top pink frame, induced genes in nadC+ lines plus quinolinate. Bottom green frame, repressed genes in nadC+ lines plus quinolinate. Expression of the 163 genes was analysed in T-DNA mutants deficient in glutathione reductase (*gr1*), catalase (*cat2*), or both (*cat2 gr1*) (Mhamdi *et al.*, 2010b). Red bars show induced genes, green bars show repressed genes. For each bar, numbers indicate how many of the induced or repressed genes in (C) were found to be significant in each line. (See color plate section at the back of the book.)

compensate for the enhanced synthetic capacity. Another interesting question is whether pyridine nucleotide redox states and associated metabolism would be strongly affected in such lines.

C. Inducible Increases in NAD(H) are Associated with Activation of Pathogenesis-Related Pathways

To provide an initial assessment of the impact of engineering increases in NAD on plant cell function, we conducted microarray analyses of RNA isolated from the leaf discs accumulating high levels of NAD. At least at the level of gene expression, there was no generalized perturbation of genes involved in primary metabolism. Indeed, the transcriptomic signature involved only a relatively small fraction of the genome (Fig. 6C). A gene encoding NR was down-regulated, consistent with links that have been noted between nitrate assimilation and NAD status (Dutilleul *et al.*, 2005; Hager *et al.*, 2010; Kaiser *et al.*, 2000, 2002; Rachmilevitch *et al.*, 2004). However, the most striking pattern emerging from the analysis was the effect on genes known to be involved in phytohormone-mediated pathogenesis responses (Table III). Among the most strongly induced genes were several associated with SA synthesis and signalling (ICS1, FMO1, PAD4, BAP2; Bartsch *et al.*, 2006; Wildermuth *et al.*, 2001). Repressed genes included wounding and jasmonic acid (JA)-responsive genes (Table III). As SA–JA antagonism is well known to be part of plant responses to biotic stress (e.g. Koornneef *et al.*, 2008), our results suggest that increases in intracellular pyridine nucleotides are sufficient to entrain some of these responses.

Changes in redox state are an integral part of plant–pathogen interactions (Chamnongpol *et al.*, 1998; Vanacker *et al.*, 2000). Thiol-sensitive components such as NPR1 are particularly important (Mou *et al.*, 2003; Tada *et al.*, 2008). Recently, it was shown that infiltration of Arabidopsis leaves with physiological concentrations of pyridine nucleotides can induce the pathogenesis-related (*PR*) genes *PR1*, *PR2*, and *PR5* (Zhang and Mou, 2009). One obvious potential link between pyridine nucleotide status and pathogenesis responses is stimulation of NADPH oxidase activities (Torres *et al.*, 2005). However, induction of *PR* genes by exogenous pyridine nucleotides was little or not affected in mutants for two of the main leaf NADPH oxidases and was suggested to occur through an extracellular mechanism, possibly dependent on calcium signalling (Zhang and Mou, 2009). Although no *PR1* probe-set was present on the array used in our experiment, we observed no induction of the SA-associated genes *PR2* or *PR5*. Hence, only a subset of pathogenesis-associated genes appears to be affected by enhanced NAD synthesis, and

TABLE III

Examples of Transcripts Showing Significantly Altered Expression in Plants with Enhanced NAD

AGI code	Annotation	Expression levels	
		NAD	H_2O_2
REPRESSED (111 genes in total)			
At1g08830	CSD1, Cu/Zn superoxide dismutase, cytosol	−1.16	Induced
At2g28190	CSD2, Cu/Zn superoxide dismutase, chloroplast	−1.50	–
At1g12520	CCS1, SOD copper chaperone	−1.00	–
At1g77760	NIA1, nitrate reductase1	−1.71	Induced
At2g22500	DIC1/PUMP5, dicarboxylate carrier 1, plant uncoupling mitochondrial protein 5	−1.57	Repressed
At1g17420	LOX3, lipoxygenase3	−1.99	Repressed
At1g72520	Putative lipoxygenase	−1.89	Repressed
At2g34600	JAZ7, jasmonate/ZIM domain 7	−1.14	Repressed
At5g53750	Early wounding-induced jasmonic acid-dependent gene	−1.64	Repressed
At5g47230	AtEFR5, ethylene-response element binding factor 5	−1.06	–
At2g44840	AtERF13, ethylene-response element binding factor 13	−1.39	–
At3g02910	AVIG-like, avirulence-induced gene-like	−1.03	–
At1g65390	Phloem protein 2A-5 (PP2A-5), innate immune response	−1.11	–
INDUCED (62 genes in total)			
At1g74710	ICS1, isochorismate synthase1, salicylic acid synthesis	+2.68	–
At1g33960	AIG1 (AVRRPT2-INDUCED GENE 1)	+2.55	–
At1g19250	FMO1, flavin-dependent monoxygenase 1	+1.84	–
At2g45760	BAP2 (BON ASSOCIATION PROTEIN 2), negative regulator of PCD	+1.20	–
At3g52430	PAD4 (PHYTOALEXIN DEFICIENT 4)	+1.48	–
At1g31580	ECS1, cell wall protein, defence response	+2.24	–
At4g25100	FSD1 (FE SUPEROXIDE DISMUTASE 1), chloroplast	+1.50	Repressed
At3g20340	Paraquat-repressed gene	+1.19	–
At3g50470	Hypersensitive response protein 3 (HR3)	+1.13	–

AGI, Arabidopsis gene identifier code. Expression levels are given as $\log_2$ treatment/control ratio (0 = no expression change). Significant induction or repression by short-term exposure to intracellular H_2O_2 in a catalase-deficient mutant (Mhamdi *et al.*, 2010b) is indicated for comparison. For experimental details, see legend of Fig. 6.

these are not necessarily those whose expression is modified by exogenous pyridine nucleotides (Zhang and Mou, 2009). Further work is required to elucidate this point.

D. LINKS BETWEEN NAD, ROS, AND THIOL STATUS?

Enhanced NAD^+ or NADH could affect redox homeostasis through a number of mechanisms. As well as affecting NADPH oxidases, increased NADPH could modify thiol status by altering glutathione reductase (GR) or TRX reductase activities. To assess whether the transcriptome changes we report were linked to changes in intracellular oxidative stress or thiol-disulfide status, we first compared the microarray data with results recently obtained using the same array in plants deficient in catalase (*cat2*) and/or GR (*gr1*) (Mhamdi *et al.*, 2010b). The comparison revealed only a partial overlap, with the greatest overlap observed for double *cat2 gr1* mutants lacking both antioxidative enzymes. Overlap was most evident for JA-associated genes, whose expression has been linked to glutathione status (Koornneef *et al.*, 2008; Mhamdi *et al.*, 2010b). Partial overlap between the two studies could reflect the activation of ROS-dependent and ROS-independent signalling pathways by increases in NAD synthesis. Activation of multiple pathways might be caused by simultaneous increases in NAD^+, NADH, $NADP^+$, and NADPH (Fig. 6B) as these forms have been shown-to impact signalling pathways differentially when supplied exogenously to Arabidopsis leaves (Zhang and Mou, 2009). Direct analyses of thiol redox status should throw further light on this question and are underway.

Current concepts suggest that apoplastic exchange of pyridine nucleotides between plant cells does not occur at very high rates under most conditions. However, it has been proposed that these compounds could appear in the extracellular milieu at significant concentrations when cellular integrity is breached during pathogen attack or wounding, and that this might function as an important signal (Zhang and Mou, 2009). Although we cannot exclude the possibility that enhanced intracellular NAD contents cause increases in extracellular NAD(P), it is noteworthy that expression of genes associated with wounding and JA signalling were repressed, not induced, in our experiments (Table III). These genes are known to be affected by the intracellular redox state (Koornneef *et al.*, 2008; Mhamdi *et al.*, 2010b) and were repressed in response to both NAD enrichment and increased intracellular H_2O_2 (Table III). Despite this, it is evident that enhanced NAD does not merely recapitulate an oxidative stress response (Fig. 6). In some cases, increased NAD produced an antagonistic effect to those observed in *cat2*. Genes that showed this behaviour included two superoxide dismutases and a gene of unknown function known to be repressed by the superoxide generator, paraquat (Table III). An important point here could be that different ROS such as superoxide, singlet O_2 and H_2O_2 produce distinct transcriptomics signatures in plants (Gadjev *et al.*, 2006). Enhanced NAD and increased H_2O_2 also affected the expression of *NIA1*, encoding an isoform of NR, in opposite ways. The effects of enhanced NAD on NR expression could be related to nitric oxide production (Rockel *et al.*, 2002). It is also interesting that although neither NaPRT was among the significant genes affected by increased NAD, one of these genes (At2g23420) was among those induced by oxidative stress in *cat2 gr1* (Mhamdi *et al.*, 2010b).

In conclusion, enhanced NAD synthesis triggers induction of specific genes, a large number of which are associated with pathogenesis responses. Although some interaction with oxidative stress responses is apparent, a second major factor may be modification of processes related to the non-redox roles of NAD (Fig. 3). The concept of "oxidative stress" is a vague one that is still poorly defined in plants (Foyer and Noctor, 2011). In particular, the way in which enhanced ROS availability impacts on different redox couples such as NAD, NADP, and glutathione remains to be elucidated (Noctor, 2006). Model systems such as the one we describe here, which allow inducible enhancement of cellular pyridine nucleotide pools, could be useful in such studies.

VII. CONCLUSIONS AND PERSPECTIVES

Interest in pathways of NAD synthesis, traditionally driven by the vitamin function of niacin, has been greatly stimulated by the recognition that NAD plays important roles in cell signalling. Some of these roles involve cleavage of the NAD molecule and could therefore be influenced by the activity of NAD synthesis pathways. Several NAD-cleaving reactions, while not themselves involving redox exchange, appear to be tightly intertwined with redox homeostasis and signalling processes.

Surprising gaps in our knowledge of NAD synthesis and metabolism remain. This is as true in plants as it is in other organisms. While ongoing studies are likely to plug some of these gaps in knowledge quite quickly, it has become apparent that NAD metabolism is complex and shows diversity both between organisms and between different tissues of the same organism. Recent data, including those shown here, clearly point to potential interactions between NAD and plant defence signalling. These interactions could involve a number of processes, including stress-related PARP activity and calcium signalling.

Recent analyses of NUDIX hydrolases in plants suggest that these components play critical roles in oxidative stress and pathogenesis responses. The influence of NAD and NADP availability on NADPH oxidases and other ROS-producing enzymes also remains to be established, as does the relationship between these nucleotides and intracellular thiols. The availability of protocols that are able to provide information on the redox state and concentration of individual components in specific compartments will be useful here.

Our data suggest that it is feasible to manipulate the pathway of pyridine nucleotide synthesis to achieve considerable increases in tissue NAD. A key development in the systems biology era is the increasing erosion of notions of compartmentalized housekeeping versus signalling functions, according to which basic energetic processes were viewed as essential but of little importance in cell signalling functions. This has been replaced by an increasingly sophisticated appreciation of the integrated, networking nature of cells and tissues. Will manipulating a factor as central to cell function as NAD(H) produce desirable effects on plant niacin contents, plant performance or stress resistance? Can any such positive effects be separated from unwanted side effects? It should now be possible to begin to answer this and related questions.

ACKNOWLEDGEMENTS

We thank Jean-Pierre Renou (URGV Evry, France) for microarray analyses, Bertrand Gakière (IBP Orsay, France) for discussions and advice, and Fanta Ouedraogo (IBP Orsay, France) for technical assistance. Funding from the ANR-Genoplante project "Redoxome" is gratefully acknowledged by J. H. and G. N. We are also grateful to the Université de Paris sud 11 and the French Embassy in China for funding towards the studies of S. L.

REFERENCES

Adams-Phillips, L., Briggs, A. G. and Bent, A. F. (2010). Disruption of poly(ADP-ribosyl)ation mechanisms alters responses of *Arabidopsis* to biotic stress. *Plant Physiology* **152**, 267–280.

Ahlfors, R., Lang, S., Overmyer, K., Jaspers, P., Brosché, M., Tauriainen, A., Kollist, H., Tuominen, H., Belles-Boix, E., Piippo, M., Inzé, D., Palva, E. T. and Kangasjärvi, J. (2004). Arabidopsis RADICAL-INDUCED CELL DEATH1 belongs to the WWE protein–protein interaction domain protein family and modulates abscisic acid, ethylene, and methyl jasmonate responses. *The Plant Cell* **16**, 1925–1937.

Anderson, R. M., Bitterman, K. J., Wood, J. G., Medvedik, O. and Sinclair, D. A. (2003). Nicotinamide and PNC1 govern lifespan extension by calorie restriction in *Saccharomyces cerevisiae*. *Nature* **423**, 181–185.

Arditti, J. and Tarr, J. B. (1979). Niacin biosynthesis in plants. *American Journal of Botany* **66**, 1105–1113.

Ashihara, H., Stasolla, C., Yin, Y., Loukanina, N. and Thorpe, T. A. (2005). *De novo* and salvage biosynthetic pathways of pyridine nucleotides and nicotinic acid conjugates in cultured plant cells. *Plant Science* **169**, 107–114.

Ashihara, H., Luit, B., Belmonte, M. and Stasolla, C. (2008). Metabolism of nicotinamide, adenine, and inosine in developing microspore-derived canola (*Brassica napus*) embryos. *Plant Physiology and Biochemistry* **46**, 752–759.

Bartsch, M., Gobbato, E., Bednarek, P., Debey, S., Schultze, J. L., Bautor, J. and Parker, J. E. (2006). Salicylic acid-independent ENHANCED DISEASE SUSCEPTIBILITY1 signaling in Arabidopsis immunity and cell death is regulated by the monooxygenase FMO1 and the nudix hydrolase NUDT7. *The Plant Cell* **18**, 1038–1051.

Belenky, P., Chistensen, K. C., Gazzaniga, F., Pletnev, A. A. and Brenner, C. (2009). Nicotinamide riboside and nicotinic acid riboside salvage in fungi and mammals. Quantitative basis for Urh1 and purine nucleoside phosphorylase function in NAD^+ metabolism. *The Journal of Biological Chemistry* **284**, 158–164.

Berglund, T. L., Ohlsson, A. B., Rydström, J., Jordan, B. R. and Strid, A. (1993). Effect of nicotinamide on gene expression and glutathione levels in tissue cultures of *Pisum sativum*. *Journal of Plant Physiology* **142**, 676–684.

Berglund, T. L., Kalbin, G., Strid, A., Rydström, J. and Ohlsson, A. B. (1996). UV-B- and oxidative stress-induced increase in nicotinamide and trigonelline and inhibition of defensive metabolism induction by poly(ADP-ribose) polymerase inhibitor in plant tissue. *FEBS Letters* **380**, 188–193.

Berríos-Rivera, S. J., San, K. Y. and Bennett, G. N. (2002). The effect of NAPRTase overexpression on the total levels of NAD, the $NADH/NAD^+$ ratio, and the distribution of metabolites in *Escherichia coli*. *Metabolic Engineering* **4**, 238–247.

Bessman, M. J., Frick, D. N. and O'Handley, S. F. (1996). The MutT proteins or "Nudix" hydrolases, a family of versatile, widely distributed, "housecleaning" enzymes. *The Journal of Biological Chemistry* **271**, 25059–25062.

Bick, J. A., Setterdahl, A. T., Knaff, D. B., Chen, Y., Pitcher, L. H., Zilinskas, B. and Leustek, T. (2001). Regulation of the plant-type 5′-adenylylsulfate reductase by oxidative stress. *Biochemistry* **40**, 9040–9048.

Bieganowski, P. and Brenner, C. (2004). Discoveries of nicotinamide riboside as a nutrient and conserved NRK genes establish a Preiss–Handler independent route to NAD^+ in fungi and humans. *Cell* **117**, 495–502.

Bogan, K. L. and Brenner, C. (2008). Nicotinic acid, nicotinamide, and nicotinamide riboside: A molecular evaluation of NAD^+ precursor vitamins in human nutrition. *Annual Review of Nutrition* **28**, 115–130.

Bykova, N. V., Rasmusson, A. G., Igamberdiev, A. U., Gardeström, P. and Møller, I. M. (1999). Two separate transhydrogenase activities are present in plant mitochondria. *Biochemical and Biophysical Research Communications* **265**, 106–111.

Chai, M. F., Chen, Q. J., An, R., Chen, Y. M., Chen, J. and Wang, X. C. (2005). NADK2, an *Arabidopsis* chloroplastic NAD kinase, plays a vital role in both chlorophyll synthesis and chloroplast protection. *Plant Molecular Biology* **59**, 553–564.

Chai, M. F., Wei, P. C., Chen, Q. J., An, R., Chen, J., Yang, S. and Wang, X. C. (2006). NADK3, a novel cytoplasmic source of NADPH, is required under conditions of oxidative stress and modulates abscisic acid responses in Arabidopsis. *The Plant Journal* **47**, 665–674.

Chamnongpol, S., Willekens, H., Moeder, W., Langebartels, C., Sandermann, H., Van Montagu, M., Inzé, D. and Van Camp, W. (1998). Defense activation and enhanced pathogen tolerance induced by H_2O_2 in transgenic tobacco. *Proceedings of the National Academy of Sciences of the United States of America* **95**, 5818–5823.

Cicchillo, R. M., Tu, L., Stromberg, J. A., Hoffart, L. M., Krebs, C. and Booker, S. J. (2005). *Escherichia coli* quinolinate synthetase does indeed harbour a [4Fe–4S] cluster. *Journal of the American Chemical Society* **127**, 7310–7311.

Day, D. A. and Wiskich, J. T. (1981). Glycine metabolism and oxaloacetate transport by pea leaf mitochondria. *Plant Physiology* **68**, 425–429.

De Block, M., Verduyn, C., De Brouwer, D. and Cornelissen, M. (2005). Poly(ADP-ribose) polymerase in plants affects energy homeostasis, cell death and stress tolerance. *The Plant Journal* **41**, 95–106.

Delumeau, O., Renard, M. and Montrichard, F. (2000). Characterization and possible redox regulation of the purified calmodulin-dependent NAD kinase from *Lycopersicon pimpinellifolium*. *Plant, Cell and Environment* **23**, 1267–1273.

Di Martino, C. and Pallotta, M. L. (2011). Mitochondria-localized NAD biosynthesis by nicotinamide mononucleotide adenylyltransferase in Jerusalem artichoke (*Helianthus tuberosus* L.) heterotrophic tissues. *Planta,* doi: 10.1007/s00425-011-1428-6.

Douce, R. and Neuburger, M. (1989). The uniqueness of plant mitochondria. *Annual Review of Plant Physiology and Plant Molecular Biology* **40**, 371–414.

Draczynska-Lusiak, B. and Brown, O. R. (1992). Protein A of quinolinate synthetase is the site of oxygen poisoning of pyridine nucleotide coenzyme synthesis in *Escherichia coli*. *Free Radical Biology and Medicine* **13**, 689–693.

Dulyaninova, N. G., Podlepa, E. M., Toulokhonova, L. V. and Bykhovsky, V. Y. (1998). Salvage pathway for NAD biosynthesis in *Brevibacterium ammoniagenes*: Regulatory properties of triphosphate-dependent nicotinate phosphoribosyltransferase. *Biochimica et Biophysica Acta* **1478**, 211–220.

Dutilleul, C., Driscoll, S., Cornic, G., De Paepe, R., Foyer, C. H. and Noctor, G. (2003a). Functional mitochondrial complex I is required for optimal photosynthetic performance in photorespiratory conditions and during transients. *Plant Physiology* **131**, 264–275.

Dutilleul, C., Garmier, M., Noctor, G., Mathieu, C. D., Chétrit, P., Foyer, C. H. and De Paepe, R. (2003b). Leaf mitochondria modulate whole cell redox homeostasis, set antioxidant capacity and determine stress resistance through altered signaling and diurnal regulation. *The Plant Cell* **15**, 1212–1226.

Dutilleul, C., Lelarge, C., Prioul, J. L., De Paepe, R., Foyer, C. H. and Noctor, G. (2005). Mitochondria-driven changes in leaf NAD status exert a crucial influence on the control of nitrate assimilation and the integration of carbon and nitrogen metabolism. *Plant Physiology* **139**, 64–78.

Elvehjem, C. A., Madden, R. J., Strong, F. M. and Woolley, D. W. (1938). The isolation and identification of the anti-black tongue factor. *The Journal of Biological Chemistry* **123**, 137–149.

Evans, J., Wang, T. C., Heyes, M. P. and Markey, S. P. (2002). LC/MS analysis of NAD biosynthesis using stable isotope pyridine precursors. *Analytical Biochemistry* **306**, 197–203.

Flachmann, R., Kunz, N., Seifert, J., Gütlich, M., Wientjes, F. J., Läufer, A. and Gassen, H. G. (1988). Molecular biology of pyridine nucleotide biosynthesis in *Escherichia coli*. Cloning and characterization of quinolinate synthesis genes nadA and nadB. *European Journal of Biochemistry* **175**, 221–228.

Foyer, C. H. and Noctor, G. (2009). Redox regulation in photosynthetic organisms: Signaling, acclimation, and practical implications. *Antioxidants and Redox Signaling* **11**, 861–905.

Foyer, C. H. and Noctor, G. (2011). Ascorbate and glutathione: The heart of the redox hub. *Plant Physiology* **155**, 2–18.

Foyer, C. H., Bloom, A. J., Queval, G. and Noctor, G. (2009). Photorespiratory metabolism: Genes, mutants, energetics, and redox signaling. *Annual Review of Plant Biology* **60**, 455–484.

Frost, G. M., Yang, K. S. and Waller, G. R. (1967). Nicotinamide adenine dinucleotide as a precursor of nicotine in *Nicotiana rustica* L. *The Journal of Biological Chemistry* **242**, 887–888.

Fuller, L. (1980). Determination of nicotinamide. *Methods in Enzymology* **66**, 3–4.

Gadjev, I., Vanderauwera, S., Gechev, T. S., Laloi, C., Minkov, I. N., Shulaev, V., Apel, K., Inzé, D., Mittler, R. and Van Breusegem, F. (2006). Transcriptomic footprints disclose specificity of reactive oxygen species signaling in Arabidopsis. *Plant Physiology* **141**, 436–445.

Gallais, S., Pou de Crescenzo, M. A. and Laval-Martin, D. L. (2000a). Changes in soluble and membrane-bound isoforms of calcium-calmodulin-dependent and -independent NAD^+ kinase, during the culture of after-ripened and dormant seeds of *Avena sativa*. *Australian Journal of Plant Physiology* **27**, 649–658.

Gallais, S., Pou de Crescenzo, M. A. and Laval-Martin, D. L. (2000b). Evidence of active $NADP^+$ phosphatase in dormant seeds of *Avena sativa* L. *Journal of Experimental Botany* **51**, 1389–1394.

Garavaglia, S., Galizzi, A. and Rizzi, M. (2003). Allosteric regulation of *Bacillus subtilis* NAD kinase by quinolinic acid. *Journal of Bacteriology* **185**, 4844–4850.

Gardeström, P. (1987). Adenylate ratios in the cytosol, chloroplasts and mitochondria of barley leaf protoplasts during photosynthesis at different carbon dioxide concentrations. *FEBS Letters* **212**, 114–118.

Gardner, P. R. and Fridovich, I. (1991). Quinolinate synthetase: The oxygen-sensitive site of *de novo* $NAD(P)^+$ biosynthesis. *Archives of Biochemistry and Biophysics* **284**, 106–111.

Gazzaniga, F., Stebbins, R., Chang, S. Z., McPeek, M. A. and Brenner, C. (2009). Microbial NAD metabolism: Lessons from comparative genomics. *Microbiology and Molecular Biology Reviews* **73**, 529–541.

Ge, X., Li, G. J., Wang, S. B., Zhu, H., Zhu, T., Wang, X. and Xia, Y. (2007). AtNUDT7, a negative regulator of basal immunity in Arabidopsis, modulates two distinct defense response pathways and is involved in maintaining redox homeostasis. *Plant Physiology* **145**, 204–215.

Graziani, G., Battaini, F. and Zhang, J. (2005). PARP-1 inhibition to treat cancer, ischemia, inflammation. *Pharmacological Research* **52**, 1–4.

Guse, A. H. and Lee, H. C. (2008). NAADP: A universal Ca^{2+} trigger. *Science Signaling* **1**, 10.

Gutierres, S., Sabar, M., Lelandais, C., Chétrit, P., Diolez, P., Degand, H., Boutry, M., Vedel, F., De Kouchkovsky, Y. and De Paepe, R. (1997). Lack of mitochondrial and nuclear-encoded subunits of complex I and alteration of the respiratory chain in *Nicotiana sylvestris* mitochondrial deletion mutants. *Proceedings of the National Academy of Sciences of the United States of America* **94**, 3436–3441.

Haferkamp, I., Schmitz-Esser, S., Linka, N., Urbany, C., Collingro, A., Wagner, M., Horn, M. and Neuhaus, H. E. (2004). A candidate NAD^+ transporter in an intracellular bacterial symbiont related to Chlamydiae. *Nature* **432**, 622–625.

Hagedorn, P. H., Flyvbjerg, H. and Møller, I. M. (2007). Modelling NADH turnover in plant mitochondria. *Physiologia Plantarum* **120**, 370–385.

Hager, J., Pellny, T., Mauve, C., Lelarge-Trouverie, C., Prioul, J. L., De Paepe, R., Foyer, C. H. and Noctor, G. (2010). Conditional modulation of NAD levels and metabolite profiles in *Nicotiana sylvestris* by mitochondrial electron transport and carbon/nitrogen supply. *Planta* **231**, 1145–1157.

Hara, N., Yamada, K., Shibata, T., Osago, H., Hashimoto, T. and Tsuchiya, M. (2007). Elevation of cellular NAD levels by nicotinic acid and involvement of nicotinic acid phosphoribosyltransferase in human cells. *The Journal of Biological Chemistry* **282**, 24574–24582.

Harbinson, J., Genty, B. and Foyer, C. H. (1990). Relationship between photosynthetic electron transport and stromal enzyme activity in pea leaves. Toward an understanding of the nature of photosynthetic control. *Plant Physiology* **94**, 545–553.

Harding, S. A., Oh, S. H. and Roberts, D. A. (1997). Transgenic tobacco expressing a foreign calmodulin gene shows an enhanced production of active oxygen species. *The EMBO Journal* **16**, 1137–1144.

Hashida, S. N., Takahashi, H., Kawai-Yamada, M. and Uchimaya, H. (2007). *Arabidopsis thaliana* nicotinate/nicotinamide mononucleotide adenyltransferase (AtNMNAT) is required for pollen tube growth. *The Plant Journal* **49**, 694–703.

Hashida, S. N., Takahashi, H. and Uchimaya, H. (2009). The role of NAD biosynthesis in plant development and stress response. *Annals of Botany* **103**, 819–824.

Hashida, S. N., Itami, T., Takahashi, H., Takahara, K., Nagano, M., Kawai-Yamada, M., Shoji, K., Goto, F., Yoshihara, T. and Uchimaya, H. (2010a). Nicotinate/nicotinamide mononucleotide adenyltransferase-mediated regulation of NAD biosynthesis protects guard cells from reactive oxygen species in ABA-mediated stomatal movement in Arabidopsis. *Journal of Experimental Botany* **61**, 3813–3825.

Hashida, S. N., Takahara, K., Hirabayashi, T., Kawai-Yamada, M., Shoji, K., Yoshihara, T. and Uchimaya, H. (2010b). Overexpression of nicotinamide adenine dinucleotide synthetase causes early ageing in *Arabidopsis thaliana*. Meeting Abstract, 21st International Conference on *Arabidopsis* Research.

Hayaishi, O. and Ueda, K. (1977). Poly(ADP-ribose) and ADP-ribosylation of proteins. *Annual Review of Biochemistry* **46**, 95–116.

Hayashi, M., Takahashi, H., Tamura, K., Huang, J., Yu, L. H., Kawai-Yamada, M., Tezuka, T. and Uchimiya, H. (2005). Enhanced dihydroflavonol-4-reductase activity and NAD homeostasis leading to cell death tolerance in transgenic rice. *Proceedings of the National Academy of Sciences of the United States of America* **102**, 7020–7025.

Heineke, D., Riens, B., Grosse, H., Hoferichter, P., Peter, U., Flügge, U. I. and Heldt, H. W. (1991). Redox transfer across the inner chloroplast envelope membrane. *Plant Physiology* **95**, 1131–1137.

Henderson, L. M. (1983). Niacin. *Annual Review of Nutrition* **3**, 289–307.

Henderson, L. M., Someroski, J. F., Rao, D. R., Wu, P. H. L., Griffith, T. and Byerrum, R. U. (1959). Lack of a tryptophan–niacin relationship in corn and tobacco. *The Journal of Biological Chemistry* **234**, 93–95.

Heuser, F., Schroer, K., Lutz, S., Bringer-Meyer, S. and Sahm, H. (2007). Enhancement of the NAD(P)(H) pool in *Escherichia coli* for biotransformation. *Engineering in Life Sciences* **7**, 343–353.

Holtgrefe, S., Gohlke, J., Starmann, J., Druce, S., Klocke, S., Altmann, B., Wojtera, J., Lindermayr, C. and Scheibe, R. (2007). Regulation of plant cytosolic glyceraldehyde 3-phosphate dehydrogenase isoforms by thiol modifications. *Physiologia Plantarum* **133**, 211–218.

Horwitt, M. K., Harper, A. E. and Henderson, L. M. (1981). Niacin–tryptophan relationships for evaluating niacin equivalents. *The American Journal of Clinical Nutrition* **34**, 423–427.

Hosokawa, Y., Mitchell, E. and Gholson, R. K. (1983). Higher plants contain L-aspartate oxidase, the first enzyme of the *Escherichia coli* quinolinate synthetase system. *Biochemical and Biophysical Research Communications* **111**, 188–193.

Hossain, M. A. and Asada, K. (1985). Monodehydroascorbate reductase from cucumber is a flavin adenine dinucleotide enzyme. *The Journal of Biological Chemistry* **260**, 12920–12926.

Hothorn, M., Wachter, A., Gromes, R., Stuwe, T., Rausch, T. and Scheffzek, K. (2006). Structural basis for the redox control of plant glutamate cysteine ligase. *The Journal of Biological Chemistry* **281**, 27557–27565.

Huang, L., Sun, Q., Qin, F., Li, C., Zhao, Y. and Zhou, D. X. (2007). Down-regulation of a *SILENT INFORMATION REGULATOR2*-related histone deacetylase gene, *OsSRT1*, induces DNA fragmentation and cell death in rice. *Plant Physiology* **144**, 1508–1519.

Hunt, L. and Gray, J. (2009). The relationship between pyridine nucleotides and seed dormancy. *New Phytologist* **181**, 62–70.

Hunt, L., Lerner, F. and Ziegler, M. (2004). NAD—New roles in signalling and gene regulation in plants. *New Phytologist* **163**, 31–44.

Hunt, L., Holdsworth, M. J. and Gray, J. E. (2007). Nicotinamidase activity is important for germination. *The Plant Journal* **51**, 341–351.

Igamberdiev, A. U. and Gardeström, P. (2003). Regulation of NAD- and NADP-dependent isocitrate dehydrogenases by reduction levels of pyridine nucleotides in mitochondria and cytosol of pea leaves. *Biochimica et Biophysica Acta* **1606**, 117–125.

Imai, S. (2009). Nicotinamide phosphoribosyltransferase (Nampt): A link between NAD biology, metabolism, and diseases. *Current Pharmaceutical Design* **15**, 20–28.

Imai, S., Armstrong, C. M., Kaeberlein, M. and Guarente, L. (2000). Transcriptional silencing and longevity protein Sir2 is an NAD-dependent histone deacetylase. *Nature* **403**, 795–800.

Ishikawa, K., Ogawa, T., Hirosue, E., Nakayama, Y., Harada, K., Fukusaki, E., Yoshimura, K. and Shigeoka, S. (2009). Modulation of the poly(ADP-ribosyl)ation reaction via the Arabidopsis ADP-ribose/NADH pyrophosphohydrolase, AtNUDX7, is involved in the response to oxidative stress. *Plant Physiology* **151**, 741–754.

Ishikawa, K., Yoshimura, K., Harada, K., Fukusaki, E., Ogawa, T., Tamoi, M. and Shigeoka, S. (2010). AtNUDX6, an ADP-ribose/NADH pyrophosphorylase in Arabidopsis, positively regulates NPR1-dependent salicylic acid signaling. *Plant Physiology* **152**, 2000–2012.

Jackson, B. J. (2003). Proton translocation by transhydrogenase. *FEBS Letters* **545**, 18–24.

Jambunathan, N. and Mahalingam, R. (2006). Analysis of Arabidopsis growth factor gene 1 (GFG1) encoding a nudix hydrolase during oxidative signaling. *Planta* **224**, 1–11.

Jambunathan, N., Penaganti, A., Tang, Y. and Mahalingam, R. (2010). Modulation of redox homeostasis under suboptimal conditions by Arabidopsis nudix hydrolase 7. *BMC Plant Biology* **10**, 173.

Journet, E.-P., Neuburger, J. and Douce, R. (1981). Role of glutamate–oxaloacetate transaminase and malate dehydrogenase in the regeneration of NAD^+ for glycine oxidation by spinach leaf mitochondria. *Plant Physiology* **67**, 467–469.

Kaiser, W. M., Kandlbinder, A., Stoimenova, M. and Glaab, J. (2000). Discrepancy between nitrate reduction in intact leaves and nitrate reductase activity in leaf extracts: What limits nitrate reduction in situ? *Planta* **210**, 801–807.

Kaiser, W. M., Stoimenova, M. and Man, H. M. (2002). What limits nitrate reduction in leaves? *In* Photosynthetic Nitrogen Assimilation and Associated Carbon and Respiratory Metabolism, (C. H. Foyer and G. Noctor, eds.), pp. 63–70. Kluwer Academic, Dordrecht, The Netherlands.

Kasimova, M. R., Grigiene, J., Krab, K., Hagedorn, P. H., Flyvbjerg, H., Andersen, P. E. and Møller, I. M. (2006). The free NADH concentration is kept constant in plant mitochondria under different metabolic conditions. *The Plant Cell* **18**, 688–698.

Katahira, R. and Ashihara, H. (2009). Profiles of the biosynthesis and metabolism of pyridine nucleotides in potatoes (*Solanum tuberosum* L.). *Planta* **231**, 35–45.

Katoh, A. and Hashimoto, T. (2004). Molecular biology of pyridine and nicotine biosynthesis. *Frontiers in Bioscience* **9**, 1577–1586.

Katoh, A., Uenohara, K., Akita, M. and Hashimoto, T. (2006). Early steps in the biosynthesis of NAD in Arabidopsis start with aspartate and occur in the plastid. *Plant Physiology* **141**, 851–857.

Kelly, G. J. and Gibbs, M. (1973). Nonreversible D-glyceraldehyde 3-phosphate dehydrogenase of plant tissues. *Plant Physiology* **52**, 111–118.

Koornneef, A., Leon-Reyes, A., Ritsema, T., Verhage, A., Den Otter, F. C., Van Loon, L. C. and Pieterse, C. M. J. (2008). Kinetics of salicylate-mediated suppression of jasmonate signaling reveal a role for redox modulation. *Plant Physiology* **147**, 1358–1368.

Koslowsky, S., Riegler, H., Bergmüller, E. and Zrenner, R. (2008). Increased biomass accumulation by increasing phosphoribosylpyrophosphate synthetase activity in *Arabidopsis thaliana* and *Nicotiana tabacum*. *Plant Biotechnology Journal* **6**, 281–294.

Kurnasov, O., Goral, V., Colabroy, K., Gerdes, S., Anantha, S., Osterman, A. and Begley, T. P. (2003). NAD biosynthesis: Identification of the tryptophan to quinolinate pathway in bacteria. *Chemistry & Biology* **10**, 1195–1204.

Lahély, S., Bergaentzlé, M. and Hasselmann, C. (1999). Fluorimetric determination of niacin in foods by high-performance liquid chromatography with post-column derivatization. *Food Chemistry* **65**, 129–133.

Lancien, M., Gadal, P. and Hodges, M. (2000). Enzyme redundancy and the importance of 2-oxoglutarate in higher plant ammonium assimilation. *Plant Physiology* **123**, 817–824.

Lin, S. J., Defossez, P. A. and Guarante, L. (2000). Requirement of NAD and SIR2 for life-span extension in *Saccharomyces cerevisiae*. *Science* **289**, 2126–2128.

Mann, D. F. and Byerrum, R. U. (1974a). Quinolinic acid phosphoribosyltransferase from castor bean endosperm. I. Purification and characterization. *The Journal of Biological Chemistry* **249**, 6817–6823.

Mann, D. F. and Byerrum, R. U. (1974b). Activation of the *de novo* pathway for nucleotide biosynthesis prior to ricinine biosynthesis in castor beans. *Plant Physiology* **53**, 603–609.

Marinoni, I., Nonnis, S., Monteferrante, C., Heathcote, P., Härtig, E., Böttger, L. H., Trautwein, A. X., Negri, A., Albertini, A. M. and Tedeschi, G. (2008). Characterization of L-aspartate oxidase and quinolinate synthase from *Bacillus subtilis*. *FEBS Journal* **275**, 5090–5107.

Matsui, A. and Ashihara, H. (2008). Nicotinate riboside salvage in plants: Presence of nicotinate riboside kinase in mungbean seedlings. *Plant Physiology and Biochemistry* **46**, 104–108.

McLennan, A. G. (2006). The nudix hydrolase superfamily. *Cellular and Molecular Life Sciences* **63**, 123–143.

Mhamdi, A., Mauve, C., Gouia, H., Saindrenan, P., Hodges, M. and Noctor, G. (2010a). Cytosolic NADP-dependent isocitrate dehydrogenase contributes to redox homeostasis and the regulation of pathogen responses in *Arabidopsis* leaves. *Plant, Cell and Environment* **33**, 1112–1123.

Mhamdi, A., Hager, J., Chaouch, S., Queval, G., Han, Y., Taconnat, Y., Saindrenan, P., Issakidis-Bourguet, E., Gouia, H., Renou, J. P. and Noctor, G. (2010b). *Arabidopsis* GLUTATHIONE REDUCTASE 1 is essential for the metabolism of intracellular H_2O_2 and to enable appropriate gene expression through both salicylic acid and jasmonic acid signaling pathways. *Plant Physiology* **153**, 1144–1160.

Micheli, V. and Sestini, S. (1997). Determining NAD synthesis in erythrocytes. *Methods in Enzymology* **280**, 211–221.

Minard, K. I. and McAlister-Henn, L. (2010). Pnc1p supports increases in cellular NAD(H) levels in response to internal or external oxidative stress. *Biochemistry* **49**, 6299–6301.

Mou, Z., Fan, W. and Dong, X. (2003). Inducers of plant systemic acquired resistance regulate NPR1 function through redox changes. *Cell* **113**, 935–944.

Murthy, U. M. N., Ollagnier-de-Choudens, S., Sanakis, Y., Abdel-Ghany, S. E., Rousset, C., Ye, H., Fontecave, M., Pilon-Smits, E. A. H. and Pilon, M. (2007). Characterization of *Arabidopsis thaliana* SufE2 and SufE3. Functions in chloroplast iron–sulfur cluster assembly and NAD synthesis. *The Journal of Biological Chemistry* **282**, 18254–18264.

Navazio, L., Bewell, M. A., Siddiqua, A., Dickinson, G., Galione, A. and Sanders, D. (2000). Calcium release from the endoplasmic reticulum of higher plants elicited by the NADP metabolite nicotinic acid adenine dinucleotide phosphate. *Proceedings of the National Academy of Sciences of the United States of America* **97**, 8693–8698.

Neuburger, M. and Douce, R. (1983). Slow diffusion of NAD^+ between intact isolated plant mitochondria and suspending medium. *Biochemical Journal* **216**, 443–450.

Neuburger, M., Day, D. A. and Douce, R. (1984). The regulation of malate oxidation in plant mitochondria by the redox state of endogenous pyridine nucleotides. *Physiologie Végétale* **22**, 571–580.

Neuburger, M., Day, D. A. and Douce, R. (1985). Transport of NAD^+ in percoll-purified potato tuber mitochondria. Inhibition of NAD^+ influx and efflux by *N*-4-azido-2-nitrophenyl-4-aminobutyryl-3′-NAD^+. *Plant Physiology* **78**, 405–410.

Noctor, G. (2006). Metabolic signalling in defence and stress: The central roles of soluble redox couples. *Plant, Cell and Environment* **29**, 409–425.

Noctor, G., Strohm, S., Jouanin, L., Kunert, K.-J., Foyer, C. H. and Rennenberg, H. (1996). Synthesis of glutathione in leaves of transgenic poplar overexpressing γ-glutamylcysteine synthetase. *Plant Physiology* **112**, 1071–1078.

Noctor, G., Arisi, A. C. M., Jouanin, L. and Foyer, C. H. (1998). Manipulation of glutathione and amino acid biosynthesis in the chloroplast. *Plant Physiology* **118**, 471–482.

Noctor, G., Queval, G. and Gakiere, B. (2006). NAD(P) synthesis and pyridine nucleotide cycling in plants and their potential importance in stress conditions. *Journal of Experimental Botany* **57**, 1603–1620.

Ogawa, T., Ueda, Y., Yoshimura, K. and Shigeoka, S. (2005). Comprehensive analysis of cytosolic nudix hydrolases in *Arabidopsis thaliana*. *The Journal of Biological Chemistry* **280**, 25277–25283.

Ogawa, T., Ishikawa, K., Harada, K., Fukusaki, E., Yoshimura, K. and Shigeoka, S. (2009). Overexpression of an ADP-ribose pyrophosphatase, AtNUDX2, confers enhanced tolerance to oxidative stress in Arabidopsis plants. *The Plant Journal* **57**, 289–301.

Ollagnier de Choudens, S., Loiseau, L., Sanakis, Y., Barras, F. and Fontecave, M. (2005). Quinolinate synthetase, an iron–sulfur enzyme in NAD biosynthesis. *FEBS Letters* **579**, 3737–3743.

Palmieri, F., Rieder, B., Ventrella, A., Blanco, E., Do, P. T., Nunes-Nesi, A., Trauth, A. U., Fiermonte, G., Tjaden, J., Agrimi, G., Kirchberger, S., Paradies, E., Fernie, A. R. and Neuhaus, H. E. (2009). Molecular identification and functional characterization of *Arabidopsis thaliana* mitochondrial and chloroplastic NAD^+ carrier proteins. *The Journal of Biological Chemistry* **284**, 31249–31259.

Pellny, T. K., Van Aken, O., Dutilleul, C., Wolff, T., Groten, K., Bor, M., De Paepe, R., Reyss, A., Van Breusegem, F., Noctor, G. and Foyer, C. H. (2008). Mitochondrial respiratory pathways modulate nitrate sensing and nitrogen-dependent regulation of plant architecture in *Nicotiana sylvestris*. *The Plant Journal* **54**, 976–992.

Pellny, T. K., Locato, V., Vivancos, P. D., Markovic, J., De Gara, L., Pallardo, F. V. and Foyer, C. H. (2009). Pyridine nucleotide cycling and control of intracellular redox state in relation to poly (ADP-ribose) polymerase activity and nuclear localization of glutathione during exponential growth of Arabidopsis cells in culture. *Molecular Plant* **2**, 442–456.

Pineau, B., Mathieu, C., Gérard-Hirne, C., De Paepe, R. and Chétrit, P. (2005). Targeting the NAD7 subunit to mitochondria restores a functional complex I and a wild type phenotype in the *Nicotiana sylvestris* CMSII mitochondrial mutant lacking *nad7*. *The Journal of Biological Chemistry* **280**, 25994–26001.

Pollak, N., Niere, M. and Ziegler, M. (2007). NAD kinase levels control the NADPH concentration in human cells. *The Journal of Biological Chemistry* **282**, 33562–33571.

Preiss, J. and Handler, P. (1958). Biosynthesis of diphosphopyridine nucleotide. II. Enzymatic aspects. *The Journal of Biological Chemistry* **233**, 493–500.

Priault, P., Tcherkez, T., Cornic, G., De Paepe, R., Naik, R., Ghashghaie, J. and Streb, P. (2006). The lack of mitochondrial complex I in a CMSII mutant of *Nicotiana sylvestris* increases photorespiration through an increased internal resistance to CO_2 diffusion. *Journal of Experimental Botany* **57**, 3195–3207.

Pugin, A., Frachisse, J. M., Tavernier, E., Bligny, R., Gout, E., Douce, R. and Guern, J. (1997). Early events induced by the elicitor cryptogein in tobacco cells: Involvement of a plasma membrane NADPH oxidase and activation of glycolysis and the pentose phosphate pathway. *The Plant Cell* **9**, 2077–2091.

Queval, G. and Noctor, G. (2007). A plate-reader method for the measurement of NAD, NADP, glutathione and ascorbate in tissue extracts. Application to redox profiling during Arabidopsis rosette development. *Analytical Biochemistry* **363**, 58–69.

Queval, G., Jaillard, D., Zechmann, B. and Noctor, G. (2011). Increased intracellular H_2O_2 availability preferentially drives glutathione accumulation in vacuoles and chloroplasts. *Plant, Cell & Environment* **34**, 21–32.

Rachmilevitch, S., Cousins, A. B. and Bloom, A. J. (2004). Nitrate assimilation in plant shoots depends on photorespiration. *Proceedings of the National Academy of Sciences of the United States of America* **101**, 11506–11510.

Raghavendra, A. S. and Padmasree, K. (2003). Beneficial interactions of mitochondrial metabolism with photosynthetic carbon assimilation. *Trends in Plant Science* **8**, 546–553.

Ramsey, K. M., Yoshino, J., Brace, C. S., Abrassart, D., Kobayashi, Y., Marcheva, B., Hong, H. K., Chong, J. L., Buhr, E. D., Lee, C., Takahashi, J. S., Imai, S. and Bass, J. (2009). Circadian clock feedback cycle through NAMPT-mediated NAD^+ biosynthesis. *Science* **324**, 651–654.

Rasmusson, A. G. and Møller, I. M. (1990). NADP-utilizing enzymes in the matrix of plant mitochondria. *Plant Physiology* **94**, 1012–1018.

Rasmusson, A. G., Soole, K. L. and Elthon, T. E. (2004). Alternative NAD(P)H dehydrogenases of plant mitochondria. *Annual Review of Plant Biology* **55**, 23–39.

Reumann, S., Heupel, R. and Heldt, H. W. (1994). Compartmentation studies on spinach leaf peroxisomes. II. Evidence for the transfer of reductant from the cytosol to the peroxisomal compartment via a malate shuttle. *Planta* **193**, 167–173.

Rius, S. P., Casati, P., Iglesias, A. A. and Gomez-Casati, D. F. (2006). Characterization of an *Arabidopsis thaliana* mutant lacking a cytosolic non-phosphorylating glyceraldehyde-3-phosphate dehydrogenase. *Plant Molecular Biology* **61**, 945–957.

Rockel, P., Strube, F., Rockel, A., Wildt, J. and Kaiser, W. M. (2002). Regulation of nitric oxide (NO) production by plant nitrate reductase *in vivo* and *in vitro*. *Journal of Experimental Botany* **53**, 103–110.

Roje, S. (2007). Vitamin B synthesis in plants. *Phytochemistry* **68**, 1904–1921.

Rose-Sallin, C., Blake, C. J., Genoud, D. and Tagliaferri, E. G. (2001). Comparison of microbiological and HPLC-fluorescence detection methods for determination of niacin in fortified food products. *Food Chemistry* **73**, 473–480.

Rousset, C., Fontecave, M. and Ollagnier de Choudens, S. (2008). The [4Fe–4S] cluster of quinolinate synthase from *Escherichia coli*. Investigation of cluster ligands. *FEBS Letters* **582**, 2937–2944.

Ryrie, I. J. and Scott, K. J. (1969). Nicotinate, quinolinate and nicotinamide as precursors in the biosynthesis of nicotinamide-adenine dinucleotide in barley. *Biochemical Journal* **115**, 679–685.

Sabar, M., De Paepe, R. and De Kouchkovsky, Y. (2000). Complex I impairment, respiratory compensations, and photosynthetic decrease in nuclear and mitochondrial male sterile mutants of *Nicotiana sylvestris*. *Plant Physiology* **124**, 1239–1249.

Sánchez, J. P., Duque, P. and Chua, N. H. (2004). ABA activates ADPR cyclase and cADPR induces a subset of ABA-responsive genes in Arabidopsis. *The Plant Journal* **38**, 381–395.

Saunders, A. H. and Booker, S. J. (2008). Regulation of the activity of *Escherichia coli* quinolinate synthase by reversible disulfide-bond formation. *Biochemistry* **47**, 8467–8469.

Saunders, A. H., Griffiths, A. E., Lee, K. H., Cicchillo, R. M., Tu, L., Stromberg, J. A., Krebs, C. and Booker, S. J. (2008). Characterization of quinolinate synthases from *Escherichia coli, Mycobacterium tuberculosis*, and *Pyrococcus horikoshii* indicates that [4Fe–4S] clusters are common cofactors throughout this class of enzymes. *Biochemistry* **47**, 10999–11012.

SCF (Scientific Committee on Food) (1993). Nutrient and energy intakes for the European Community. *In* Reports of the Scientific Committee for Food **Series 31**, Commission of the European Communities, Luxembourg.

Scheibe, R., Backhausen, J. E., Emmerlich, V. and Holtgrefe, S. (2005). Strategies to maintain redox homeostasis during photosynthesis under changing conditions. *Journal of Experimental Botany* **56**, 1481–1489.

Schippers, J. H. M., Nunes-Nesi, A., Apetrei, R., Hille, J., Fernie, A. R. and Dijkwel, P. P. (2008). The *Arabidopsis onset of leaf death5* mutation of quinolinate synthase affects nicotinamide adenine dinucleotide biosynthesis and causes early ageing. *The Plant Cell* **20**, 2909–2925.

Shaul, O. and Galili, G. (1992). Increased lysine synthesis in tobacco plants that express high levels of bacterial dihydropicolinate synthase in their chloroplasts. *The Plant Journal* **2**, 203–209.

Shen, W., Wei, Y., Dauk, M., Zheng, Z. and Zou, J. (2003). Identification of a mitochondrial glycerol-3-phosphate dehydrogenase from *Arabidopsis thaliana*: Evidence for a mitochondrial glycerol-3-phosphate shuttle in plants. *FEBS Letters* **536**, 92–96.

Shen, W., Wei, Y., Dauk, M., Tan, Y., Taylor, D. C., Selvaraj, G. and Zou, J. (2006). Involvement of a glycerol-3-phosphate dehydrogenase in modulating the NADH/NAD^+ ratio provides evidence of a mitochondrial glycerol-3-phosphate shuttle in Arabidopsis. *The Plant Cell* **18**, 422–441.

Sinclair, S. J., Murphy, K. J., Birch, C. D. and Hamill, J. D. (2000). Molecular characterization of quinolinate phosphoribosyl transferase (QPRTase) in *Nicotiana*. *Plant Molecular Biology* **44**, 603–617.

Smythe, G. A., Braga, O., Brew, B. J., Grant, R. S., Guillemin, G. J., Kerr, S. J. and Walker, D. W. (2002). Concurrent quantification of quinolinic, picolinic, and nicotinic acids using electron-capture negative-ion gas chromatography–mass spectrometry. *Analytical Biochemistry* **301**, 21–26.

Sorci, L., Martynowski, D., Rodlonov, D. A., Eyobo, Y., Zogaj, X., Klose, K. E., Nikolaev, E. V., Magni, G., Zhang, H. and Osterman, A. L. (2009). Nicotinamide mononucleotide synthetase is the key enzyme for an alternative route of NAD biosynthesis in *Francisella tularensis*. *Proceedings of the National Academy of Sciences of the United States of America* **106**, 3083–3088.

Stitt, M., McLilley, R. and Heldt, H. W. (1982). Adenine nucleotide levels in the cytosol, chloroplasts, and mitochondria of wheat leaf protoplasts. *Plant Physiology* **70**, 971–977.

Strand, M. K., Stuart, G. R., Longley, M. J., Graziewicz, M. A., Dominick, O. C. and Copeland, W. C. (2003). *POS5* gene of *Saccharomyces cerevisiae* encodes a mitochondrial NADH kinase required for stability of mitochondrial DNA. *Eukaryotic Cell* **2**, 809–820.

Szal, B., Dabrowska, Z., Malmberg, G., Gardeström, P. and Rychter, A. (2008). Changes in energy status of leaf cells as a consequence of mitochondrial genome arrangement. *Planta* **227**, 697–706.

Tada, Y., Spoel, S. H., Pajerowska-Mukhtar, K., Mou, Z., Song, J., Wang, C., Zuo, J. and Dong, X. (2008). Plant immunity requires conformational changes of NPR1 via *S*-nitrosylation and thioredoxins. *Science* **321**, 952–956.

Takahama, U., Shimizu-Takahama, M. and Heber, U. (1981). The redox state of the NADP system in illuminated chloroplasts. *Biochimica et Biophysica Acta* **637**, 530–539.

Takahara, K., Kasajima, I., Takahashi, H., Hashida, S., Itami, T., Onodera, H., Toki, S., Yanagisawa, S., Kawai-Yamada, M. and Uchimaya, H. (2010). Metabolome and photochemical analysis of rice plants overexpressing Arabidopsis NAD kinase gene. *Plant Physiology* **152**, 1863–1873.

Takahashi, H., Takahara, K., Hashida, S. N., Hirabayashi, T., Fujimori, T., Kawai-Yamada, M., Yamaya, T., Yanagisawa, S. and Uchimaya, H. (2009). Pleiotropic modulation of carbon and nitrogen metabolism in Arabidopsis plants overexpressing the *NAD kinase2* gene. *Plant Physiology* **151**, 110–113.

Tarr, J. B. and Arditti, J. (1982). Niacin biosynthesis in seedlings of *Zea mays*. *Plant Physiology* **69**, 553–556.

Todisco, S., Agrimi, G., Castegna, A. and Palmieri, F. (2006). Identification of the mitochondrial NAD^+ transporter in *Saccharomyces cerevisiae*. *The Journal of Biological Chemistry* **281**, 1524–1531.

Tong, L., Lee, S. and Denu, J. M. (2009). Hydrolase regulates NAD^+ metabolites and modulates cellular redox. *The Journal of Biological Chemistry* **284**, 11256–11266.

Torres, M. A., Jones, J. D. G. and Dangl, J. L. (2005). Pathogen-induced, NADPH oxidase-derived reactive oxygen intermediates suppress spread of cell death in *Arabidopsis thaliana*. *Nature Genetics* **37**, 1130–1134.

Trentmann, O., Jung, B., Neuhaus, H. E. and Haferkamp, I. (2008). Nonmitochondrial ATP/ADP transporters accept phosphate as third substrate. *The Journal of Biological Chemistry* **283**, 36486–36493.

Turner, W. L., Waller, J. C., Vanderbeld, B. and Snedden, W. A. (2004). Cloning and characterization of two NAD kinases from Arabidopsis. Identification of a calmodulin-binding isoform. *Plant Physiology* **135**, 1243–1255.

Turner, W. L., Waller, J. C. and Snedden, W. A. (2005). Identification, molecular cloning and functional characterization of a novel NADH kinase from *Arabidopsis thaliana* (thale cress). *Biochemical Journal* **385**, 217–223.

Vanacker, H., Carver, T. L. W. and Foyer, C. H. (2000). Early H_2O_2 accumulation in mesophyll cells leads to induction of glutathione during the hypersensitive response in the barley–powdery mildew interaction. *Plant Physiology* **123**, 1289–1300.

Vanderauwera, S., De Block, M., Van de Steene, N., De Cottet, B. V., Metzlaff, M. and Van Breusegem, F. (2007). Silencing of poly(ADP-ribose) polymerase in plants alters abiotic stress signal transduction. *Proceedings of the National Academy of Sciences of the United States of America* **104**, 15150–15155.

Vrablik, T. L., Huang, L., Lange, S. E. and Hanna-Rose, W. (2009). Nicotinamidase modulation of NAD^+ biosynthesis and nicotinamide levels separately affect reproductive development and cell survival in *C. elegans*. *Development* **136**, 3637–3646.

Wagner, R., Feth, F. and Wagner, K. G. (1986). The pyridine-nucleotide cycle in tobacco: Enzyme activities for the recycling of NAD. *Planta* **167**, 226–232.

Waller, G. R., Yang, K. S., Gholson, R. K., Hadwiger, L. A. and Chaykin, S. (1966). The pyridine nucleotide cycle and its role in the biosynthesis of ricinine by *Ricinus cummunis* L.. *The Journal of Biological Chemistry* **241**, 4411–4418.

Waller, J. C., Dhanoa, P. K., Schumann, U., Mullen, R. T. and Snedden, W. A. (2010). Subcellular and tissue localization of NAD kinases from *Arabidopsis*: Compartmentalization of de novo NADP biosynthesis. *Planta* **231**, 305–317.

Wang, G. and Pichersky, E. (2007). Nicotinamidase participates in the salvage pathway of NAD in Arabidopsis. *The Plant Journal* **49**, 1020–1029.

Wildermuth, M. C., Dewdney, J., Wu, G. and Ausubel, F. M. (2001). Isochorismate synthase is required to synthesize salicylic acid for plant defence. *Nature* **414**, 562–565.

Windahl, K. L., Trenerry, V. C. and Ward, C. M. (1998). The determination of niacin in selected foods by capillary electrophoresis and high performance liquid chromatography: Acid extraction. *Food Chemistry* **65**, 263–270.

Winter, H., Robinson, G. and Heldt, H. W. (1993). Subcellular volumes and metabolite concentrations in barley leaves. *Planta* **191**, 180–190.

Winter, H., Robinson, G. and Heldt, H. W. (1994). Subcellular volumes and metabolite concentrations in spinach leaves. *Planta* **193**, 530–535.

Wubbolts, M. G., Terpstra, P., Van Beilen, J. B., Kingma, J., Meesters, H. A. R. and Witholt, B. (1990). Variation of cofactor levels in *Escherichia coli.* Sequence analysis and expression of the pncB gene encoding nicotinic acid phosphoribosyltransferase. *The Journal of Biological Chemistry* **265**, 17665–17672.

Xie, W., Xu, A. and Yeung, E. S. (2009). Determination of NAD^+ and NADH in a single cell under hydrogen peroxide stress by capillary electrophoresis. *Analytical Chemistry* **81**, 1280–1284.

Yamada, K., Hara, N., Shibata, T., Osago, H. and Tsuchiya, M. (2006). The simultaneous measurement of nicotinamide adenine dinucleotide and related compounds by liquid chromatography/electrospray ionization tandem mass spectrometry. *Analytical Biochemistry* **352**, 282–285.

Ying, W., Garnier, P. and Swanson, R. A. (2003). NAD^+ repletion prevents PARP-1-induced glycolytic blockade and cell death in cultured mouse astrocytes. *Biochemical and Biophysical Research Communications* **308**, 809–813.

Zhang, X. and Mou, Z. (2009). Extracellular pyridine nucleotides induce PR gene expression and disease resistance in Arabidopsis. *The Plant Journal* **57**, 302–312.

Zheng, X., Hayashibe, E. and Ashihara, H. (2005). Changes in trigonelline (*N*-methylnicotinic acid) content and nicotinic acid metabolism during germination of mungbean (*Phaseolus aureus*) seeds. *Journal of Experimental Botany* **56**, 1615–1623.

Zimmermann, P., Heinlein, C., Orendi, G. and Zentgraf, U. (2006). Senescence-specific regulation of catalase in *Arabidopsis thaliana* (L.) Heynh. *Plant, Cell and Environment* **29**, 1049–1056.

Pantothenate Biosynthesis in Higher Plants

MICHAEL E. WEBB*[,1] AND ALISON G. SMITH[†]

*School of Chemistry and Astbury Centre for Structural Molecular Biology, University of Leeds, Leeds, United Kingdom
[†]Department of Plant Sciences, University of Cambridge, Cambridge, United Kingdom

[1]Corresponding author: E-mail: m.e.webb@leeds.ac.uk

Advances in Botanical Research, Vol. 58

0065-2296/11 $35.00
DOI: 10.1016/B978-0-12-386479-6.00001-9

ABSTRACT

Pantothenate (vitamin B_5) is an essential metabolite for all organisms because it is the precursor to the 4′-phosphopantetheine moiety of coenzyme A (CoA) and acyl carrier protein. Pantothenate must be obtained from the diet by animals, but plants, bacteria, and fungi can synthesise it *de novo* by the condensation of β-alanine with pantoate, which is synthesised in two steps from α-ketoisovalerate, an intermediate in branched chain amino acid biosynthesis. In plants, the first and the last enzymes in this pathway, ketopantoate hydroxymethyltransferase and pantothenate synthetase, have been identified and characterised, but the enzyme responsible for the intermediate step of reduction of ketopantoate to pantoate has not been identified. Similarly, the source of β-alanine for pantothenate biosynthesis in plants has not been established; we suggest that catabolism of the polyamine, spermidine, is the most likely source of this metabolite. In contrast, all five plant enzymes required to convert pantothenate into CoA have been identified and characterised. Strains of bacteria producing increased quantities of pantothenate have been generated by metabolic engineering, and knowledge of the bacterial pathway has been used in the first attempts to engineer plants with similarly increased pantothenate production.

I. BIOLOGICAL FUNCTION AND REQUIREMENTS

A. INTRODUCTION

Pantothenate (Fig. 1A), or vitamin B_5, is one of the structurally simplest of the vitamins and is ubiquitous in nature. In all organisms, pantothenate is the precursor for the 4′-phosphopantetheine moiety of both coenzyme A (CoA) and the covalently linked phosphopantetheine prosthetic group of acyl carrier proteins (ACPs; Alberts and Vagelos, 1966). These cofactors play a vital role in central carbohydrate and lipid metabolism, as well as being involved in many secondary metabolite biosynthetic pathways. Pantothenate was first identified by Williams *et al.* (1933) as an essential nutrient for yeast, *Saccharomyces cerevisiae*. This small molecule was later found to be identical to the factors subsequently shown by Elvehjem and Koehn (1935) to prevent dermatitis in chicks and by Lythgoe *et al.* (1940) to be important for rat nutrition. Hydrolysis of purified pantothenate was shown to produce β-alanine (Weinstock *et al.*, 1939), and structural analysis of the resulting pantoyl lactone (Stiller *et al.*, 1940a) led to the determination of the structure of pantothenate (Williams and Major, 1940) and two subsequent total

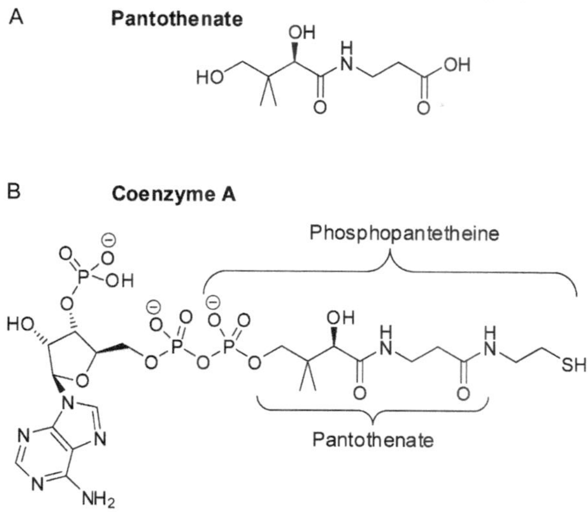

Fig. 1. Structure of pantothenate and coenzyme A.

syntheses (Stiller *et al.*, 1940b; Williams *et al.*, 1940). The pathway to pantothenate, via the condensation of pantoate with β-alanine, is best characterised in bacteria (Fig. 2; Webb *et al.*, 2004). In this review, we consider in detail what is known of the synthesis of pantothenate in plants, its uptake and subsequent conversion to the active cofactors in humans and other organisms, and the prospects for engineering the pathway in plants.

B. BIOLOGICAL FUNCTION OF PANTOTHENATE

Coenzyme A (CoA, Fig. 1B) is so called because it was identified by Lipmann *et al.* (1947) as the heat-stable cofactor for acetylation reactions, the A standing for acetylation. The active part of the molecule is the terminal thiol group, which is covalently linked via a thioester bond to acyl groups such as acetate, or longer chain fatty acids. The CoA derivative is more soluble in the aqueous environment of the cell and is said to be activated because the ΔG of hydrolysis of the thioester linkage is large and negative (e.g. -31.5 kJ mol^{-1} for acetyl CoA). This then facilitates the formation of covalent bonds, such as citrate from acetyl CoA and oxaloacetate in the Krebs' cycle. CoA is involved in innumerable reactions of central metabolism (e.g. fatty acid oxidation, and biosynthesis of glycerolipids and sterols) as well as secondary metabolic pathways, including those for polyketides, non-ribosomal protein synthesis, flavonoids, and lignin. In *Escherichia coli*, it has been estimated that approximately 100 enzymes (over 4% of the total) use either CoA or a CoA ester as substrate (Begley *et al.*, 2001).

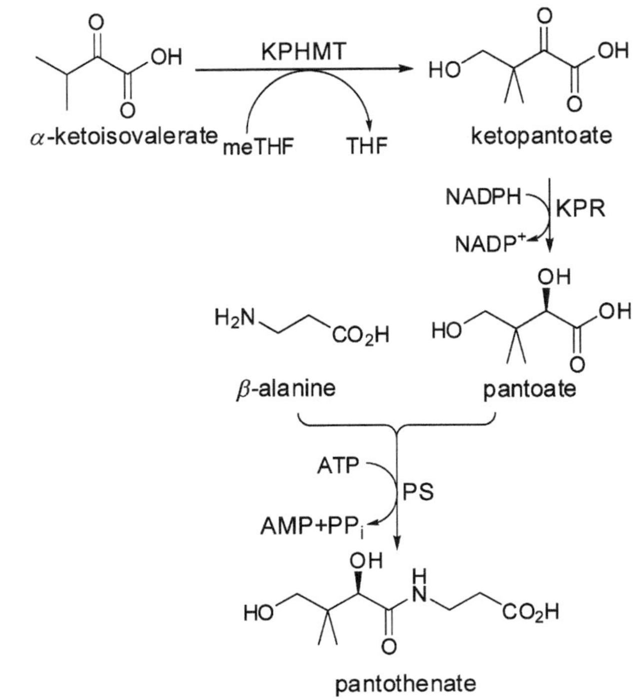

Fig. 2. Biosynthetic pathway for pantothenate. Pantothenate is formed by the condensation of β-alanine and pantoate catalysed by the ATP-dependent pantothenate synthetase (PS). Pantoate is synthesis**ed in two steps from α-ketoisovalerate (α-KIVA), the oxo-acid of valine, by ketopantoate hydroxymethyltransferase (KPHMT), which has a 5,10-methylene tetrahydrofolate cofactor, followed by NADPH-reduction by ketopantoate reductase (KPR).

ACPs have a much more restricted, although equally important, role in fatty acid synthesis, and in *E. coli*, ACP is the most abundant soluble protein constituting about 0.25% of the total soluble protein (Magnuson *et al.*, 1993). Again the acyl groups are attached via a thioester link to the terminal thiol. Transthioesterification is readily achieved and this reactivity is central to the chemistry of these thioesters. The pK_a of the alpha proton is also reduced by thioesterification, enabling Claisen ester condensation chemistry to occur readily in pathways of fatty acid biosynthesis.

C. EFFECTS OF PANTOTHENATE DEFICIENCY

Pantothenate is widespread in all foodstuffs—indeed the name is derived from the Greek '*panto*' then, which means from everywhere—and so deficiency is not observed in the general population. A 'burning foot syndrome' exhibited by malnourished prisoners of war in the Second World War was attributed to pantothenate deficiency although this was not confirmed (Bender, 1999). However, in poultry chicks, dermatitis around the eyes and beak was found to be alleviated by administering pantothenate. In adult mice, pantothenate deficiency is associated with lower levels of total CoA and an associated inability to maintain normal levels of glycogen (Smith *et al.*, 1987). In rats, mild pantothenate deficiency is associated with short-term elevation in plasma triglyceride and free fatty acid levels before significant weight loss occurs (Wittwer *et al.*, 1990). These final symptoms are similar to those observed after feeding with the L-enantiomer of the naturally occurring vitamin D-pantothenate. The L-enantiomer has no biological activity but can antagonise the incorporation of D-pantothenate when in excess (Sarett and Barboriak, 1963). Moreover, maternal pantothenate deficiency led to reduced food intake in neonatal pups and thereby to change in lipid composition (Rajalakshmi and Nakhasi, 1975). Finally, rats fed on a pantothenate-deficient diet exhibited loss of fur colour (Morgan and Simms, 1940): this lends a weak pseudo-scientific basis for the widespread inclusion of pantothenol into hair care products, although the surfactant properties are probably more important in this case (Hoezel *et al.*, 1998).

A small number of clinical studies of pantothenate deficiency in man have been performed with the antimetabolites ω-methylpantothenate (Fry *et al.*, 1976) and homopantothenate (pantoyl-γ-aminobutyric acid; Noda *et al.*, 1988). In both cases, a number of generalised symptoms were observed including depression, frequent infection, fatigue, and minor neurological disorders including numbness and the 'burning foot' syndrome described in earlier case studies.

Organism level effects of deficiency in a wide variety of animals of economic importance have been defined (Smith and Song, 1996 and references therein). These organisms include pigs, birds, and fish. In all cases, pantothenate deficiency leads to loss of appetite and an associated decreased growth rate. Additional symptoms include demyelination of motor neurons in chicks associated with ataxia, extensive colitis and ulcers in swine, adrenal lesions in rats and fused gill lamellae in fish (see Table I, reproduced from Smith and Song, 1996). As a consequence, pantothenate is routinely added to animal feeds and is the major market for the manufactured vitamin. The effect on weight gain and maintenance in all organisms means that it is possible to define minimum dietary requirements for a wide range of animals (see Table II, reproduced from Smith and Song, 1996); however, no routine test for pantothenate nutritional status has been identified. And in practice, the ubiquity of pantothenate (either free or bound in CoA) in natural foodstuffs means that no minimum dietary allowance is recommended.

TABLE I
Symptoms of Pantothenate Deficiency

Species	Visible signs	Histopathological abnormalities
Humans	Depression; personality changes; frequent infection; fatigue; sleep disturbances	Cardiac instability; abdominal pains; neurological disorders; Reye-like syndrome; muscle weakness
Chicks	Dermatitis at corners of beak and eyelids; rough feather; retarded growth; ataxia	Thymus involution; distended gall bladder; demyelination of motor neurons; moderate duodenal ulcerations
Swine	Loss of hair; reddening of skin; excess nasal secretion; changes in tongue; diarrhea; GI bleeding; locomotor abnormalities in gait	Extensive colitis and small ulcers; inflammatory changes in large intestine; chromatolysis of dorsal root ganglion cells; demyelination of peripheral nerves
Rats/ mice	Skin lesions; greying of hair and bald patches; red 'porphyrin' whiskers; paralysis of hind legs	Adrenal lesions; lipid depleted, shrunk, and dying cells
Fish	Clubbed gills; intralamellar proliferative lesions; listless	Fused gill lamallae covered with exudate; fatty liver; kidneys deposited with glycogen and hyaline bodies; clumped mitochondria

Reproduced in part from Smith and Song (1996). In fact, given the widespread distribution of pantothenate in most foodstuffs, this is confined to animals on artificial diets, or to experimentally induced treatments, such as with the antimetabolites ω-methylpantothenate (Fry *et al.*, 1976), and homopantothenate (pantoyl-γ-aminobutyric acid; Noda *et al.*, 1988).

D. DIETARY ABSORPTION AND METABOLISM IN MAMMALS

Pantothenate is absorbed at low concentrations through the intestinal wall via Na-dependent active transport (Fenstermacher and Rose, 1986; Prasad *et al.*, 1999), while at higher concentrations, it passively diffuses through the cell membrane (Shibata *et al.*, 1983). All forms of the vitamin act as dietary nutrients; CoA can be cleaved by a non-specific nucleotide pyrophosphatase (Skrede, 1973) and lysosomal acid phosphatase (Bremer *et al.*, 1972) to generate pantetheine. Pantetheine can then be degraded to generate pantothenate and cysteamine by pantetheinase (Dupre *et al.*, 1973). Pantetheine is also absorbed through the intestine; in fact, this process occurs more rapidly than pantothenate transport (Ono *et al.*, 1974; Turner and Hughes, 1962).

The protein responsible for this pantetheinase activity in mammals has been identified. Using the N-terminal sequence of the protein isolated from pig kidney, Maras *et al.* (1999) determined that this protein was found to be

TABLE II
Recommended Dietary Requirements of Pantothenate

Species	Minimum requirement ($mg\ kg^{-1}$ diet)	Determinant
Humans		
Adults	4–7[a]	Usual intake; urinary excretion
Infants	2–3[a]	Human milk composition
Chickens		
Chicks	7.8–10	Weight gain; feed efficiency; gain in fat
Laying hens	1.9–8.8	Egg production; maintenance of weight
Turkey		
Poults (7–23 days)	10.5	Weight gain; lack of dermatitis
Hens (1–3 years)	16.0	Fertile eggs; hatchability
Swine		
Breeding sows	11.9–12.5	Litter birth weight; litter weanling weights
Rats	8.0–10.0	Weight gain; acetylation of sulfanilamide
Mice	6.0–30.0	Weight gain; absence of symptoms
Catfish	10.0–15.0	Weight gain; lack of clubbed gills

[a] $mg\ day^{-1}$.

Reproduced in part from Smith and Song (1996).

similar to the GPI-anchored protein vanin-1. This protein and its human homologues VNN1 and VNN2 are involved in lymphocyte migration. Subsequent characterisation by Pitari *et al.* (2000) has shown that depletion of the enzyme leads to lack of free cysteamine in tissue and has a role in combating oxidative damage.

Once taken up, pantothenate is converted rapidly to phosphopantetheine and CoA (Fig. 3). Pantothenate kinase (PanK) first phosphorylates pantothenate to generate 4′-phosphopantothenate. This undergoes CTP- or ATP-dependent ligation with cysteine to generate phosphopantothenoylcysteine, which is then decarboxylated by phosphopantethenoylcysteine decarboxylase (PPCDC) to produce 4′-phosphopantetheine. This intermediate can be incorporated directly into ACPs by their cognate phosphopantetheinyl transferases (Walsh *et al.*, 1997). Alternatively, transfer of an adenylyl group from ATP, catalysed by phosphopantetheine adenylyltransferase (PPAT), leads to dephosphocoenzyme A, and the final active form of the cofactor is produced by dephosphocoenzyme A kinase (DPCK). The pathway from pantothenate to CoA is conserved in all organisms, with the exception of some intracellular parasites, such as Mycoplasma, Rickettsia, and Chlamydia, which appear to encode orthologues only of the terminal gene, for DPCK. This suggests that

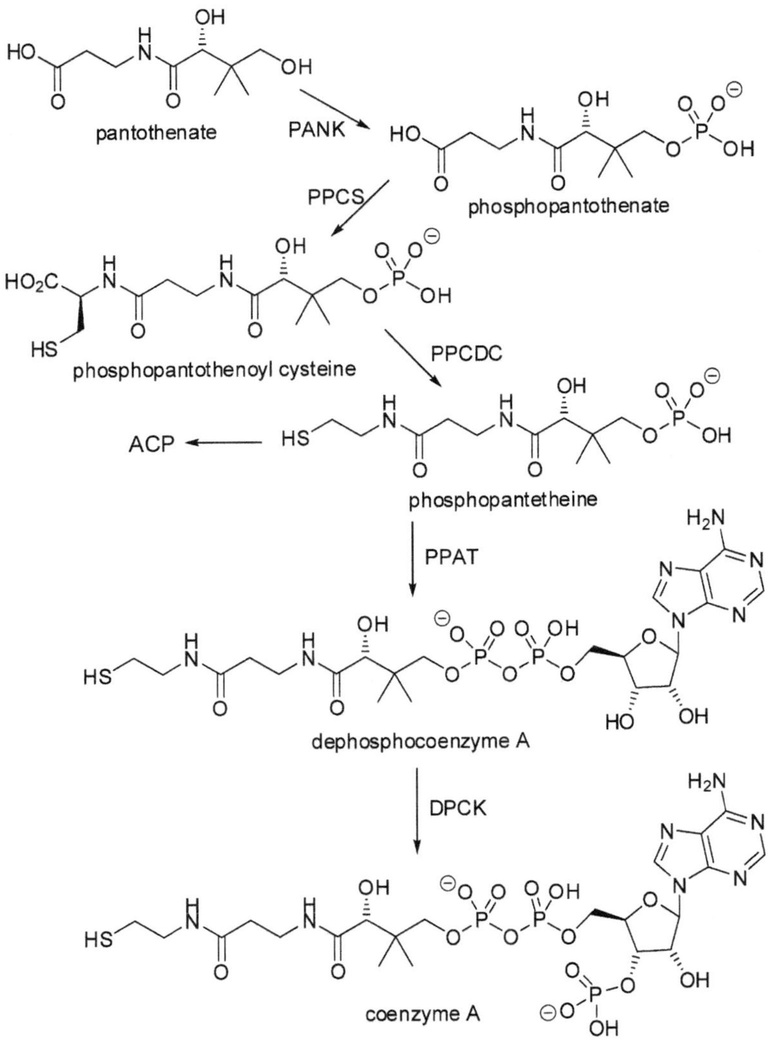

Fig. 3. Biosynthetic pathway from pantothenate to CoA. The conversion of pantothenate to CoA takes place in five steps. Firstly, pantothenate is phosphorylated by pantothenate kinase (PANK), and then it is ligated to cysteine by 4′-phosphopantethenoylcysteine synthetase (PPCS). Decarboxylation by 4′-phosphopantethenoylcysteine decarboxylase (PPCDC) produces the 4′-phosphopantetheine moiety, which is either ligated to a serine residue in acyl carrier protein (ACP), or to adenosine monophosphate by 4′-phosphopantetheine adenylyltransferase (PPAT) to produce dephosphocoenzyme A. The last step is phosphorylation of the 3′-hydroxyl group by dephosphocoenzyme A kinase (DPCK).

they have evolved an uptake system for CoA or dephosphoCoA. Defects in the CoA biosynthesis genes are generally lethal due to the essential nature of the products (as discussed in Section IV for plants). Moreover, for the majority of organisms, neither CoA nor the intermediates can be taken up by cells, and so auxotrophs are only known that are facultative, for example, carrying temperature-sensitive mutations.

Metabolism of pantothenate is regulated chiefly at the level of the enzyme PanK. In mammals, there are five isoforms of this enzyme with distinct metabolic functions, and defects in each have different consequences. Two of these isoforms, PanK1α and PanK1β, are products of a single gene as a result of alternative splicing (Rock *et al.*, 2002a) and are chiefly expressed in hepatic tissue. They are associated with the metabolic transition from glucose utilisation/fatty acid synthesis to gluconeogenesis and oxidation of fatty acids, as demonstrated by accumulation of lipid droplets in the livers of *PanK1*$^{-/-}$ mice (Leonardi *et al.*, 2010). In contrast, PanK2 is chiefly expressed in the mitochondria of neuronal tissues and point mutations in this gene are associated with the condition PanK-associated neurodegenerative disorder (formerly known as Hallervorden–Spatz syndrome), which is a rapidly progressing autosomal recessive disorder leading to dystonia, parkinsonism, and iron accumulation in the brain, and ultimately death (Hayflick *et al.*, 2003). In a mouse model of this condition, mutations were associated with retinal degeneration and arrest of spermiogenesis, but the iron accumulation and neurodegenerative symptoms were not observed (Kuo *et al.*, 2005). Biochemical characterisation of the third isoform, PanK3 which is also highly expressed in the liver (Zhou *et al.*, 2001) shows that in contrast to PanK1β, it is strongly feedback regulated by both acetyl CoA and long chain acyl-CoAs (Zhang *et al.*, 2005) and this is consistent with a role in glycolysis and fatty acid synthesis. Little is known about the final isoform, PanK4, other than that it is expressed ubiquitously but is chiefly found in muscle tissue (Zhou *et al.*, 2001).

II. DISTRIBUTION IN PLANTS

Because plants synthesise pantothenate *de novo*, they are a good dietary source of the vitamin, although the highest levels are found in meat, dairy products, and mushrooms (USDA, 2005). In order to measure bioavailable pantothenate, it is first necessary to liberate it from the cofactors CoA and ACP, and numerous studies have investigated the optimal way of doing this. Since pantothenate is sensitive to both low and high pH, it is not possible to do this by alkaline or acid hydrolysis, and instead enzymes are used, with pantetheinase being the most suitable (Gonthier *et al.*, 1998). Pantothenate can then be quantified using a bioassay, in which growth of a bacterial auxotroph such as *Lactobacillus plantarum* (Wyse *et al.*, 1985) or *E. coli* (Chakauya *et al.*, 2008) is found to be related to pantothenate concentration. However, it is essential to construct careful standard curves to determine the linear range of sensitivity. Using these approaches, the estimated levels of pantothenate level in most common vegetables and fruits have been determined and range between 2 and 25 μM, equivalent to 5–25 nmol g^{-1} fresh weight (FW; Friedrich, 1988; USDA, 2005). The highest levels are found in potatoes, grains and cereals, broccoli, and tomatoes. Sunflower kernels have the highest pantothenate level measured to date in plants (301 nmol g^{-1} FW), while in cauliflower and broccoli florets, levels are 31 and 26 nmol g^{-1} FW, respectively; since these are consumed in larger amounts, they are probably the best sources of dietary vitamin B5. However, processing (such as freezing, canning, or refining) leads to a reduction in both free and bound pantothenate. For example, Rychlik (2000) found that rice grains contain 95 nmol g^{-1} FW total pantothenate compared to 20 nmol g^{-1} FW in polished rice. Interestingly, the free pantothenate was less affected with 25 nmol g^{-1} FW in the whole grain compared to 18 nmol g^{-1} FW in polished rice. Most studies have concentrated on measuring pantothenate in the edible parts of plants, but Rathinasabapathi and Raman (2005) quantified pantothenate in young leaves from several higher plants and found a 10-fold variation in amount from 0.73 nmol g^{-1} FW in *Rosa sinensis* to 7.3 nmol g^{-1} FW in *Citrus* but in all cases much lower than that detected in the edible fruits, flowers, and seeds.

A similar picture was seen in *Brassica napus*, in which levels in different plant tissues were compared (Chakauya *et al.*, 2008). Free pantothenate in leaves was measured at 11 ± 1 nmol g^{-1} FW compared with 15 and 41 nmol g^{-1} FW found in flowers and mature seeds, respectively. Similarly, that in green seeds was half that of the pods (fruit and seed). This might indicate that the vitamin comes from the fruit, possibly from biosynthesis or remobilisation from other tissues. Assimilates for seed storage can be the result of one of two physiological processes: utilisation of assimilates from current photosynthesis, and remobilisation of substances accumulated pre-anthesis (Prasad *et al.*, 2002). There are reported cases of mobilisation of carbon from flag leaves and stems to the grain in chickpeas and wheat (Davies *et al.*, 2000). If this is the case then pantothenate might be synthesised during an early stage of plant growth and recycled throughout the plant to areas of high demand. When combined, the results seem to suggest that pantothenate might be mobilised from photosynthetic parts of the plant including older leaves and the hull to nourish the developing seed. This is shown by the increase in free pantothenate as the seed develops to the same level as in a fully imbibed seed (Chakauya *et al.*, 2008).

However, consideration of pantothenate levels in plants is complicated by the incorporation of pantothenate into CoA and ACP. A metabolic profile analysis of *Arabidopsis thaliana* looked at acetyl CoA levels and found that there was several fold difference between seeds and leaves (Gibon *et al.*, 2002). They estimated that early developing seeds had about 25–30 nmol g^{-1} dry weight falling to about 5 nmol g^{-1} dry weight at maturity. Using HPLC analysis, Tumaney *et al.* (2004) estimated that early-mid-maturing *B. napus* seeds have 25 nmol g^{-1} FW falling to 12.5 nmol g^{-1} at late maturing, compared to just 5 nmol g^{-1} FW in leaves acetyl CoA levels in *Arabidopsis* and spinach leaves were within the same range, 5 and 6.8 nmol g^{-1} FW, respectively. It is likely that changes in the acetyl CoA pattern reflects altered activity in lipid metabolism in the different tissues, with higher levels required in seeds for the demands of triacylglyceride production or mobilisation in development or germination, respectively. However, what is not certain is whether the ratio of free to bound pantothenate remains the same in all tissues.

At the subcellular level, acetyl CoA is found in all subcellular compartments in plants, including plastids, mitochondria, peroxisomes, and cytosol where it is involved in several metabolic functions (Fatland *et al.*, 2004; Tumaney *et al.*, 2004). The concentration has been measured at 30–35 μM in spinach stroma by Post-Beittenmiller *et al.* (1992), who estimated that 75% of cellular acetyl CoA was in the chloroplast. However, it is widely accepted that membranes are impermeable to CoA and its thioesters, implying that either CoA is synthesised in all cellular compartments or made in one compartment and then transported to others during biosynthesis. An example of such a CoA transporter is the *Arabidopsis* ATP-binding cassette (ABC) transporter encoded by the *PXA1* gene (Zolman *et al.*, 2001). This transporter is involved in the import of long chain acyl-CoAs into the peroxisomes for β-oxidation. Further consideration of the subcellular distribution of pantothenate and CoA is given in Section IV.F describing the distribution of the biosynthetic enzymes.

III. BIOSYNTHETIC PATHWAY AND LOCATION OF THE PATHWAY

In all prototrophic organisms studied to date, pantothenate is synthesised from 3-methyl-2-oxobutyric acid (α-ketoisovalerate—α-KIVA) and β-alanine (Fig. 2; Webb *et al.*, 2004). The former compound is an intermediate in valine biosynthesis, whereas β-alanine can be derived from a number of sources including decarboxylation of L-aspartate; degradation of uracil; and degradation of the polyamines, spermine, and spermidine (discussed in detail in Section V.B). The first enzyme of the pathway is ketopantoate hydroxymethyltransferase (KPHMT), which converts α-KIVA into ketopantoate. This is reduced to D-pantoate, which is then condensed with β-alanine to generate D-pantothenate, catalysed by pantothenate synthetase (PS). It is only these two enzymes KPHMT and PS that have been characterised in plants. Putative enzymes responsible for the remaining steps are discussed in Sections III.C and V.B.

A. KETOPANTOATE HYDROXYMETHYLTRANSFERASE

KPHMT is responsible for the first step in the pathway, the transfer of a formaldehyde equivalent onto an enolate ion-derived from α-KIVA to generate ketopantoate (Fig. 4). The cellular formaldehyde equivalent is provided by methylene tetrahydrofolate. This activity was first observed in bacteria by McIntosh *et al.* (1957), who demonstrated the presence of the

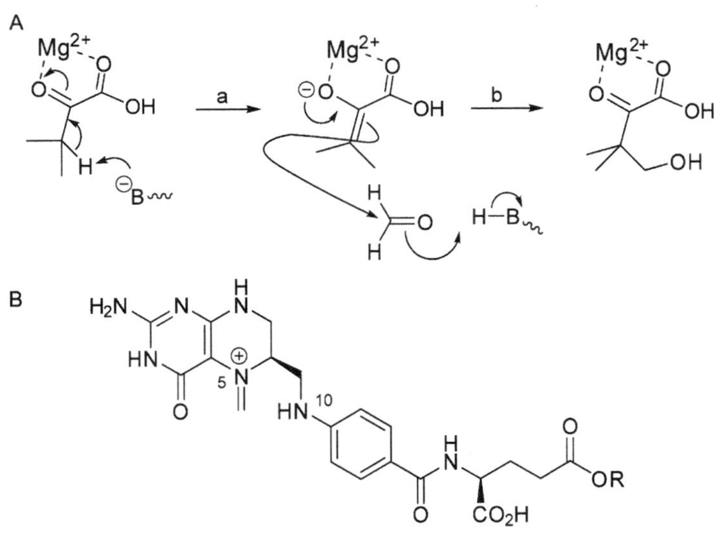

Fig. 4. Outline mechanism of ketopantoate hydroxymethyltransferase (KPHMT). (A) KPHMT catalyses the transfer of a formaldehyde equivalent from methylene tetrahydrofolate (B) to α-ketoisovalerate. (a) α-Ketoisovalerate is bound to an active-site magnesium ion which promotes formation of an enolate via deprotonation using an active-site base. (b) The enolate nucleophile attacks formaldehyde with concerted proton transfer to generate ketopantoate. (B) Structure of the ring-open iminium form of methylene tetrahydrofolate.

activity in *E. coli*. Teller *et al.* (1976) purified this enzyme and characterised its catalytic properties (Powers and Snell, 1976). The gene was subsequently isolated by functional complementation (Jones *et al.*, 1993), and this eventually led to the structural solution of the enzyme from both *E. coli* (von Delft *et al.*, 2003) and *Mycobacterium tuberculosis* (Chaudhuri *et al.*, 2003). The enzyme is a homodecameric structure with each protomer forming a $(\beta\alpha)_8$ barrel (Fig. 5A). The active site lies in a deep pocket formed by the C-termini of the β-strands (Fig. 5B). The catalytic properties of the bacterial enzyme have been extensively characterised by Sugantino *et al.* (2003). Intriguingly,

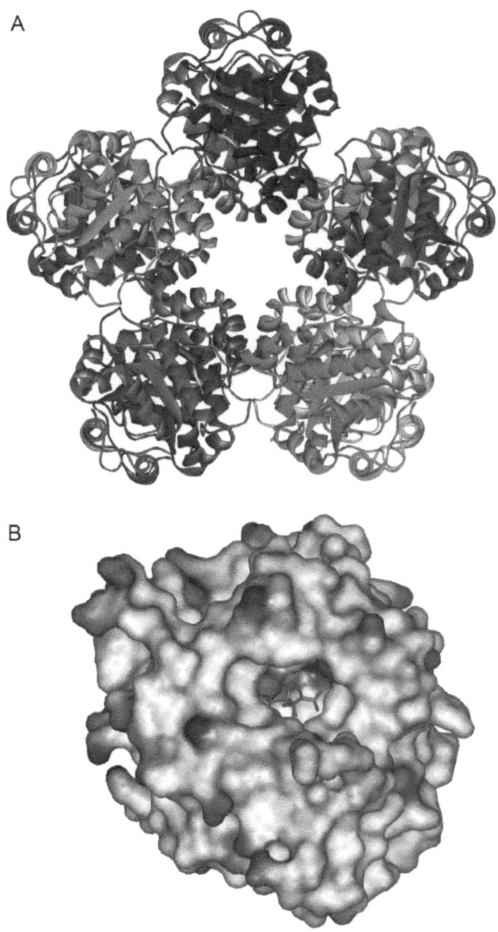

Fig. 5. Crystal structure of bacterial ketopantoate hydroxymethyltransferase. (A) *E. coli* KPHMT exists as a decamer of 10 $(\beta\alpha)_8$-barrel protomers (pdb 1m3u; von Delft *et al.*, 2003). (B) Pantoate is bound to a magnesium ion within the 'barrel' of each promoter.

the enzyme from this source does not have an absolute requirement for a tetrahydrofolate cofactor for activity but will accept free formaldehyde as an electrophilic partner for hydroxymethylation.

The existence of this enzymatic step in plants was first suggested in an auxotrophic strain of *Datura innoxia* (thorn-apple), which required supplementation with pantothenate, pantoate, or ketopantoate for survival (Savage *et al.*, 1979). This suggested the presence of KPHMT in the plant, though no activity could be measured in cell-free extracts (Sahl *et al.*, 1988). KPHMT activity in plants was then unambiguously demonstrated by Jones *et al.* (1994) who fed ^{14}C-labelled valine to pea-leaf discs, followed by HPLC analysis of an extract. Radiolabel was found in peaks corresponding to α-KIVA, ketopantoyl-lactone, and pantoyl-lactone. The acid treatment during extraction would have degraded any pantothenate in the sample.

Ottenhof *et al.* (2004) used the amino acid sequence of the *E. coli* enzyme to identify two homologues of the enzyme in the complete genome sequence of *Arabidopsis*, termed KPHMT1 (At2g46110) and KPHMT2 (At3g61530). The coding sequence for both genes could be amplified from an *Arabidopsis* cDNA library indicating that they are both transcribed, and both genes were shown to encode functional enzymes from the fact that they complemented the *E. coli* Hfr3000 YA139 *panB* mutant. Two gene products were also identified by BLAST searching of the rice genome (2674.t00011 and 4685.t00006), and in indeed, in all plant genomes publically available, there are two *PanB* genes detectable. The predicted protein sequence of the plant enzymes have 30% identity with the *E. coli* KPHMT protein with all the residues conserved in bacteria being conserved in plants (Fig. 6). Both isoforms of the enzyme are synthesised with N-terminal extensions relative to the *E. coli* protein. When these sequences from *Arabidopsis* KPHMT1 and KPHMT2 were fused to GFP, they were able to target GFP to the mitochondria indicating that the KPHMT isoforms would similarly be located in this organelle. This observation allowed the detection of enzyme activity in plant extracts for the first time; mitochondria from both peas and *Arabidopsis* were isolated, and KPHMT activity could then be directly assayed in extracts of these organelles. This was achieved using a coupled assay with the next enzyme in the bacterial pathway, ketopantoate reductase (KPR), and monitoring the oxidation of NADPH (Ottenhof *et al.*, 2004).

By overexpressing the recombinant *Arabidopsis* enzymes in *E. coli*, both isoforms KPHMT2 (Coxon, 2006) and KPHMT1 (Webb, unpublished results) have been shown to be catalytically active. In the case of KPHMT2, the methylene tetrahydrofolate cofactor was shown to be required for catalytic activity. This suggests that the precise enzymatic mechanism of the plant enzyme may be different to that of the bacterial enzyme for which formaldehyde can be used as a substrate.

```
A. thaliana 1   MASSLTRN-CSRFSKAIS-VRFMSNLPENTVYGGPKP--QNPNQRVTLTHLRQKHRRGEP  56
A. thaliana 2   MAASLVRSCCSRASRAITTVRFMSNVPEDTVYGGPKP--QNSNQRVTLTQLRQKHRKGEP  58
E. coli         ----------------------MKP--------------------TTISLLQKYKQEKKR  18
M. tuberculosis ----------------------MS---EQTIYGANTPGGSGPRTKIRTHHLQRWKADGHK  35
B. subtilis     ----------------------MK---------------------TKLDFLKMKESEEP-  16
C. glutamicum   ----------------------MS---GIDAK------------KIRTRHFREAKVNGQK  23
S. cerevisiae   --------------------MNIMKRQLCTSSKRFFSTAKNVVKYNTIQDIRNKYFTGTP  40

A. thaliana 1   ITVVTAYDYPSAVHLDTAGIDVCLVGDSASMVVHGHDTTLPISLDEMLVHCRAVARGAKR 116
A. thaliana 2   ITMVTAYDYPSAVHIDTAGIDVCLVGDSAAMVVHGYDTTLPISLEEMLVHCRAVARGAKR 118
E. coli         FATITAYDYSFAKLFADEGLNVMLVGDSLGMTVQGHDSTLPVTVADIAYHTAAVRRGAPN  78
M. tuberculosis WAMLTAYDYSTARIFDEAGIPVLLVGDSAANVVYGYDTTVPISIDELIPLVRGVVRGAPH  95
B. subtilis     IVMLTAYDYPAAKLAEQAGVDMILVGDSLGMVVLGLDSTVGVTVADMIHHTKAVKRGAPN  76
C. glutamicum   VSVLTSYDALSARIFDEAGVDMLLVGDSAANVVLGRDTTLSITLDEMIVLAKAVTIATKR  83
S. cerevisiae   LSMCTAYDFITATWVNKANCDLLLVGDSLAMTSLGYDSTITLSLNEFKYHVASVCRAEGS 100

A. thaliana 1   PLLVGDLPFGTYESSSSQAVDTAVRVLKE-GGMDAIKLEGGS----ASRITAAKAIVEAG 171
A. thaliana 2   PLLVGDLPFGTYESSTNQAVDTAVRVLKE-GGMDAIKLEGGS----PSRITAAKSIVEAG 173
E. coli         CLLLADLPFMAYATP-EQAFENAATVMR--AGANMVKIEGG-----EWLVETVQMLTERA 130
M. tuberculosis ALVVADLPFGSYEAGPTAALAAATRFLKD-GGAHAVKLEGG-----ERVAEQIACLTAAG 149
B. subtilis     TFIVTDMPFMSYHLSKEDTLKNAAAIVQE-SGADALKLEGG-----EGVFESIRALTLGG 130
C. glutamicum   ALVVVDLPFGTYEVSPNQAVESAIRVMRE-TGAAAVKIEGG-----VEIAQTIRRIVDAG 157
S. cerevisiae   SMVVVDMPFGTFESGISDGLKNAIDIMKLDSKVTSVKVEVGSYTKDKYAMKFIEELCSRG 160

A. thaliana 1   IAVIGHVGLTPQAISVLGGFRPQGRN-IASAVKVVETAMALQEAGCFSVVLECVPPPVAA 230
A. thaliana 2   IAVMGHVGLTPQAISVLGGFRPQGKN-IASAVKVVETAMALQEAGCFSVVLECVPPPVAA 232
E. coli         VPVCGHLGLTPQSVNIFGGYKVQGRG-DEAGDQLLSDALALEAAGAQLLVLECVPVELAK 189
M. tuberculosis IPVMAHIGFTPQSVNTLGGFRVQGR--GDAAEQTIADAIAVAEAGAFAVVMEMVPAELAT 207
B. subtilis     IPVVSHLGLTPQSVGVLGGYKVQGKD-EQSAKKLIEDSIKCEEAGAMMLVLECVPAELTA 189
C. glutamicum   IPVVGHIGYTPQSEHSLGGHVVQGR--GASSGKLIADARALEQAGAFAVVLEMVPAEAAR 215
S. cerevisiae   IPVMAHIGLTPQKVHSLGGYKVQGSKSLLQMQELYETAMQLQKIGCWSILIECVPHKMAQ 220

A. thaliana 1   AATSALKIPTIGIGAGPFCSGQVLVYHDLLGMMQHPHHAKVTPKFCKQYANVGEVINKAL 290
A. thaliana 2   AATSALNIPTIGIGAGPFCSGQVLVYHDLLGMMQHPHHAKVTPKFCKQYAQVGEVINKAL 292
E. coli         RITEALAIPVIGIGAGNVTDGQILVMHDAFGITGG-----HIPKFAKNFLAETGDIRAAV 244
M. tuberculosis QITGKLTIPTVGIGAGPNCDGQVLVWQDMAGFSGA-----KTARFVKRYADVGGELRRAA 262
B. subtilis     KIAETLSIPVIGIGAGVKADGQVLVYHDIIGHGVE-----RTPKFVKQYTRIDETIETAI 244
C. glutamicum   EVTEDLSITTIGIGAGNGTDGQVLVWQDAFGLNRG-----KKPRFVREYATLGDSLHDAA 270
S. cerevisiae   FITSKLSVPTIGIGAGNGTSGQVLVISDLLGMQGD-----SVPKFVKQAVNMTDIATQGL 275

A. thaliana 1   MEYKEEVSKKVFPGPSHSPYKITASELDGFLTELQKLGFDKAASAAALAAENMEPSK--- 347
A. thaliana 2   LEYKEEVSKHLFPGPSHSPYKISSSDLDGFLSELQKLGLDKAASDAAASAEKMDHSDSPS 352
E. coli         RQYMAEVESGVYPGEEHS-----------FH----------------------------- 264
M. tuberculosis MQYAQEVAGGVFPADEHS-----------F------------------------------ 281
B. subtilis     SGYVQDVRHRAFPEQKHS-----------FQMNQTVLDGLYGGK---------------- 277
C. glutamicum   QAYIADIHAGTFPGEAES-----------F------------------------------ 289
S. cerevisiae   KEYIASVEDRTFPERGTHTFKVKEDLWNEFLSSINEK----------------------- 312

A. thaliana 1   --
A. thaliana 2   SQ 354
E. coli         --
M. tuberculosis --
B. subtilis     --
C. glutamicum   --
S. cerevisiae   --
```

Fig. 6. Sequence alignment of ketopantoate hydroxymethyltransferases. The two *Arabidopsis* KPHMTs are aligned with those from yeast (*S. cerevisiae*) and four representative bacterial sequences. Conserved residues are highlighted.

B. PANTOTHENATE SYNTHETASE

PS catalyses the ATP-dependent ligation of pantoate and β-alanine to form pantothenate via a pantoyl adenylate intermediate (Fig. 7). PS activity was first reported by Wieland and Moller (1941) in crude lysates of yeast. The activity was subsequently identified in extracts of *E. coli* by Maas in 1952,

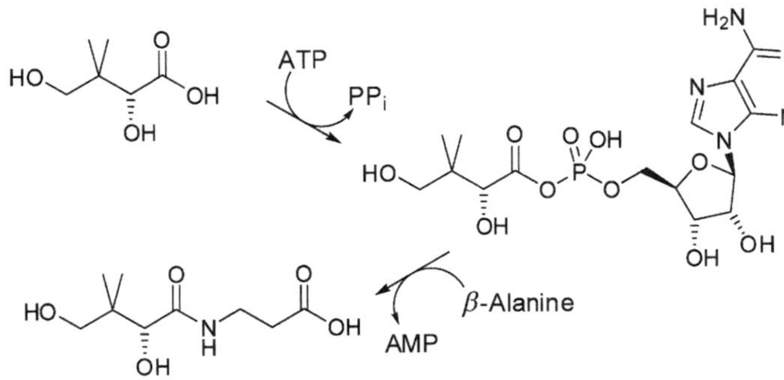

Fig. 7. Mechanism of pantothenate synthetase. Pantothenate synthetase catalyses formation of a peptide bond between pantoate and β-alanine by forming a pantoyl adenylate intermediate from ATP and pantoate, generating pyrophosphate, followed by displacement of AMP by β-alanine.

and the protein responsible for this activity purified 50-fold by Pfleiderer *et al.* (1960) and 300-fold by Kreiling *et al.* (1962) before purification to homogeneity (2100-fold purification) by Miyatake *et al.* (1976). The protein from both *E. coli* (von Delft *et al.*, 2001) and *M. tuberculosis* (Zheng and Blanchard, 2001) was subsequently overexpressed and purified and the crystal structures were determined (von Delft *et al.*, 2001; Wang and Eisenberg, 2003).

The protein is dimeric (Fig. 8A) and each protomer comprises two domains. The N-terminal domain has a Rossmann fold with the active site at the interface formed between the N- and C-terminal domains. The protein is structurally very similar to class I aminoacyl-tRNA synthetases and is thus a member of the cytidylyltransferase superfamily. By analogy with these enzymes, a hinge-bending mechanism has been proposed in which the C-terminal domain opens and closes over the active site during the enzymatic reaction, because the catalytically essential residues are spatially separated into two domains in the apo structure of the *E. coli* enzyme. In contrast, structures for both the apo and pantoyl adenylate complex of the *M. tuberculosis* enzyme show little or no conformational change between them, being essentially closed structures (Wang and Eisenberg, 2003). Recently, the structures of both the apoenzyme and a pantoyl adenylate complex of *Staphylococcus aureus* PS have been solved (Satoh *et al.*, 2010). Structural analysis indicated that the apoenzyme adopts an open and relatively mobile structure, while the complex structure is closed and entirely rigid (Fig. 8B). Structural comparison of the apoenzyme and the complex revealed how *S. aureus* PS undergoes open/close conformational change and also determined the key interactions with the adenine ring of PA (Fig. 8C) for a hinge-bending domain closure.

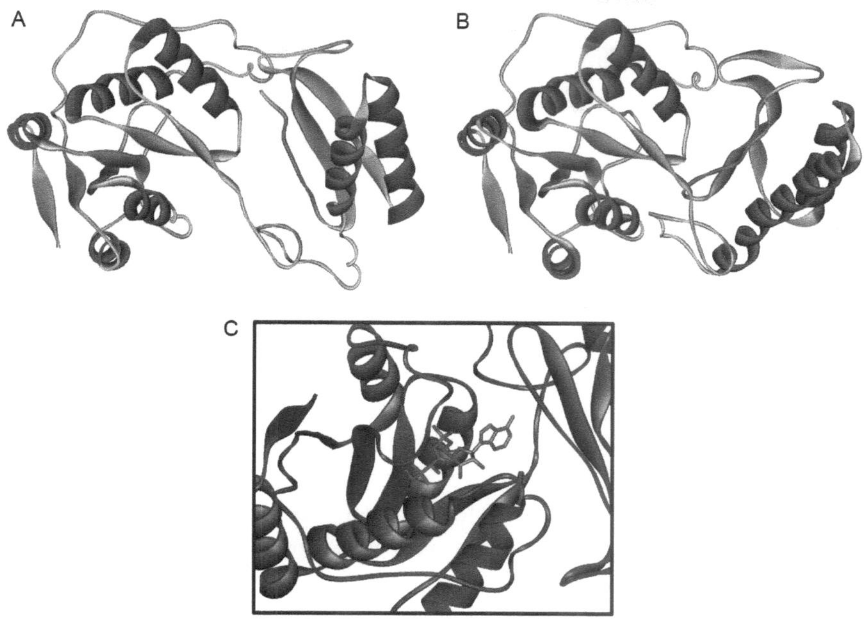

Fig. 8. Structure of pantothenate synthetase. (A) Structure of the apoform of pantothenate synthetase from *S. aureus* (pdb 3ag5; Satoh *et al.*, 2010). The N-terminal domain (left hand side) has a Rossmann fold, whilst the active site is at the interface between this domain and the C-terminal part of the enzyme. (B) Structure of the same enzyme in which the pantoyl adenylate intermediate is bound (pdb 3ag6), showing that the conformation change in which the C-terminus has closed over the active site. (C) Close-up of the active site with pantoyl adenylate structure.

The first report of PS in higher plants was that of Genschel *et al.* (1999) who used functional complementation of an *E. coli panC* mutant AT1371 to identify the genes for PS from *Lotus japonicus*, *Oryza sativa*, and also from yeast, *S. cerevisiae*. By comparison of the plant enzyme with the bacterial homologues, the protein appears to not to have an N-terminal extension characteristic of a transit peptide for targeting to organelles, suggesting that it would be located in the plant cytosol. The protein from *L. japonicus* was overexpressed in *E. coli* and purified to 95% homogeneity. This *L. japonicus* PS exhibited Michaelis–Menten kinetics with respect to ATP, β-alanine, and D-pantoate although substrate inhibition was observed at higher concentrations of D-pantoate.

Ottenhof *et al.* (2004) subsequently used this gene sequence in BLAST (sequence similarity) searches to identify the gene for PS in *Arabidopsis* (At5g48840). The catalytic activity of this protein was then once again confirmed by complementation of the *E. coli panC* mutant AT1371. Using

GFP-fusions and confocal fluorescence microscopy, the enzyme was shown to localise to the cytosol, confirming the predictions from Genschel *et al.* (1999). This protein was subsequently characterised further by Jonczyk and Genschel (2006). In this study, they sought to understand the molecular basis for the substrate inhibition by pantoate observed for the enzyme from *L. japonicus* (Genschel *et al.*, 1999). Analysis of the kinetics of the *Arabidopsis* enzyme revealed that the variation in rate for different concentrations of pantoate at fixed concentrations of ATP could not be accounted for by a simple model for uncompetitive substrate inhibition (i.e. formation of a dead-end S.E.S complex). The data did, however, fit well to an empirical equation of the form $v = (\alpha_1[S] + \alpha_2[S]^2)/(1 + \beta_1[S] + \beta_2[S]^2)$, and this equation was therefore used to interpret their data. This equation has the same form as the analytical equation for partial uncompetitive inhibition (i.e. the S.E.S complex is still able to form product but at a reduced rate), and the data are consistent with the inhibited rate being ca. 10% of the uninhibited rate. β-Alanine utilisation was also negatively cooperative at high concentrations of pantoate.

Substrate inhibition of this type is not observed for the bacterial PSs, and the molecular basis for the difference has been studied (Jonczyk and Genschel, 2006). The simplest origin of substrate inhibition would be allosteric communication between the two subunits of PS. Sequence alignment of plant and bacterial PSs revealed an amino acid insertion in the plant protein N-terminal to the dimerisation region of the protein (Fig. 9). In order to investigate the potential role for this part of the protein in allosteric substrate inhibition, two mutated forms of the protein (named AtPTS) were generated. The first had a single E132A alteration at end of the insertion, which led to a reduction in substrate inhibition. A complete loss of substrate inhibition by pantoate was seen in a second mutant in which residues P97-T133 were replaced with P93-T107 of the *E. coli* protein, so that the entire insertion had been removed. Intriguingly, the loss of substrate inhibition by pantoate revealed increasing substrate inhibition by ATP which had previously been masked.

The observation of substrate inhibition in this system may not be directly relevant *in vivo*, since the total flux through the plant biosynthetic pathway (discussed in Section VI) is low. The cellular concentration of pantoate is therefore always below the concentration at which substrate inhibition occurs. Jonczyk and Genschel (2006) proposed that instead allosteric communication between the two sites actually serves to enhance substrate binding at low substrate concentrations, and since higher substrate concentrations are not encountered *in planta* there is no evolutionary pressure to reduce the negative effect of substrate inhibition.

```
O. sativa        ----MAAPREPEVIRDKAAMRAWSRRRRAEGKTVAVVPTMGYLHQGHLSLISAAAAAASA  56
L. japonicus     --------MAPMVISDKDEMRKWSRSMRSQGKLIALVPTMGFLHEGHLSLVRDAHNHAD-  51
A. thaliana      --------MEPEVIRDKDSMRKWSRAMRSQGKTIGLVPTMGYLHEGHLSLVRQSLALTD-  51
S. cerevisiae    -------MKIFHTVEEVVQWRTQELRETRFRETIGFVPTMGCLHSGHASLISQSVKENTY  53
B. subtilis      ----------MRQITDISQLKEAIKQYHSEGKSIGFVPTMGFLHEGHLTLADKAR-QEND  49
M. tuberculosis  MTIPAFHPGELNVYSAPGDVADVSRALRLTGRRVMLVPTMGALHEGHLALVRAAKRVPGS  60
E. coli          ----------MLIIETLPLLRQQIRRLRMEGKRVALVPTMGNLHDGHMKLVDEAKARAD-  49
C. glutamicum    ----------MQVATTKQALIDA----LLHHKSVGLVPTMGALHSGHASLVKAAR-AEND  45

O. sativa        DPVAIVVTIYVNPSQFAPS---EDLATYPSDFAGDLRKLAST-GVVDAVFNP--PDLYVR 110
L. japonicus     ---LVAVSIYVNPGQFSPT---EDLSAYPSDFQGDLQKLMSVPGGVDVVFHP--HNLYDY 103
A. thaliana      ---VTVVSIYVNPGQFSPT---EDLSTYPSDFSGDLTKLAALSGGKVVVFNP--KNLYDY 103
S. cerevisiae    ----TVVSIFVNPSQFAPT---EDLDNYPRTLPDDIKLLESL--KVDVLFAPNAHVMYPQ 104
B. subtilis      ---AVIMSIFVNPAQFGPN---EDFEAYPRDIERDAALAENA--GVDILFTPDAHDMYPG 101
M. tuberculosis  ---VVVVSIFVNPMQFGAG---EDLDAYPRTPDDDLAQLRAE--GVEIAFTPTTAAMYPD 112
E. coli          ---VVVVSIFVNPMQFDRP---EDLARYPRTLQEDCEKLNKR--KVDLVFAPSVKEIYPN 101
C. glutamicum    ---TVVASIFVNPLQFEALGDCDDYRNYPRQLDADLALLEEA--GVDIVFAPDVEEMYPG 100

O. sativa        GAGR--RGAASGG----AISCLEEAAGDGHETWVRVERLEKGMCGASRPVFFRGVATIVS 164
L. japonicus     GGDGGDAVAECGGDG--VVSCVDRRSGFGHETWVRAEKLEKPLCGKSRPVFFRGVATIVT 161
A. thaliana      GGET-KKINDGGGNGGRVVSCVEE-GGLGHETWIRVERLEKGFCGKSRPVFFRGVATIVT 161
S. cerevisiae    GIPLDIEEQ--------------------KGPFVSVLGLSEKLEGKTRPNFFRGVATVVT 144
B. subtilis      EKNV---------------------------TIH-VERRTDVLCGRSREGHFDGVAIVLT 133
M. tuberculosis  GLRT---------------------------TVQ-PGPLAAELEGGPRPTHFAGVLTVVL 144
E. coli          GTET--------------------------HTYVDVPGLSTMLEGASRPGHFRGVSTIVS 133
C. glutamicum    GLPL---------------------------VWARTGSIGTKLEGASRPGHFDGVATVVA 133

O. sativa        KLFNIIEPDVAVFGKKDYQQWRVICRMVRDLDFAIEIIGSEIVREADGLAMSSRNVHLSR 224
L. japonicus     KLFNIVEPDVAVFGKKDYQQWKIIQRMVRDLDFSIKVIGSEVIREKDGLAMSSRNVYLSP 221
A. thaliana      KLFNIVEPDVALFGKKDYQQWRIIQRMVRDLNFGIEIVGSDIAREKDGLAMSSRNVRLSD 221
S. cerevisiae    KLFNIVMADVAYFGQKDIQQFIVLQCMVDELFVNTRLQMMPIVRNNNGLALSSRNKYLCP 204
B. subtilis      KLFNLVKPTRAYFGLKDAQQVAVVDGLISDFFMDIELVPVDTVREEDGLAKSSRNVYLTA 193
M. tuberculosis  KLLQIVRPDRVFFGEKDYQQLVLIRQLVADFNLDVAVVGVPTVREADGLAMSSRNRYLDP 204
E. coli          KLFNLVQPDIACFGEKDFQQLALIRKMVADMGFDIEIVGVPIMRAKDGLALSSRNGYLTA 193
C. glutamicum    KLFNLVRPDRAYFGQKDAQQVAVIRRLVADLDIPVEIRPVPIIRGADGLAESSRNQRLSA 193

O. sativa        EEREKALSISRSLVDARTGA----LKGNTDCKQIKNKIVQTLT----ETGGQVDYVEIVE 276
L. japonicus     EEREKAVSINKSLFRAKSAA----EDGQIHCEKLINLVVQSIT----EAGGRIDYAEIVD 273
A. thaliana      EERQRALSISRSLAMAKASV----AEGKTNCAELKDMIIQQVV----GSAGRVDYVEIVD 273
S. cerevisiae    ESLKISENLYRGLKAAENAIRRLAPGGRLSRSEIIDTVTQIWAPYVDSHDFKIDYVSLAD 264
B. subtilis      EERKEAPKLYRALQTSAELV----QAGERDPEAVIKAAKDIIET---TSGT-IDYVELYS 245
M. tuberculosis  AQRAAAVALSAALTAAAHAA----TAG---AQAALDAARAVLDA---APGVAVDYLELRD 254
E. coli          EQRKIAPGLYKVLSSIADKL----QAGERDLDEIITIAGQELN----EKGFRADDIQIRD 247
C. glutamicum    DQRAQALVLPQVLSGLQRRK----AAG---EALDIQGARDTLAS---ADGVRLDHLEIVD 243

O. sativa        QESLVPVEQIDGPV-----VICVAAWFGKV-------RLIDNIEIDTRS----------- 313
L. japonicus     QNNLEKVEWIKGPV-----VFCVSAWFGKA-------RLIDNIEINL------------- 308
A. thaliana      QETLEGVEEIKSGV-----VICVAAWFGTV-------RLIDNIEINVSL----------- 310
S. cerevisiae    FKTLDELSDVENTSEQQPIVISCAVYVTDREKPDTVVRLIDNIVI--------------- 309
B. subtilis      YPELEPVNEIAGKM-----ILAVAVAFSKA-------RLIDNIIIDIREMERI------- 286
M. tuberculosis  IG-LGPMP-LNGSG-----RLLVAARLGTT-------RLLDNIAIEIGTFAGTDRPDGYR 300
E. coli          ADTLLEVSETSKRA-----VILVAAWLGDA-------RLIDNKMVELA------------ 283
C. glutamicum    PATLEPLE-IDG-------LLTQPALVVGA-IFVGPVRLIDNIEL--------------- 279

O. sativa        ---------
L. japonicus     ---------
A. thaliana      ---------
S. cerevisiae    ---------
B. subtilis      ---------
M. tuberculosis  AILESHWRN 309
E. coli          ---------
C. glutamicum    ---------
```

Fig. 9. Sequence alignment of representative bacterial and plant sequences for pantothenate synthetase. Highlighted residues (black) are conserved throughout bacterial and plant sequences. The inserted sequence in plants and the corresponding *E. coli* sequence identified by Jonczyk and Genschel (2006) as contributing to allosteric regulation of catalytic activity are highlighted in grey.

Jonczyk *et al.* (2008) further investigated the *in planta* function of PS. Two T-DNA insertional mutants of At5g48840 were identified from the Salk collection (Alonso *et al.*, 2003), Salk_101909 (*pts-1*) and Salk_594477 (*pts-2*). Heterozygous mutant plants exhibited no phenotype, but it was not possible to isolate homozygous plants of either knockout line. Examination of the developing seeds in the siliques of heterozygous plants revealed that a proportion were arrested at a pre-globular stage of development, and the numbers of heterozygous and wild-type (WT) plants from the remaining seeds were consistent with this hypothesis. The addition of exogenous pantothenate to growing heterozygous plants was sufficient to abolish the arrest of seed development, and all seeds produced by these plants were able to germinate on agar supplemented with pantothenate. Homozygous mutants were able to grow into fertile plants on nutrient agar, although they were not able to survive on soil. This was hypothesised to be due to a lack of active transport into the roots of plants.

Jonczyk *et al.* (2008) used a combination of published transcript data (Meyers *et al.*, 2004) and promoter:GUS fusion assays to investigate the expression patterns of *Arabidopsis PanC* gene during vegetative growth and flower development. Both sets of data showed constitutive expression of the gene across vegetative tissue types with very little variation. During flower development, both the promoter:GUS fusion assays and microarray data (Schmid *et al.*, 2005) showed expression in the sepals and carpels with less in the petals and stamen.

C. REDUCTION OF KETOPANTOATE

The identification and characterisation of KPR and the corresponding *panE* gene in bacteria lagged behind that of the other three enzymes, in part because of the presence of another enzyme able to catalyse the same reaction. Acetohydroxyacid isomerase (AHIR, EC 1.1.1.86, encoded by *ilvC*), responsible for the reductive rearrangement of 2-acetolactate to yield 3-hydroxy-3-methyl-2-oxobutyrate in branched chain amino acid biosynthesis (Fig. 10), has been shown also to catalyse reduction of ketopantoate (Primerano, 1983), and indeed, an *ilvC panE* double mutation is required in order to generate a pantothenate auxotroph (Frodyma and Downs, 1998a). In the industrially important *Corynebacterium glutamicum*, *panE* is absent and only AHIR catalyses this transformation (Merkamm *et al.*, 2003).

The *panE* gene encoding KPR was eventually identified in *Salmonella typhimurium*, where a gene previously called *apbA* was found to encode a protein with KPR activity (Frodyma and Downs, 1998b). The enzyme preferentially uses NADPH rather than NADH as hydrogen donor (Frodyma

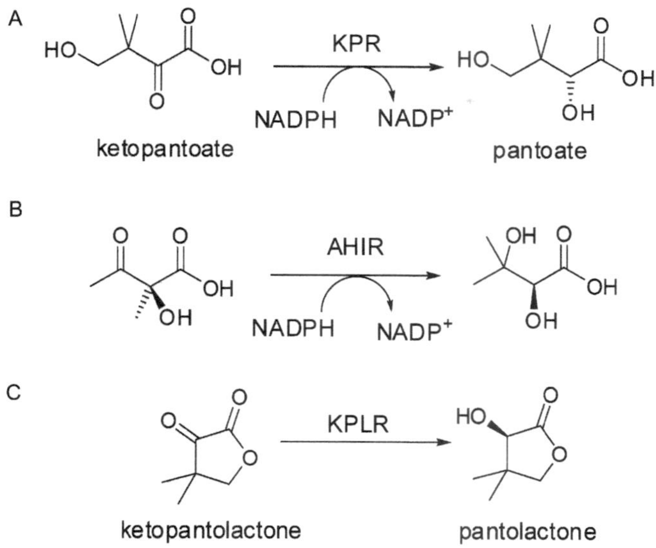

Fig. 10. Alternative methods for ketopantoate reduction. Three distinct routes for ketopantoate reduction have been identified: (A) Ketopantoate reductase (KPR) catalyses the NADPH-dependent reduction of ketopantoate to generate pantoate in *E. coli.* (B) Acetohydroxyacid isomeroreductase (AHIR, encoded by *ilvC*) can also catalyse the NADPH-dependent reduction of ketopantoate to pantoate in *E. coli* and is the sole pathway in *C. glutamicum.* (C) Reduction of ketopantolactone to pantolactone has also been proposed a possible route for this reduction in plants.

and Downs, 1998a). Thereafter, the *E. coli panE* gene was isolated, the enzyme protein overexpressed and purified, and the X-ray crystal structure of the apoprotein solved (Fig. 11; Matak-Vinkovic *et al.*, 2001). Several other structures have been determined, including that of the ternary complex with products $NADP^+$ and pantoate bound (Ciulli *et al.*, 2007).

To date, no dedicated enzyme for ketopantoate reduction has been characterised in plants. The sequence conservation between KPRs, even in bacteria, is low with only 16 residues completely conserved, all these lining the active site (Fig. 11), so it was not surprising that sequence-based searches of plant genomes have not identified a clear candidate. Instead there are 138 short-chain oxidoreductases, characterised by a Rossmann fold at the N-terminus. On the other hand, a KPR homologue was readily identifiable in yeast (Ottenhof *et al.*, 2004). Ottenhof *et al.* (2004) used a structure-led search of the complete *Arabidopsis* proteome with the *E. coli* KPR structure a query. One putative candidate (At5g34780) was identified, but the catalytic activity of this protein has not been confirmed.

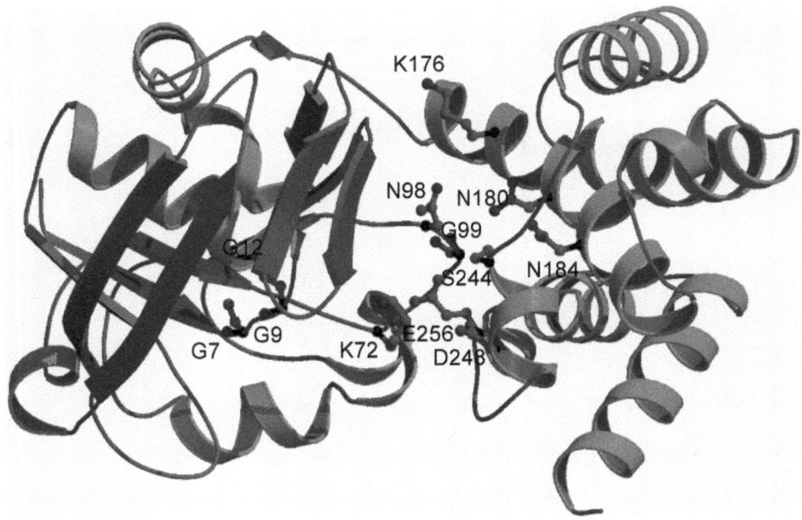

Fig. 11. Structure of ketopantoate reductase (KPR) from *E. coli*. At a sequence level, KPR has very few conserved residues and all of these are found close to the enzyme active site.

The branched chain amino acid biosynthetic pathway is well characterised in plants (Binder, 2010). In *Arabidopsis*, acetohydroxyacid isomerase (or ketolacid reductoisomerase—KARI) is encoded by At3g59610. The plant enzymes were originally characterised from spinach (Dumas *et al.*, 1992) and barley (Durner *et al.*, 1993), and the structure of enzymes from both spinach (Biou *et al.*, 1997) and rice (Leung and Guddat, 2009) has subsequently been elucidated—both proteins crystallised as dimers and showed an unusual knot structure. Though the bacterial enzyme has been shown to catalyse the transformation, the corresponding experiment has not yet been performed with the plant enzyme.

A third possibility, reduction of ketopantolactone to form pantolactone (Fig. 10), was postulated by Julliard (1994) who purified an ~34-kDa monomeric protein from spinach leaf chloroplasts. This protein was able to catalyse the NADPH-dependent reduction of ketopantolactone in a manner similar to two bacterial enzymes isolated by King *et al.* (1974). However, it also accepted a range of other substrates, including the structurally unrelated isatin and bornanedione (which also have higher affinity for the enzyme) suggesting that this enzyme is a broad spectrum reductase rather than being associated with pantothenate biosynthesis (Hata *et al.*, 1989). Moreover, plant PSs, in common with their bacterial counterparts, are unable to utilise pantolactone, and a further enzyme would be required to convert it to the open chain form. Although ketopantolactone readily ring-opens at room temperature and neutral pH ($t_{1/2}$ 4 min), pantolactone is much more stable. There is also an issue over compartmentation, discussed further in Section IV.F.

D. SOURCE OF β-ALANINE

In bacteria, β-alanine is synthesised by the catalytic decarboxylation of L-aspartate by the pyruvoyl-dependent aspartate α-decarboxylase (Williamson and Brown, 1979). This enzyme, which is discussed in more detail in Section V.B, is homotetrameric with a double-ψ β-barrel fold (Fig. 12; Albert *et al.*, 1998). Ottenhof *et al.* (2004), using both sequence- and structure-based searches were unable to find a homologue of ADC present in the *Arabidopsis* genome and proteome, nor in other plant genomes or yeast, and it is generally considered that this enzyme is confined to prokaryotes. Numerous other routes to β-alanine have been described, including the catabolism via oxidation of spermine and spermidine (White *et al.*, 2001) and uracil (West *et al.*, 1985; described below in Section V.B). These are the principle pathways operating in *S. cerevisiae* and *Drosophila melanogaster*, respectively. Both pathways known to operate in plants but which is responsible for β-alanine formation for pantothenate biosynthesis have not been established. As discussed in Section V.B, polyamine degradation is currently the most likely candidate source for this metabolite in plants.

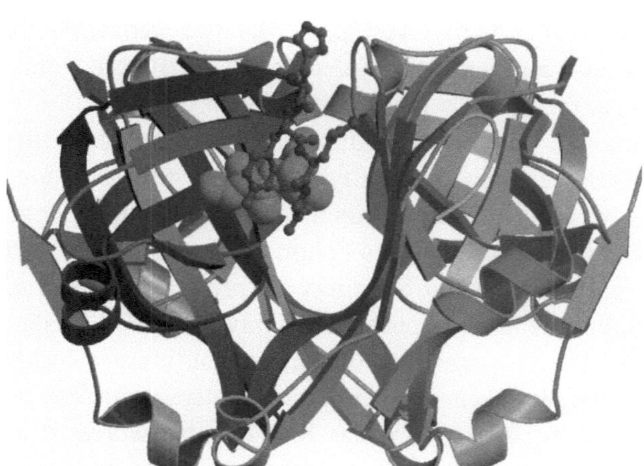

Fig. 12. Structure of aspartate α-decarboxylase (ADC). Aspartate α-decarboxylase forms a tightly bound tetramer of double-ψ β-barrels (Albert *et al.*, 1998), which auto-catalytically cleave to generate the pyruvoyl group required for catalysis.

IV. REGULATION, TURNOVER, AND METABOLISM

As in all organisms, in plants pantothenate is rapidly converted to 4′-phosphopantetheine in CoA and ACP by an identical series of reactions (Fig. 3). All five enzymes have now been identified in plants and the pathway reconstituted *in vitro* (Kupke *et al.*, 2003). Several of the enzymes have also been studied in greater detail.

A. PANTOTHENATE KINASE

PanK catalyses the ATP-dependent phosphorylation of pantothenate to form phosphopantothenate and has been shown to be the major point of regulation in CoA biosynthesis. Three distinct classes of PanKs are observed in nature: Type I PanKs exemplified by *E. coli* CoaA are found exclusively in prokaryotes, as are Type III PanKs (Brand and Strauss, 2005). The Type II PanKs are generally only found in eukaryotes and as discussed in Section I.D some are characterised by strong feedback inhibition by CoA and its esters (Rock *et al.*, 2002b). PanK activity in plants was first described by Falk and Guerra (1993) who measured it in the stroma of spinach chloroplasts. This activity could be chromatographically separated into two peaks suggesting that distinct isoforms of the enzyme were present in the plant. Kupke *et al.* (2003) identified two putative genes for this enzyme in *Arabidopsis* on the basis of sequence similarity to other eukaryotic enzymes and demonstrated that at least one of these proteins, AtPANK1 (At1g60440), was catalytically active. Tilton *et al.* (2006) investigated the two genes at a genetic and physiological level and demonstrated that both AtPANK1 and AtPANK2 (At4g32180) are functionally active in plants.

The N-terminal domain of AtPANK2 is homologous to full-length AtPANK1 (50% sequence similarity); however, the gene for AtPANK2 also encodes an additional C-terminal domain of similar length, which is also found in mammalian PANK4 (Zhou *et al.*, 2001). The domain is also found as a single protein in a variety of prokaryotes, and although its function is unknown, the presence of a highly conserved cluster of charged residues led to the proposal that it might constitute a metal-binding site. Expression of the plant genes was investigated using real-time PCR analysis (Tilton *et al.*, 2006). AtPANK1 was found to be expressed in all tissues examined, but expression was highest in the early stages of seed development (5 days after flowering), whereas AtPANK2 transcript levels were highest during the later stages of seed development (10–11 days). This is consistent with a role for CoA in fatty acid synthesis for storage lipids in the seeds.

The functional role of the two proteins in plants was then investigated using T-DNA insertion lines from the Salk institute collection (Alonso *et al.*, 2003). The full-length transcript levels of homozygous plants from both *pank1:pank1* and *pank2:pank2* lines were shown to be reduced to $< 0.5\%$ of WT levels by real-time PCR, although low levels (9%) of a truncated transcript for AtPANK2 could be detected. No phenotype could be observed for either homozygous knockout, but it was not possible to isolate homozygous double knockout *pank1:pank1/pank2:pank2* plants following crossing. A more detailed examination of developing seeds in the siliques of *pank1: pank1/PANK2:pank2* mutant plants revealed that approximately 25% of seeds did not develop (as shown by shrivelled seed remnants). This suggests that both homologues function *in vivo* to supply CoA. This observation was further supported by analysis of fatty acid and CoA levels; only minor decreases (10–18%) in either were observed in single mutants suggesting complementary and non-redundant roles for the two PanK isoforms.

B. PHOSPHOPANTETHENOYLCYSTEINE SYNTHETASE

The second step of CoA biosynthesis, ligation of phosphopantothenate and cysteine to form phosphopantethenoylcysteine (PPC), is catalysed by PPC synthetase. The *Arabidopsis* gene for this enzyme AtCoaB (At1g12350) was identified by Kupke *et al.* (2003) on the basis of sequence similarity to the human monofunctional protein (Daugherty *et al.*, 2002). In bacteria, both this step and the subsequent decarboxylation are catalysed by a single bifunctional Dfp protein (Kupke *et al.*, 2000), but both the plant and mammalian enzymes are monofunctional. Following identification of the candidate proteins, Kupke *et al.* (2003) overexpressed AtCoaB as an N-terminal MBP fusion and demonstrated that this protein catalysed the ligation of phosphopantothenate to cysteine *in vitro*. However, in contrast to the bacterial enzyme, which uses CTP to form a 4′-phosphopantothenoyl-CMP intermediate (Kupke, 2002), the plant enzyme was active with ATP. The human enzyme, the structure of which has been determined (Manoj *et al.*, 2003, 1p9o), has similarly been shown to use ATP, although it is considered that both eukaryotic enzymes go via the same intermediate.

C. PHOSPHOPANTETHENOYLCYSTEINE DECARBOXYLASE

PPCDC, which catalyses the decarboxylation of PPC to form phosphopantetheine is the most extensively characterised enzyme in the plant CoA biosynthetic pathway. Genes for two forms of this protein, AtHAL3A and AtHAL3B, were initially identified by Espinosa-Ruiz *et al.* (1999) due to their sequence similarity to yeast HAL3, which is involved in tolerance to salt stress through inhibition of the cell cycle-linked PPZ type-1 protein phosphatase. Similarity to a large family of prokaryotic flavoproteins was noted at this time, although the function of these proteins had not yet been established. The three-dimensional structure of this protein was then determined by Albert *et al.* (2000) who demonstrated that the trimeric protein was likely to catalyse the α,β-dehydrogenation of a peptidylcysteine (similar to that observed in PPC).

Following the identification of the role of these proteins in bacterial CoA synthesis (Kupke *et al.*, 2000), one of the two homologues, AtHAL3A, was shown to catalyse the decarboxylation of 4′-phosphopantothenoylcysteine to generate 4′-phosphopantetheine (Kupke *et al.*, 2001). AtHAL3A was overexpressed and purified and catalytic activity established via an HPLC-based assay. Mutation of a single histidine residue (H90N) conserved between AtHAL3A and the bacterial homologue, Dfp, was sufficient to abolish decarboxylation. Similarly, mutation of a conserved cysteine residue (C158A) in Dfp was sufficient to abolish its decarboxylation activity.

These site-directed mutagenesis studies were considerably expanded in the subsequent study of Hernández-Acosta *et al.* (2002), who characterised 11 site-directed mutants of AtHAL3A. Three of these mutations (N142D, M145L, and C175S) led to loss of activity, whereas mutation of Lys34 (K34N, K34Q, and K34R) led to formation of an additional product that was identified as the dephosphorylated product, pantetheine, demonstrating that Lys34 is likely to be involved in recognition of the phosphate group of 4′-PPC. This product was also observed in smaller amounts for all mutants suggesting that pantothenoylcysteine is also a substrate. This was therefore incubated with mutants and this led to observation of another product that was identified as pantothenoyl aminoethenethiol. On the basis of this observation, they proposed that AtHAL3A catalyses the decarboxylation of PPC via a two-step mechanism of oxidation (by FMN to form $FMNH_2$) of phosphopantothenoylcysteine to generate a thioaldehyde, which then spontaneously decarboxylates to generate a phosphopantothenoylaminoethenthiol intermediate that is reduced (by $FMNH_2$ to form FMN). This latter oxidation is dependent upon Cys175. Subsequent structural characterisation of AtHAL3A-C175S confirmed the existence of this intermediate: oxidised pantothenoylcysteamine (pantothenoylaminoethenethiol) was resolved crystallographically (Steinbacher *et al.*, 2003) supporting the proposed mechanism.

The *in vivo* function of both isoforms of PPCDC, AtHAL3A, and AtHAL3B, has been investigated by Rubio *et al.* (2006) via characterisation of three T-DNA knockout mutants. Two T-DNA insertion mutants of

AtHal3a, *hal3a-1* and *hal3a-2*, were identified in the Max-Planck Institute (Cologne, Germany) collection via PCR-based screening. The insertions were in the intron and first exon, respectively, after the ATG start codon and in neither case could a functional transcript be identified in homozygous seedlings. For AtHal3b, a T-DNA insertion mutant from the Salk collection (SALK_045607) was identified. The expression level of the HAL3A and HAL3B transcripts was then assessed by quantitative RT-PCR which confirmed that a fourfold higher expression level for HAL3A over HAL3B. The expression of both transcripts was not detectable in the corresponding *hal3a* and *hal3b* mutant plants, and the abundance of the other transcript was not affected.

Neither the *hal3a-1* nor the *hal3b* mutant plants exhibited a phenotype with respect to growth rate, salt tolerance or response to osmotic stress, despite the observations of Espinosa-Ruiz *et al.* (1999) who reported that transformation with a full-length antisense HAL3A cDNA construct led to plants with delayed growth rate and reduced tolerance of osmotic stress. Rubio *et al.* (2006) suggested that this phenotype may have been due to knockdown of both HAL3A and HAL3B transcripts in these plants, since the genes are very similar at the nucleotide level and this was not assayed in the original study. Rubio *et al.* were unable to generate homozygous *hal3a-1 hal3b* mutants, although they were able to identify *hal3a-1/hal3a-1:HAL3B/hal3b* (*aaBb*) mutants. Examination of the seeds of these plants indicated that approximately a quarter were arrested during embryogenesis. Analysis of the seedlings resulting from the remaining seeds revealed *aaBB* and *aaBb* genotype in a 1:2 ratio as expected. This demonstrates that the *aabb* genotype was embryo-lethal as expected but also that a low-abundance transcript for AtHAL3B alone is sufficient to provide sufficient CoA for growth.

In the absence of a double homozygous mutant, Rubio *et al.* (2006) investigated the *aaBb* mutant further. Whereas the single *hal3a-1* and *hal3b* mutant plants did not exhibit a phenotype, the *aaBb* mutants exhibited delayed growth and reduced tolerance to osmotic stress. Sucrose-depleted media was able to support the germination of the seeds of these plants but seedling establishment was impaired, as shown by stunted plants with short roots. This phenotype was attributed to a shortage in the supply of CoA to allow the β-oxidation of storage fatty acids to support growth—direct measurement of the lipid levels in *aaBb* seedlings grown in the absence of sucrose was consistent with this hypothesis. The growth defect could be chemically rescued by the addition of exogenous pantetheine but not pantothenate, providing further support that the defect in this step was responsible for the observed phenotype.

D. 4′-PHOSPHOPANTETHEINE ADENYLYLTRANSFERASE

The gene encoding 4′-PPAT, which catalyses the penultimate step for CoA biosynthesis, the adenylation of 4′-phosphopantetheine to generate dephosphocoenzyme A, was identified in the *Arabidopsis* genome by Kupke *et al.* (2003) using the sequence of the human protein (Daugherty *et al.*, 2002). Unlike the mammalian enzyme, the plant PPAT is a monofunctional protein. The identity of the *Arabidopsis* enzyme protein was verified by overexpression as an N-terminal MBP fusion protein, and characterisation of the reaction products by HPLC. Together with the previous three enzymes in the pathway, the purified protein catalysed the formation of dephosphocoenzyme A from pantothenate. With the addition of the fifth protein (AtCoaE discussed below), CoA was formed. This demonstrated that the protein was a functional 4′-PPAT.

Further biochemical analysis of the overexpressed enzyme was reported by Rubio *et al.* (2008). This analysis was dependent upon measurement of the kinetics of the reverse reaction from dephosphoCoA to generate phosphopantetheine and ATP, which was assayed by coupling to the activity of hexokinase and glucose 6-phosphate dehydrogenase. The activity was shown to be inhibited by CoA but not acetyl CoA, suggesting a regulatory role for PPAT in plants.

Rubio *et al.* (2008) also characterised the role of PPAT in plants by analysis of a T-DNA insertion mutant from the Salk collection (SALK_093728) designated *ppat-1*, which has a T-DNA insert in the second intron. Analysis of transcript abundance in this line by quantitative RT-PCR showed a decrease of 90%. Homozygous *ppat-1* seedlings were viable, although only 30% of germinated seeds survived transplanting. The seedlings initially grew on medium both with and without sucrose supplementation, the phenotype being less severe than that observed for the *aaBb hal3* seedlings previously described (Rubio *et al.*, 2006). Transformation of these plants with a copy of the PPAT cDNA under the control of the 35S promoter led to rescue of this phenotype, demonstrating that a lack of PPAT expression was responsible for the observed phenotype. More detailed analysis of the overexpressing lines revealed increased levels of CoA and improved salt and osmotic stress tolerance.

E. DEPHOSPHO-COA KINASE

The last required step for CoA biosynthesis is the phosphorylation of the 2′-hydroxyl of the ribose moiety in dephosphocoenzyme A by DPCK. The protein responsible for this step in *Arabidopsis* (AtCoaE) was identified by

Kupke *et al.* (2003) on the basis of similarity to putative monofunctional CoaE enzymes from other eukaryotes and the monofunctional DPCK from *E. coli* (Mishra *et al.*, 2001). The function of the protein in the transformation of dephosphocoenzyme A to CoA was confirmed by analysis of the reconstituted biosynthetic pathway by HPLC as described above for PPAT, but it has not been studied further.

F. CELLULAR AND TISSUE PATTERNS IN PLANTS

Although the complete pantothenate biosynthetic pathway has not been elucidated, knowledge of the subcellular distribution of the first and last enzymes of the pathway enables considerable insight into the organisation of pantothenate biosynthesis within plant cells (Fig. 13). KPHMT is confined

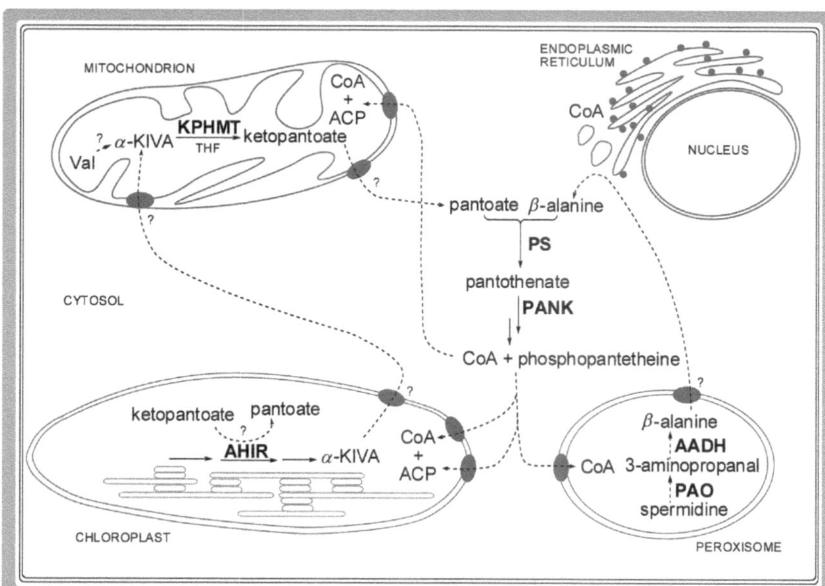

Fig. 13. Schematic of a plant cell indicating the known location of the enzymes of pantothenate and CoA biosynthesis, and the possible transporters. KPHMT is mitochondrial, whereas PS is found in the cytosol. Thus either pantoate or ketopantoate must be exported from the mitochondrion. The source of β-alanine is unknown but polyamine degradation occurs in the peroxisome and transport to the cytosol is therefore required. PANK is most likely cytosolic, and although the location of the other CoA biosynthesis enzymes have not been studied directly, there is no evidence that they are synthesised with transit peptides, so they too are likely to be in this compartment. This means that there must be CoA transporters on the membranes of the peroxisomes/ glyoxysomes, mitochondria, and chloroplasts. There may also be another transporter for 4′phosphopantetheine on the chloroplast for the production of ACP.

to plant mitochondria, which are also the site of the last steps of synthesis of tetrahydrofolate (Neuburger *et al.*, 1996; Ravanel *et al.*, 2001), the essential cofactor for this enzyme. On the other hand, PS is located in the cytosol, so either ketopantoate or pantoate needs to be exported across the mitochondrial membranes. Pantoate and β-alanine can both be taken up by plant tissue (Rathinasabapathi and Raman, 2005), as can pantothenate itself (Jonczyk *et al.*, 2008), indicating that transporters are present, but currently there is no further biochemical or molecular data available.

Some insights have come from the extensive study of transport of pantothenate across membranes in bacteria, where pantothenate permease, encoded by panF, was first identified in *E. coli* (Jackowskii and Alix, 1990), and shown to catalyse the Na^+-dependent import of pantothenate across the plasma membrane. The protein has 12 membrane-spanning domains, a similar topological profile to the superfamily of cation-dependent carriers. The mammalian pantothenate transporter, which was first cloned from rabbit intestine by Prasad *et al.* (1999), is a member of the same group of proteins, although the intestinal protein from rabbits has been shown to transport biotin in addition to pantothenate. In contrast, in budding yeast (*S. cerevisiae*), the pantothenate transporter in the plasma membrane was identified as FEN2, a protein with no sequence similarity to either the bacterial or the mammalian carriers (Stolz and Sauer, 1999). Moreover, it was shown to be a H^+-dependent symporter, rather than using Na^+. A homologue is found in fission yeast, *Schizosaccharomyces pombe*, and intriguingly, the malarial parasite, *Plasmodium falciparum*, also has a H^+-pantothenate transporter (Saliba and Kirk, 2001), thus offering a potential target for anti-malarial therapeutics. However, there have been no studies on pantothenate transporters in higher plants. Sequence similarity searches with the rabbit Na^+-pantothenate transporter as query identify At5g45380, annotated as a high-affinity urea-sodium-symporter (Liu *et al.*, 2003), as the best hit but with only 19% identical (39% similar) residues. In contrast, there are no convincing homologues of the yeast FEN2 protein.

The first committed step towards CoA synthesis is catalysed by PanK, activity of which was found to be predominantly localised in chloroplasts from spinach (Falk and Guerra, 1993). This would necessitate transport of pantothenate from its site of synthesis into the chloroplast. However, molecular studies of the two *Arabidopsis* PANK isoforms suggest that they are in fact localised to the cytosol, since neither is synthesised with N-terminal extensions suggestive of transit peptides (Kupke *et al.*, 2003). This has been confirmed for AtPANK1 by GFP-fusion experiments (H. M. Whitney and A. G. Smith, unpublished observations). CoA itself is found in all cellular compartments, and ACP for *de novo* fatty acid synthesis is in plastids. This multicompartmentation of enzymes involved in pantothenate and CoA synthesis in plant cells implies that there must be transporters present, responsible for shuttling intermediates of these pathways between the different compartments. A transporter for CoA into mitochondria has been reported for potato mitochondria (Neuburger *et al.*, 1984). A complete understanding of pantothenate metabolism in plants will therefore require the study of these transporters.

V. MAJOR DIFFERENCES TO EUBACTERIA AND OTHER PROTOTROPHIC ORGANISMS

The chief differences between plant pantothenate biosynthesis and that in other organisms rest in the enzymes responsible for ketopantoate reduction and for the synthesis of β-alanine.

A. KETOPANTOATE REDUCTASE

Ketopantoate reductase is responsible for the reversible reduction of ketopantoate to generate D-pantoate in the eubacterial pantothenate pathway. This small monomeric protein is a prototypical short-chain ketoreductase, using NADPH as a cofactor (Frodyma and Downs, 1998a; Shimizu *et al.*, 1988). The mechanisms of the enzyme from *E. coli* (Zheng and Blanchard, 2000a, 2000b, 2003) have been well characterised and the three-dimensional structure elucidated (Matak-Vinkovic *et al.*, 2001) as described in Section III.C. No close homologue of this enzyme exists in plants, and as yet no enzyme has been shown to unambiguously carry out the required catalytic activity (see Section III.C). Nonetheless, [^{14}C]-valine feeding experiments to pea-leaf disks, which resulted in the incorporation of radiolabel into ketopantolactone and pantolactone (Jones *et al.*, 1993) clearly demonstrates that plants are able to catalyse this transformation. Nonetheless, given the fact that other enzymes have been characterised that can catalyse this reaction (described in detail in Section III.C), it remains possible that a direct homologue of bacterial KPR is not present in plants.

B. β-ALANINE SYNTHESIS

Five distinct pathways for β-alanine synthesis have been described in the literature. In bacteria, this molecule is synthesised via the action of the enzyme L-aspartate α-decarboxylase, encoded by *panD* (Fig. 14 route (i); Williamson and Brown, 1979). This enzyme is unusual in that it is a member

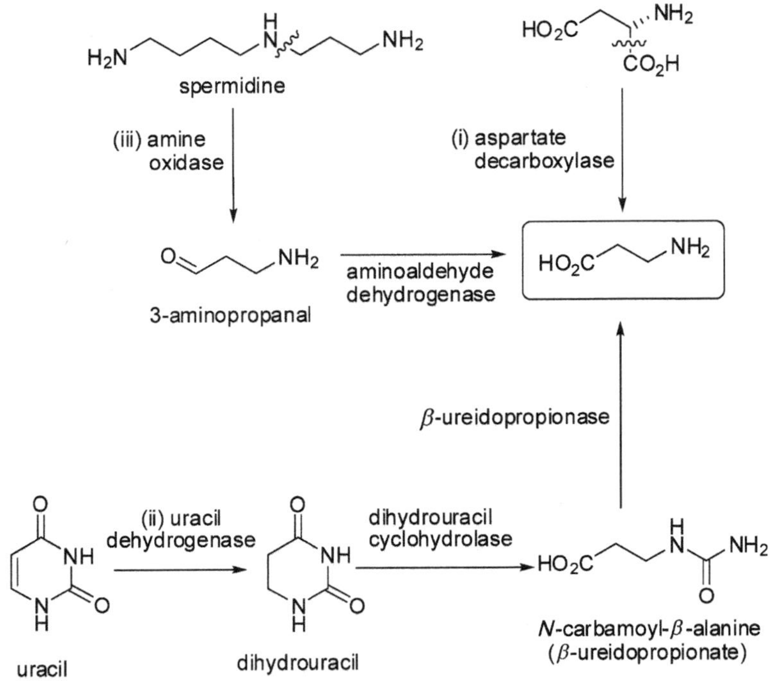

Fig. 14. β-Alanine synthesis routes. Three major routes to β-alanine have been characterised. Decarboxylation of aspartate (i) to generate β-alanine is required to support pantothenate biosynthesis in bacteria in the absence of supplementation of growth media with uracil. Uracil can be catabolised to generate β-alanine via a three-step pathway starting with uracil dehydrogenase (ii); however, this pathway is unable to support prototrophic growth. The third pathway, the FAD-dependent oxidation of spermidine (iii) to generate 3-aminopropanal, was identified by White *et al.* (2003) in yeast, and is the most likely candidate for formation of β-alanine in plants.

of the small class of pyruvoyl-dependent decarboxylases (Van Poelje and Snell, 1990). This class of enzymes, which includes *S*-adenosylmethionine decarboxylase and phosphatidylserine decarboxylase, contains a covalently bound pyruvoyl group that is involved in catalysis, rather than the more widespread pyridoxal- or FMN-dependent decarboxylases. This is formed via the intramolecular rearrangement of a serine residue to form an ester intermediate (Ramjee *et al.*, 1997; Schmitzberger *et al.*, 2003) before chain cleavage yields a dehydroalanyl residue that is then hydrolysed to form the pyruvoyl group. The pyruvoyl group binds to the amino acid substrate to form an iminium intermediate, which can then stabilise the negative charge formed as a result of decarboxylation. The enzyme is observed in all eubacteria but is not present in any eukaryotic organisms.

In the yeast, *S. kluyveri* and other organisms including *D. melanogaster* (which uses β-alanine in cuticle formation) β-alanine is made from the degradation of pyrimidines. This three-step pathway leads from uracil via dihydrouracil to β-ureidopropionate and finally β-alanine and urea (Fig. 14 route (ii)). As described above this pathway is also found in *E. coli* but without supplementation with excess pyrimidine bases, the flux through the pathway is insufficient to support prototrophic growth by *panD* mutant cells. The enzymes for both the synthesis and degradation of pyrimidines to β-alanine are present in plants (Zrenner *et al.*, 2006), and this pathway has recently been characterised (Zrenner *et al.*, 2009). Walsh *et al.* (2001) have reported the overexpression and functional characterisation of the last enzyme required for this pathway in *Arabidopsis*, β-ureidopropionase. Walsh and coworkers showed that the protein was functionally active, suggesting that plants are able to synthesise β-alanine via this route. Zrenner *et al.* (2006) report a detailed consideration of the three enzymes noting that the enzymes are all predicted to be localised to different organelles and therefore (as for pantothenate biosynthesis) there is a requirement for membrane transport of intermediates. They also considered the expression pattern of these genes and showed that they were highest in senescent leaves. However, homozygous knockout mutants of all three genes are able to grow and show no phenotypic differences to WT plants (Zrenner *et al.*, 2009). Given the severe growth defects observed as a result of mutation in the pathway from pantothenate to CoA, this suggests that pyrimidine degradation is not essential for β-alanine, and thus pantothenate biosynthesis in plants.

In *S. cerevisiae*, an alternative pathway via oxidation of spermine and spermidine to generate 3-aminopropanal is observed (Fig. 14 route (iii)). White *et al.* (2001) demonstrated that a putative amine oxidase, FMS1 was rate-limiting for both β-alanine and pantothenate accumulation in this organism. Deletion of this protein leads to auxotrophs requiring either β-alanine or pantothenate to support growth, demonstrating that this enzyme catalyses the only pathway for β-alanine in this organism. In contrast, overexpression of the protein led to excretion of pantothenate into the growth media. In fact, deletion of any of the four genes for the polyamine biosynthetic pathway also leads to auxotrophs requiring pantothenate or β-alanine supplementation for growth. Landry and Sternglanz (2003) subsequently characterised the recombinant protein and showed that it was a flavoprotein binding FMN in a 1:1 stoichiometry and that the enzyme oxidised spermine to generate spermidine and 3-aminopropanal. The structural basis for formation of these products rather than 1,3-diaminopropane and 4-(4-aminobutyl)-aminobutanal was resolved by Huang *et al.* (2005), who determined the structure of FMS1 in complex with spermine and FMN. The second product of this oxidation, spermidine is also required for hypusine biosynthesis (an essential modification of eukaryotic initiation factor 5A (Chattopadhyay *et al.*, 2003). Indeed, Chattopadhyay *et al.* (2008) have subsequently shown that over 50% of spermidine is used for this purpose alone in polyamine-limiting conditions.

The polyamine biosynthetic pathway in plants is distinct from that in other eukaryotes in that direct decarboxylation of arginine is the principle pathway for putrescine formation. Arginine decarboxylase forms agmatine, which is hydrolysed to form *N*-carbomoylputrescine and the precursor putrescine (Fuell *et al.*, 2010). In other eukaryotes, decarboxylation of ornithine yields putrescine directly, but the required enzyme ornithine decarboxylase is not found in *Arabidopsis*. Formation of spermidine is essential for plant survival (Imai *et al.*, 2004b), and a homozygous double knockout of the two spermidine synthase genes is embryo-lethal. It is possible that a lack of the hypusine modification on eIF5A in plants has a similar effect to that observed in other eukaryotes. In contrast, spermine synthase is not essential (Imai *et al.*, 2004a) and this therefore puts the role of polyamine degradation in formation of β-alanine for pantothenate into question. In plants there are five homologous isoforms of the polyamine oxidase FMS1 (PAO1–5) that have distinctive tissue (Takahashi *et al.*, 2010), and substrate (Fincato *et al.*, 2011) specificities. Three of these isoforms catalyse the oxidation of spermidine in addition to spermine (see Table III, reproduced from Fincato *et al.*, 2011), indeed one isoform PAO3 catalyses this oxidation preferentially over that of spermine. Fincato *et al.* also demonstrated that both PAO2 and PAO4 formed putrescine from spermidine (rather than diaminopropane) suggesting that the second product was 3-aminopropanal. This therefore supports a role for these enzymes in β-alanine biosynthesis. Of particular note is that this particular isoform is most strong expressed in flowers, in a similar pattern to that observed for other enzymes of pantothenate biosynthesis (Takahashi *et al.*, 2010). The crystal structure of the homologous polyamine oxidase from maize has recently been elucidated in complex with spermidine suggesting that this protein will function in this pathway (3kpf, 3lir, 3ku9).

The second step for β-alanine production in yeast is the oxidation of 3-aminopropanal to form β-alanine directly. White *et al.* (2001) initially assigned this function to one of the wide range of aldehyde dehydrogenases, but subsequent work demonstrated that, in fact, only two of the available dehydrogenases, Ald2p and Ald3p, are required (White *et al.*, 2003). Tylichová *et al.* (2010) reported the structural and functional characterisation of two aminoaldehyde dehydrogenases from *Pisum sativum*. These NAD^+-dependent dehydrogenases have broad substrate specificity and catalyse the oxidation of a wide range of C_3 and C_4 aminoaldehydes to the

TABLE III

Kinetic Constants of Substrate Oxidation by Recombinant Polyamine Oxidases (PAO) from Arabidopsis

	AtPAO1		AtPAO2		AtPAO3		AtPAO4	
	k_{cat} (s^{-1})	K_m (mM)	k_{cat} (s^{-1})	K_m (mM)	k_{cat} (s^{-1})	K_m (mM)	k_{cat} (s^{-1})	K_m (mM)
Spermine	2.5 ± 0.4	120 ± 20	4.2 ± 1.2	270 ± 30	1.7 ± 0.5	580 ± 40	4.6 ± 1.0	47 ± 5
Spermidine	0	0	4.6 ± 1.5	409 ± 40	3.4 ± 1.4	274 ± 50	0.1 ± 0.03	139 ± 18

The data clearly show that while spermine is the preferred substrate for isoforms 1, 2, and 4, spermidine is the preferred substrate for AtPAO3 though this can still oxidise spermine. The values given are the mean ± standard error of at least three independent experiments.Reproduced from Fincato *et al.* (2011).

TABLE IV
Kinetic Parameters of AMADH2 from Pisum sativum

	PsAMADH2		
Ligand	K_m (mM)	V_{max} (nmol s^{-1} mg^{-1})	V_{max}/K_m (relative)
4-Trimethylaminobutanal	21	140	0.35
4-Guanidinobutanal	7	78	0.59
3-Aminopropanal	10	190	1
4-Aminobutanal	29	57	0.01

Measurement of enzyme activity of *P. sativum* AMADH2 with different substrates clearly shows that it preferentially oxidises 3-aminopropanal to form β-alanine, with a threefold selectivity over 4-trimethylaminobutanal, the preferred substrate for the other isoform AMADH1. Thus AMADH2 is a good candidate for involvement in β-alanine production in plants via oxidation of polyamines (Fig. 14, route (iii)). Data reproduced from Tylichová *et al.* (2010).

corresponding ω-amino acids. Both isoforms oxidised 3-aminopropanal to generate β-alanine; however, one isoform (PMADH2) preferentially oxidised this substrate as shown in Table IV reproduced from Tylichová *et al.* (2010) suggesting that this pathway of β-alanine synthesis is fully extant in plants.

Two further pathways to β-alanine, catabolism of propionate (Kupiecki, 1957) or cyanide (Maruyama *et al.*, 2000), have also been proposed, but there is little molecular evidence to support their involvement in the plant pantothenate pathway. The conversion of propionate (via propionyl CoA) to β-alanine was first observed by Kupiecki in humans in 1957, the final step, transamination of malonic semialdehyde with γ-aminobutyric acid has subsequently been observed by Arst (1978) in *Aspergillus nidulans*. The existence of such a pathway in plants has been tested by the use of radiotracers in *Limonium latifolium* (Plumbaginaceae). This plant accumulates β-alanine betaine as an osmoprotectant. Feeding with [2-^{14}C]-propionate led to accumulation of radiolabelled β-alanine (Rathinasabapathi, 2002), but it was not established whether this was then incorporated into pantothenate. Subsequent feeding studies by Duhaze *et al.* (2003) have shown with a series of feeding experiments that the pathways of polyamine and uracil degradation can also contribute to the cellular pool of β-alanine in this organism. The final pathway, that of cyanide detoxification, is dependent upon the initial reaction of cyanide with cysteine to form cyanoalanine, which can subsequently be decarboxylated and hydrolysed to form β-alanine (Bruskova *et al.*, 1988; Maruyama *et al.*, 2000). There is no evidence to suggest that this pathway is a significant source of β-alanine in the absence of unusually high concentrations of cyanide.

In conclusion, there are a several pathways to β-alanine in different organisms, but it seems most likely that, as in yeast, polyamine degradation provides the major source of β-alanine for pantothenate biosynthesis in plants.

C. DIFFERENCES IN THE ARCHAEAL PATHWAY

The pantothenate biosynthesis pathway in the archaea is distinct from that in all other organisms. The earliest attempt to reconstruct this pathway was that of Genschel (2004) who used a sequence-database mining approach to identify candidate proteins to catalyse the formation of pantoate (KPHMT and KPR) but this was not able to identify direct homologues of either PS or PanK. Chromosomal proximity was therefore used as a marker to identify candidate proteins for these two activities. Linkage to the gene for KPHMT pointed to two conserved archaeal proteins (COG1701, a protein of unknown function and COG1829, a predicted kinase). These were also linked to the chromosomal loci for candidate bifunctional PPCS/PPCDC.

The function of the putative PPCS and PPCDC was confirmed by Kupke and Schwarz (2006) who purified the isolated candidate CoaB and CoaC domains of the archaeal Dfp protein. As expected both domains were functionally active although the archaeal protein catalyses the CTP-dependent formation of 4′-phosphopantothenoylcysteine as observed for the eubacterial proteins (rather than ATP as observed for eukaryotic homologues). The authors also demonstrated that the second domain catalysed the oxidative decarboxylation of PPC to form 4′-phosphopantetheine, although key residues implicated in the mechanism of the plant enzyme are missing in this protein, including the cysteine residue involved in the oxidation/reduction cycle.

Ronconi *et al.* (2008) performed the first characterisation of a candidate archaeal PS from *Methanosarcina mazei*. The overexpressed protein was inactive and did not exhibit PS activity on its own. Addition of *E. coli* PanK, however, led to formation of 4′-phosphopantothenate. Ronconi *et al.* suggest that this is due to strong product inhibition of their candidate PS by pantothenate, which is relieved when the production concentration is depleted by the action of PanK. On the basis of predominant formation of ADP in their assays and isotope exchange experiments, they concluded that the archaeal PS was an ADP-forming synthetase.

Yokooji *et al.* (2009) developed an alternative hypothesis following their overexpression and purification of a candidate PanK from *Thermococcus kodakaraensis*, TK2141. PanK activity could not be detected in cell-free lysates of this organism, and although TK2141 had PanK activity, it was

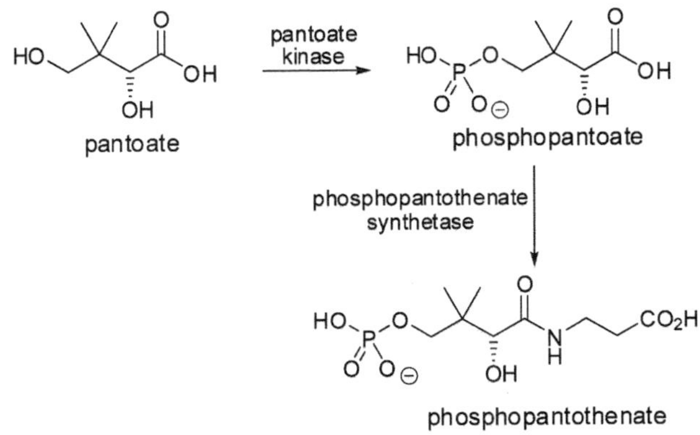

Fig. 15. Pathway to 4′-phosphopantothenate in archaea. In archaea, pantothenate synthetase and pantothenate kinase are absent. Instead, pantoate is phosphorylated by pantoate kinase to generate 4-phosphopantoate and this is then a substrate for phosphopantothenate synthetase, which generates 4′-phosphopantothenate directly.

more effective as a pantoate kinase (V_{max} sevenfold higher) suggesting that this was the true activity. Yokooji *et al.* then investigated the linked TK1686 gene and demonstrated that this protein catalysed the ATP-dependent ligation of 4-phosphopantoate and β-alanine to form 4′-phosphopantothenate directly (with accompanying formation of AMP) in contrast to the pathway in other organisms (Fig. 15). These enzymes are the homologues of those identified by Genschel (2004), and the difference in observed behaviour is therefore surprising. They hypothesised that the observation of phosphopantothenate synthesis by Ronconi *et al.* (2008) might have been due to low levels of pantoate phosphorylation by *E. coli* panK.

VI. ENGINEERING THE PATHWAY

The importance of pantothenate both as an essential nutrient and as a commodity chemical in its own right means that considerable efforts have been made to engineer the pathway to enhance pantothenate concentrations in bacteria and in plants. About 4000 tonnes of pantothenate is produced annually for cosmetics, vitamin preparations, and feed additives (Vandamme, 1992). Current production is by bulk chemical synthesis and this requires optical resolution to separate the L-isomer from the biologically active D-isomer. Since free pantothenate is unstable, most commercial preparations are in the form of calcium pantothenate (for feeds) or the more stable alcohol form (panthenol). Both panthenol and other various forms of pantothenate are absorbed by animals and then subsequently converted to the acid form. BASF (http://www.products.basf.com) reports that vitamin demand is increasing by 4% annually. To meet this demand other production routes are sought either to substitute for the chemical synthesis or to complement it.

A. ENGINEERING THE PATHWAY IN BACTERIA

Microorganisms such as *E. coli* produce 15 times more pantothenate than their cellular requirements. Because of this, there have been many attempts to tap this product through fermentation processes (Marx *et al.*, 2002; Sahm and Eggeling, 1999). Studies of the whole pathway in both *E. coli* and the industrially important *C. glutamicum* have been undertaken with a view to improving the yields of pantothenate from industrial scale fermentation. PS is highly expressed in *E. coli* and this suggests that the supply of one of its substrates is limiting. Powers and Snell (1976) proposed that pantoate was limiting, whereas evidence for limitation by β-alanine has also been reported (Cronan, 1980). Subsequently, Jackowskii and Rock (1981) demonstrated that while β-alanine was limiting for pantothenate synthesis, neither compound was limiting for CoA synthesis, due to regulation of subsequent steps. A β-alanine auxotroph was used to determine the limiting concentration (8 mM) of exogenous β-alanine required to trigger excretion of excess pantothenate and it was calculated that 15-fold more pantoate is formed than is required for either pantothenate or CoA production. Elischewski *et al.* (1999) subsequently demonstrated that overexpression of KPR leads to enhanced excretion of pantothenate suggesting that this enzyme rather than KPHMT is the limiting enzyme in this branch of the pathway.

In *C. glutamicum*, β-alanine supply is also the limiting factor for pantothenate production and supplementation with β-alanine leads to an increase in pantothenate production (Dusch *et al.*, 1999). This effect could also be obtained by transformation with *C. glutamicum panD* (though intriguingly not with *E. coli panD*, possibly due either to poor protein expression or to slow post-translational activation of this enzyme). Subsequent studies have shown that the chief limitation on the flux through the pantoate branch is competition between the pantothenate pathway and that for the branched chain amino acids (Fig. 16). Many of the enzymes in this pathway are multifunctional and function in branches for both isoleucine and valine synthesis. Overexpression of both *panB* and *panC*, both with and without overexpression of the *ilvB*, *ilvC*, *ilvN*, and *ilvD* gene products, leads to an increase in pantothenate production (Sahm and Eggeling, 1999). More

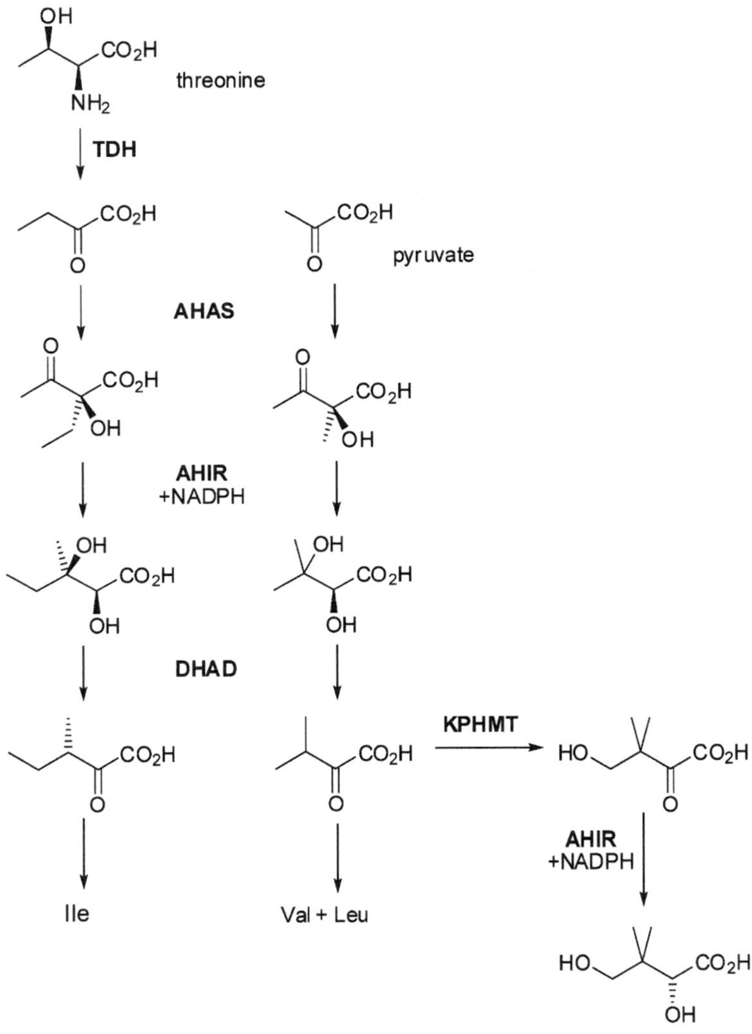

Fig. 16. Relationship between pantothenate biosynthesis and the branched chain amino acid biosynthesis pathway. The branched chain amino acid pathway uses a common set of enzymes (shown in bold) to transform pyruvate and 2-ketobutyrate (generated by threonine deaminase—TDH, encoded by *ilvA*) into valine and isoleucine respectively. In *C. glutamicum*, the second common enzyme in this pathway, acetohydroxyacid isomeroreductase (AHIR, encoded by *ilvC*) is required both to form α-ketoisovaleric acid (α-KIVA) and to reduce ketopantoate to pantoate. Competitive substrate inhibition by the other substrates reduces the flux through the pantothenate biosynthesis pathway. Deletion of TDH is therefore an effective strategy to increase the flux through the pathway. DHAD, dihydroxyacid dehydratase; AHAS, acetohydroxyacid synthase.

interestingly, deletion of *ilvA* (only in the isoleucine pathway) leads to an increase in pantothenate production possibly due to an increased cellular concentration of α-KIVA. It has subsequently been shown that AHIR (encoded by *ilvC*) is in fact the only functional enzyme for ketopantoate reduction in this organism and so this deletion would also serve to allow an increased flux through the pantothenate pathway (Merkamm *et al.*, 2003).

The cellular requirement for branched chain amino acids for protein synthesis is generally much higher than that for CoA and this is reflected by the large intrinsic difference in the fluxes of the two pathways (Chassagnole *et al.*, 2002, 2003). From α-KIVA the flux to valine is 10-fold that to pantothenate. Overproduction of pantothenate in prokaryotes is probably fundamentally limited at this branchpoint; there are limitations in the supply of the 5,10-methylene tetrahydrofolate cofactor for KPHMT, and overexpression of *panB* leads to considerable glycine accumulation (from the action of serinehydroxymethyltransferase to form 5,10-methyleneTHF). Engineering in glycine catabolising enzymes and feeding with serine may be able to overcome this limitation. Competition for AHIR between the two pathways will also limit the flux through the pathway in *C. glutamicum*. This could be overcome by expression of a dedicated KPR.

B. ENGINEERING THE PATHWAY IN PLANTS

There is increasing interest in using plants for biotransformation processes, and industrial support of vitamin phytofarming has been increasing in the past decade (Herbers, 2003). For example, BASF has already invested about €600 million in developing capacities to increase vitamin production in plants. This follows successful expression of enzymes for β-carotene (provitamin A) biosynthesis in the rice endosperm achieving a yield of about 2 $\mu g\ g^{-1}$ β-carotene to produce the 'golden rice' (Ye *et al.*, 2000). This has been followed by extensive refining and optimisation, for example, by tissue specific promoters, such that Diretto *et al.* (2007) report total carotenoid levels of 114 $\mu g\ g^{-1}$ dry weight in potato tubers. These studies have included both feeding and genetic modification-based approaches.

A range of approaches, similar to those explored in bacterial systems, have been taken to enhance the production of pantothenate in plant systems. Rathinasabapathi and Raman (2005) first investigated which branch of the pantothenate biosynthetic pathway was limiting for formation of pantothenate in *L. latifolium*, *Lycopersicon esculentum*, *Phaseolus vulgaris*, and *Citrus x paradisi* by direct measurement of pantothenate in leaves after treatment with either pantolactone or β-alanine. Only treatment with pantolactone was observed to increase concentration of total pantothenate. In order to confirm that this observation was not due to poor uptake of β-alanine, they demonstrated that ^{14}C-labelled β-alanine was accumulated by the leaves but that this accumulation was not accompanied by additional production of pantothenate.

In contrast, β-alanine appears to be limiting for pantothenate production in tobacco. Fouad and Rathinasabapathi (2006) generated transgenic tobacco in which *E. coli* L-aspartate decarboxylase was expressed constitutively. This led to a 1.2- to 4-fold increase in cellular β-alanine, an accompanying 3.2- to 4.1-fold increase in pantothenate and an accompanying increase in total amino acid concentration. This expression was also found to enhance the plants ability to cope with heat-induced stress. This work has now been extended by Fouad and Altpeter (2009) who investigated the effect of transplastomic expression on the heat-stress tolerance of these plants. In this case, the gene for *E. coli panD* was inserted into the chloroplast genome by site-specific recombination and the resultant plants were shown to have increased tolerance to high temperature stress. This phenotype was, however, attributed to β-alanine rather than to increased production of pantothenate.

In *Arabidopsis*, Jonczyk *et al.* (2008) demonstrated that while expression of the gene for PS is necessary for pantothenate biosynthesis, it is not limiting. Transformation of WT plants with a construct for constitutive cytosolic expression of *E. coli panC* did not lead to an increase in plant pantothenate biosynthesis although it did lead to a 300-fold increase in the measured PS activity. Chakauya *et al.* (2008) investigated the effect of transformation of oil seed rape (*B. napus*) with the genes encoding either PS (*panC*) or KPHMT (*panB*) under the control of the strong constitutive CaMV35S promoter. No significant differences in the concentration of pantothenate could be observed for the transgenic lines expressing either *Arabidopsis* or *E. coli* PS. Expression of *E. coli* KPHMT, however, led to a 1.5- to 2.5-fold increase in pantothenate levels in leaves, flowers, siliques, and seed. This suggests that in oil seed rape, as in the organisms studied by Rathinasabapathi and Raman (2005), the supply of pantoate is limiting for pantothenate biosynthesis. However, the limited increase in pantothenate production suggests that the supply of both substrates for pantothenate synthesis is tightly controlled in this organism.

VII. CONCLUSION

Despite being one of the simplest vitamins, with only four enzymatic steps in its biosynthesis, the pathway leading to pantothenate in plants still remains unclear. Only two of the four enzymes of the bacterial pathway, KPHMT and PS, have been shown to be unambiguously present in plants. The involvement of polyamine degradation now appears likely to be the pathway for of β-alanine synthesis: the enzyme is functionally active in *Zea mays*, and the best candidate for a competing pathway, pyrimidine degradation, is non-essential. The final remaining step required for pantothenate synthesis, the reduction of ketopantoate to pantoate can be catalysed by the eubacterial acetohydroxyacid reductoisomerase (AHIR), and plant AHIRs have now been overexpressed and characterised. If these enzymes do catalyse this transformation to any degree, then this reaction would be sufficient to support pantothenate biosynthesis.

Little is known about the regulation of the pathway to pantothenate in plants, the pathway is undeniably a 'low flux' pathway, and enzyme activities in crude cell lysates can only be detected after enrichment of the lysates by first isolating the organelles in which the enzymes are found. Nonetheless, although this makes it challenging to study the pathway, prospects for developing biotransformation systems in plants with enhanced levels of the vitamin are more straightforward, particularly by reference to studies in bacteria (Chakauya *et al.*, 2006).

REFERENCES

Albert, A., Dhanaraj, V., Genschel, U., Khan, G. L., Ramjee, M. K., Pulido, R., Sibanda, B. L., Von Delft, F., Witty, M., Blundell, T. L., Smith, A. G. and Abell, C. (1998). Crystal structure of aspartate decarboxylase at 2.2 angstrom resolution provides evidence for an ester in protein self-processing. *Nature Structural Biology* **5**, 289–293.

Albert, A., Martinez-Ripoll, M., Espinosa-Ruiz, A., Yenush, L., Cuiláñez-Macià, F. A. and Serrano, R. (2000). The X-ray structure of the FMN-binding protein AtHal3 provides the structural basis for the activity of a regulatory subunit involved in signal transduction. *Structure* **8**, 961–969.

Alberts, A. W. and Vagelos, P. R. (1966). Acyl carrier protein. 8. Studies of acyl carrier protein and coenzyme A in *Escherichia coli* pantothenate or β-alanine auxotrophs. *Journal of Biological Chemistry* **241**, 5201–5204.

Alonso, J. M., Stepanova, A. N., Leisse, T. J., Kim, C. J., Chen, H. M., Shinn, P., Stevenson, D. K., Zimmerman, J., Barajas, P., Cheuk, R., Gadrinab, C. Heller, C. *et al.* (2003). Genome-wide insertional mutagenesis of Arabidopsis thaliana. *Science* **301**, 653–657.

Arst, H. N. (1978). GABA transaminase provides an alternative route of β-alanine synthesis in *Aspergillus nidulans*. *Molecular and General Genetics* **163**, 23–27.

Begley, T. P., Kinsland, C. and Strauss, E. (2001). The biosynthesis of coenzyme A in bacteria. *Vitamins and Hormones—Advances in Research and Applications* **61**, 157–171.

Bender, D. A. (1999). Optimum nutrition: Thiamin, biotin and pantothenate. *Proceedings of the Nutrition Society* **58**, 427–433.

Binder, S. (2010). Branched-chain amino acid metabolism in *Arabidopsis thaliana*. *The Arabidopsis Book* **8**, e0137.

Biou, V., Dumas, R., Cohen-Addad, C., Douce, R., Job, D. and Pebay-Peyroula, E. (1997). The crystal structure of plant acetohydroxy acid isomeroreductase complexed with NADPH, two magnesium ions and a herbicidal transition state analog determined at 1.65 Å resolution. *The EMBO Journal* **16**, 3405–3415.

Brand, L. A. and Strauss, E. (2005). Characterization of a new pantothenate kinase isoform from *Helicobacter pylori*. *Journal of Biological Chemistry* **280**, 20185–20188.

Bremer, J., Wojtczak, A. and Skrede, S. (1972). Leakage and destruction of CoA in isolated mitochondria. *European Journal of Biochemistry* **25**, 190–197.

Bruskova, R. K., Tolosa, E. A., Goryachenkova, E. V. and Izmailov, S. F. (1988). Asparagine synthetase, β-cyanoalanine synthase, and β-cyanoalanine hydratase in white lupine seedlings. *Soviet Plant Physiology* **35**, 245–252.

Chakauya, E., Coxon, K. M., Whitney, H. M., Ashurst, J. L., Abell, C. and Smith, A. G. (2006). Pantothenate biosynthesis in higher plants: Advances and challenges. *Physiologia Plantarum* **126**, 319–329.

Chakauya, E., Coxon, K. M., Wei, M., MacDonald, M. V., Barsby, T., Abell, C. and Smith, A. G. (2008). Towards engineering increased pantothenate (vitamin B_5) levels in plants. *Plant Molecular Biology* **68**, 493–503.

Chassagnole, C., Letisse, F., Diano, A. and Lindley, N. D. (2002). Carbon flux analysis in a pantothenate overproducing *Corynebacterium glutamicum* strain. *Molecular Biology Reports* **29**, 129–134.

Chassagnole, C., Diano, A., Letisse, F. and Lindley, N. D. (2003). Metabolic network analysis during fed-batch cultivation of *Corynebacterium glutamicum* for pantothenic acid production: First quantitative data and analysis of by-product formation. *Journal of Biotechnology* **104**, 261–272.

Chattopadhyay, M. K., Tabor, C. W. and Tabor, H. (2003). Spermidine but not spermine is essential for hypusine biosynthesis and growth in *Saccharomyces cerevisiae*: Spermine is converted to spermidine in vivo by the FMS1-amine oxidase. *Proceedings of the National Academy of Sciences of the United States of America* **100**, 13869–13874.

Chattopadhyay, M. K., Park, M. H. and Tabor, H. (2008). Hypusine modification for growth is the major function of spermidine in *Saccharomyces cerevisiae* polyamine auxotrophs grown in limiting spermidine. *Proceedings of the National Academy of Sciences of the United States of America* **105**, 6554–6559.

Chaudhuri, B. N., Sawaya, M. R., Kim, C. Y., Waldo, G. S., Park, M. S., Terwilliger, T. C. and Yeates, T. O. (2003). The crystal structure of the first enzyme in the pantothenate biosynthetic pathway, ketopantoate hydroxymethyltransferase, from *M. tuberculosis*. *Structure* **11**, 753–764.

Ciulli, A., Chirgadze, D. Y., Smith, A. G., Blundell, T. L. and Abell, C. (2007). Crystal structure of *Escherichia coli* ketopantoate reductase in a ternary complex with $NADP^+$ and pantoate bound. *Journal of Biological Chemistry* **282**, 8487–8497.

Coxon, K. M. (2006). PhD thesis, University of Cambridge.

Cronan, J. E. (1980). β-Alanine synthesis in *E. coli*. *Journal of Bacteriology* **141**, 1291–1297.

Daugherty, M., Polanuyer, B., Farrell, M., Scholle, M., Lykidis, A., De Crécy-Lagard, V. and Osterman, A. (2002). Complete reconstitution of the human coenzyme A biosynthetic pathway via comparative genomics. *Journal of Biological Chemistry* **277**, 21431–21439.

Davies, S. L., Turner, N. C., Palta, J. A., Siddique, K. H. M. and Plummer, J. A. (2000). Remobilisation of carbon and nitrogen supports seed filling in chickpea subjected to water deficit. *Australian Journal of Agricultural Research* **51**, 855–866.

Diretto, G., Al-Babili, S., Tavazza, R., Papcchioli, V., Beyer, P. and Giuliano, G. (2007). Metabolic engineering of potato carotenoid content through tuber-specific overexpression of a bacterial mini-pathway. *PLoS ONE* **4**, e350.

Duhaze, C., Gagneul, D., Leport, L., Larher, F. R. and Bouchereau, A. (2003). Uracil as one of the multiple sources of β-alanine in *Limonium latifolium*, a halotolerant β-alanine betaine accumulating *Plumbaginaceae*. *Plant Physiology and Biochemistry* **41**, 993–998.

Dumas, R., Job, D., Ortholand, J. Y., Emeric, G., Greiner, A. and Douce, R. (1992). Isolation and kinetic properties of acetohydroxy acid isomeroreductase from spinach (*Spinacia oleracea)* chloroplasts overexpressed in *Escherichia coli*. *Biochemical Journal* **288**, 865–874.

Dupre, S., Rosei, M. A., Bellussi, L., Grosso, E. D. and Cavallani, D. (1973). The substrate specificity of pantetheinase. *European Journal of Biochemistry* **40**, 103–107.

Durner, J., Knörzer, O. and Böger, P. (1993). Ketol-acid reductoisomerase from barley (*Hordeum vulgare*). *Plant Physiology* **103**, 903–910.

Dusch, N., Puhler, A. and Kalinowski, J. (1999). Expression of the *Corynebacterium glutamicum panD* gene encoding L-aspartate-α-decarboxylase leads to pantothenate overproduction in *Escherichia coli*. *Applied and Environmental Microbiology* **65**, 1530–1539.

Elischewski, F., Puhler, A. and Kalinowski, J. (1999). Pantothenate production in *Escherichia coli* K12 by enhanced expression of the *panE* gene encoding ketopantoate reductase. *Journal of Biotechnology* **75**, 135–146.

Elvehjem, C. A. and Koehn, C. J. (1935). Studies on vitamin B2 (G): The non-identity of vitamin B2 and flavins. *Journal of Biological Chemistry* **108**, 709.

Espinosa-Ruiz, A., Bellés, J. M., Serrano, R. and Culiáñez-Macià, F. A. (1999). *Arabidopsis thaliana* AtHAL3: A flavoprotein related to salt and osmotic tolerance and plant growth. *The Plant Journal* **20**, 529–539.

Falk, K. L. and Guerra, D. J. (1993). Coenzyme A biosynthesis in plants: Partial purification and characterization of pantothenate kinase from spinach. *Archives of Biochemistry and Biophysics* **301**, 424–436.

Fatland, B. L., Ke, J., Anderson, M. D., Mentzen, W. I., Cui, L. W., Allred, C. C., Johnston, J. L., Nikolau, B. J. and Wurtele, E. S. (2004). Molecular characterization of a heteromeric ATP-citrate lyase that generates acetyl coenzyme A in Arabidopsis. *Plant Physiology* **130**, 740–756.

Fenstermacher, D. K. and Rose, R. C. (1986). Absorption of pantothenic acid in rat and chick intestine. *American Journal of Physiology* **250**, G155–G160.

Fincato, P., Moschou, P. N., Spedaletti, V., Tavazza, R., Angelini, R., Federico, R., Roubelakis-Angelakis, K. A. and Tavladoraki, P. (2011). Functional diversity inside the *Arabidopsis polyamine oxidase* gene family. *Journal of Experimental Botany* **62**, 1155–1168.

Fouad, W. M. and Altpeter, F. (2009). Transplastomic expression of bacterial L-aspartate-α-decarboxylase enhances photosynthesis and biomass production in response to high temperature stress. *Transgenic Research* **18**, 707–718.

Fouad, W. M. and Rathinasabapathi, B. (2006). Expression of bacterial L-aspartate-alpha-decarboxylase in tobacco increase beta-alanine and pantothenate levels and improves thermotolerance. *Plant Molecular Biology* **60**, 495–505.

Friedrich, W. (1988). Vitamins. Walter de Gruyter, Berlin, New York.

Frodyma, M. E. and Downs, D. (1998a). ApbA, the ketopantoate reductase enzyme of Salmonella typhimurium is required for the synthesis of thiamine via the alternative pyrimidine biosynthetic pathway. *Journal of Biological Chemistry* **273**, 5572–5576.
Frodyma, M. E. and Downs, D. (1998b). The *panE* gene, encoding ketopantoate reductase, maps at 10 minutes and is allelic to *apbA* in *Salmonella typhimurium*. *Journal of Bacteriology* **180**, 4757–4759.
Fry, P. C., Fox, H. M. and Tao, H. G. (1976). Metabolic response to a pantothenic acid deficient diet in humans. *Journal of Nutritional Science and Vitaminology* **22**, 339–346.
Fuell, C., Elliott, K. A., Hanfrey, C. C., Fanceschetti, M. and Michael, A. J. (2010). Polyamine biosynthetic diversity in plants and algae. *Plant Physiology and Biochemistry* **48**, 513–520.
Genschel, U. (2004). Coenzyme A biosynthesis: Reconstruction of the pathway in archaea and an evolutionary scenario based on comparative genomics. *Molecular Biology and Evolution* **21**, 1242–1251.
Genschel, U., Powell, C. A., Abell, C. and Smith, A. G. (1999). The final step of pantothenate biosynthesis in higher plants: Cloning and characterization of pantothenate synthetase from *Lotus japonicus* and *Oryza sativum* (rice). *Biochemical Journal* **341**, 669–678.
Gibon, Y., Vigeolas, H., Tiessen, A., Geigenberger, P. and Stitt, M. (2002). Sensitive and high throughput metabolite assays for inorganic pyrophosphate, ADPGlc, nucleotide phosphates, and glycolytic intermediates based on a novel enzymic cycling system. *The Plant Journal* **30**, 221–235.
Gonthier, A., Fayol, V., Viollet, J. and Hartmann, D. J. (1998). Determination of pantothenic acid in foods: Influence of the extraction method. *Food Chemistry* **63**, 287–294.
Hata, H., Shimizu, S., Hattori, S. and Yamada, H. (1989). Ketopantoyl-lactone reductase from *Candida parapsilosis*: Purification and characterization as a conjugated polyketone reductase. *Biochimica Biophysica Acta* **990**, 175–181.
Hayflick, S. J., Westaway, S. K., Levinson, B., Zhou, B., Johnson, M. A., Ching, K. H. L. and Gitschier, J. (2003). Genetic, clinical and radiographic delineation of Hallervorden-Spatz syndrome. *New England Journal of Medicine* **348**, 33–40.
Herbers, K. (2003). Vitamin production in transgenic plants. *Journal of Plant Physiology* **160**, 821–829.
Hernández-Acosta, P., Schmid, D. G., Jung, G., Culiáñez-Macià, F. A. and Kupke, T. (2002). Molecular characterization of the *Arabidopsis thaliana* flavoprotein AtHAL3a reveals the general reaction mechanism of 4′-phosphopantothenoylcysteine decarboxylases. *Journal of Biological Chemistry* **277**, 20490–20498.
Hoezel, H., Hannich, M. and Schaefer, G. (1998). Hair shampoo having gloss-enhancing properties. Patent 6,231,843. United States, Wella Aktiengesellschaft.
Huang, Q., Liu, Q. and Hao, Q. (2005). Crystal structures of Fms1 and its complex with spermine reveal substrate specificity. *Journal of Molecular Biology* **348**, 951–959.
Imai, A., Akiyama, T., Kato, T., Sato, S., Tabata, S., Yamamoto, K. T. and Takahashi, T. (2004a). Spermine is not essential for survival of Arabidopsis. *FEBS Letters* **556**, 148–152.
Imai, A., Matsuyama, T., Hanzawa, Y., Akiyama, T., Tamaoki, M., Saji, H., Shirano, Y., Kato, T., Hayashi, H., Shibata, D., Tabata, S. Komeda, Y.

et al. (2004b). Spermidine synthase genes are essential for survival of Arabidopsis. *Plant Physiology* **135**, 1565–1573.
Jackowskii, S. and Alix, J. H. (1990). Cloning, sequence, and expression of the pantothenate permease (*panF*) gene of *Escherichia coli*. *Journal of Bacteriology* **172**, 3842–3848.
Jackowskii, S. and Rock, C. O. (1981). Regulation of coenzyme A biosynthesis. *Journal of Bacteriology* **148**, 926–932.
Jonczyk, R. and Genschel, U. (2006). Molecular adaptation and allostery in plant pantothenate synthetases. *Journal of Biological Chemistry* **281**, 37435–37446.
Jonczyk, R., Ronconi, S., Rychlik, M. and Genschel, U. (2008). Pantothenate synthetase is essential but not limiting for pantothenate biosynthesis in Arabidopsis. *Plant Molecular Biology* **66**, 1–14.
Jones, C. E., Brook, J. M., Buck, D., Abell, C. and Smith, A. G. (1993). Cloning and sequencing of the *Escherichia coli PanB* gene, which encodes ketopantoate hydroxymethyltransferase, and overexpression of the enzyme. *Journal of Bacteriology* **175**, 2125–2130.
Jones, C. E., Dancer, J. E., Smith, A. G. and Abell, C. (1994). Evidence for the pathway to pantothenate in plants. *Canadian Journal of Chemistry* **72**, 261–263.
Julliard, J. H. (1994). Purification and characterization of oxopantoyl lactone reductase from higher-plants—Role in pantothenate synthesis. *Botanica Acta* **107**, 191–200.
King, H. L., Dyar, R. E. and Wilkin, D. R. (1974). Ketopantoyl lactone and ketopantoic acid reductases. Characterization of the reactions and purification of two forms of ketopantoyl lactone reductase. *Journal of Biological Chemistry* **249**, 4689–4695.
Kreiling, A., Ludwig, W. and Pfleiderer, G. (1962). Uber pantothensauresynthetase aus *E. coli* 4. Chromatographische Reinigung und Untersuchung uber die Wirkgruppe. *Biochemische Zeitschrift* **336**, 241–250.
Kuo, Y. M., Duncan, J. L., Westaway, S. K., Yang, H., Nune, G., Yujun Xu, E., Hayflick, S. J. and Gitschier, J. (2005). Deficiency of pantothenate kinase 2 (*Pank2*) in mice leads to retinal degeneration and azoospermia. *Human Molecular Genetics* **14**, 49–57.
Kupiecki, F. P. (1957). The enzymatic synthesis of β-aminoisobutyrate. A product of valine metabolism, and of β-Alanine, a product of β-hydroxypropionate metabolism. *Journal of Biological Chemistry* **231**, 743–754.
Kupke, T. (2002). Molecular characterization of the 4′-phosphopantothenoylcysteine synthetase domain of bacterial dfp flavoproteins. *Journal of Biological Chemistry* **277**, 36137–36145.
Kupke, T. and Schwarz, W. (2006). 4′-phosphopantetheine biosynthesis in archaea. *Journal of Biological Chemistry* **281**, 5435–5444.
Kupke, T., Uebele, M., Schmid, D., Jung, G., Blasesse, M. and Steinbacher, S. (2000). Molecular characterization of lantibiotic-synthesizing enzyme EpiD reveals a function for bacterial Dfp proteins in coenzyme A biosynthesis. *Journal of Biological Chemistry* **275**, 31838–31846.
Kupke, T., Hernández-Acosta, P., Steinbacher, S. and Culiáñez-Macià, F. A. (2001). *Arabidopsis thaliana* flavoprotein AtHAL3a catalyzes the decarboxylation of 4′-phosphopanthenoylcysteine to 4′-phosphopantetheine, a key step in coenzyme A biosynthesis. *Journal of Biological Chemistry* **276**, 19190–19196.

Kupke, T., Hernández-Acosta, P. and Culiáñez-Macià, F. A. (2003). 4′-phosphopantetheine and coenzyme A biosynthesis in plants. *Journal of Biological Chemistry* **278**, 38229–38237.

Landry, J. and Sternglanz, R. (2003). Yeast Fms1 is a FAD-utilizing polyamine oxidase. *Biochemical and Biophysical Research Communications* **303**, 771–776.

Leonardi, R., Rehg, J. E., Rock, C. O. and Jackowski, S. (2010). Pantothenate kinase 1 is required to support the metabolic transition from the fed to the fasted state. *PLoS ONE* **5**, e11107.

Leung, E. W. W. and Guddat, L. W. (2009). Conformational changes in a plant ketol-acid reductoisomerase upon Mg^{2+} and NADPH binding as revealed by two crystal structures. *Journal of Molecular Biology* **389**, 167–182.

Lipmann, F., Kaplan, N. O., Novelli, G. D., Tuttle, L. C. and Guirard, B. M. (1947). Coenzyme for acetylation, a pantothenic acid derivative. *Journal of Biological Chemistry* **167**, 869–870.

Liu, L. H., Ludewig, U., Frommer, W. B. and von Wiren, N. (2003). AtDUR3 encodes a type of high-affinity urea/H^+ symporter in Arabidopsis. *Plant Cell* **15**, 790–800.

Lythgoe, B., Macrae, T. F., Stanley, R. H., Todd, A. R. and Work, C. E. (1940). The vitamin B2 complex of liver: Identity of the liver filtrate factor with pantothenic acid. *Biochemical Journal* **34**, 1335–1340.

Maas, W. K. (1952). Pantothenate studies. III. Description of the extracted pantothenate-synthesizing enzyme of *Escherichia coli*. *Journal of Biological Chemistry* **198**, 23–32.

Magnuson, K., Jackowski, S., Rock, C. O. and Cronan, J. E. (1993). Regulation of fatty acid biosynthesis in *Escherichia coli*. *Microbiology Reviews* **57**, 522–542.

Manoj, N., Strauss, E., Begley, T. P. and Ealick, S. E. (2003). Structure of human phosphopantothenoylcysteine synthetase at 2.3 Å resolution. *Structure* **11**, 927–936.

Maras, B., Barra, D., Dupré, S. and Pitari, G. (1999). Is pantetheinase the actual identity of mouse and human vanin-1 proteins? *FEBS Letters* **461**, 149–152.

Maruyama, A., Ishizawa, K. and Takagi, T. (2000). Purification and characterization of beta-cyanoalanine synthase and cysteine synthases from potato tubers: Are beta-cyanoalanine synthase and mitochondrial cysteine synthase the same enzyme? *Plant Cell Physiology* **42**, 200–208.

Marx, A., Wege, C., Binder, M., Becker, U., Strand, S. and Estler, M. (2002). Scale-up of the biotechnological in-situ production of calcium pantothenate and metabolic phenotyping to improve our understanding of process characteristics. *Chemie Ingenieur Technik* **74**, 705.

Matak-Vinkovic, D., Vinkovic, M., Saldanha, S. A., Ashurst, J. L., Von Delft, F., Inoue, T., Miguel, R. N., Smith, A. G., Blundell, T. L. and Abell, C. (2001). Crystal structure of *Escherichia coli* ketopantoate reductase at 1.7 angstrom resolution and insight into the enzyme mechanism. *Biochemistry* **40**, 14493–14500.

McIntosh, E. N., Purko, M. and Wood, W. A. (1957). Ketopantoate formation by a hydroxymethylation enzyme from *Escherichia coli*. *Journal of Biological Chemistry* **228**, 499–510.

Merkamm, M., Chassagnole, C., Lindley, N. D. and Guyonvarch, A. (2003). Ketopantoate reductase activity is only encoded by *ilvC* in *Corynebacterium glutamicum*. *Journal of Biotechnology* **104**, 253–260.

Meyers, B. C., Vu, T. H., Tej, S. S., Ghazal, H., Matvienko, M., Agrawal, V., Ning, J. and Haudenschild, C. D. (2004). Analysis of the transcriptional complexity of Arabidopsis thaliana by massively parallel signature sequencing. *Nature Biotechnology* **22**, 1006–1011.

Mishra, P., Park, P. K. and Drueckhammer, D. G. (2001). Identification of *yacE* (*coaE*) as the structural gene for dephosphocoenzyme A kinase in *Escherichia coli* K-12. *Journal of Bacteriology* **183**, 2774–2778.

Miyatake, K., Nakano, Y. and Kitaoka, S. (1976). Pantothenate synthetase of *Escherichia coli* B 1. Physicochemical properties. *Journal of Biochemistry* **79**, 673–678.

Morgan, A. F. and Simms, H. D. (1940). Greying of fur and other disturbances in several species due to a vitamin deficiency. *Journal of Nutrition* **19**, 233–250.

Neuburger, M., Day, D. A. and Douce, R. (1984). Transport of coenzyme A in plant mitochondria. *Archives of Biochemistry and Biophysics* **15**, 253–258.

Neuburger, M., Rébeillé, F., Jourdain, A., Nakamura, S. and Douce, R. (1996). Mitochondria are a major site for folate and thymidylate synthesis in plants. *Journal of Biological Chemistry* **271**, 9466–9472.

Noda, S., Umezaki, H., Yamamoto, I., Araki, T., Murakami, T. and Ishil, N. (1988). Reye-like syndrome following treatment with the pantothenic acid antagonist, calcium hopantothenate. *Journal of Neurology, Neurosurgery and Psychiatry* **51**, 582–585.

Ono, S., Kameda, K. and Abiko, Y. (1974). Metabolism of panthethine in the rat. *Journal of Nutritional Science and Vitaminology* **20**, 203–213.

Ottenhof, H. H., Ashurst, J. L., Whitney, H. M., Saldanha, S. A., Schmitzberger, F., Gweon, H. S., Blundell, T. L., Abell, C. and Smith, A. G. (2004). Organisation of the pantothenate (vitamin B5) biosynthesis pathway in higher plants. *The Plant Journal* **37**, 61–72.

Pfleiderer, G., Kreiling, A. and Weiland, T. (1960). Uber Pantothensauresynthetase aus *E. coli* 1. Anreicherung mit Hilfer eines optischen Testes. *Biochemische Zeitschrift* **333**, 302–307.

Pitari, G., Malergue, F., Martin, F., Philippe, J. M., Massucci, M. T., Chabret, C., Maras, B., Duprè, S., Naquet, P. and Galland, F. (2000). Pantetheinase activity of membrane-bound Vanin-1: Lack of free cysteamine in tissues of vanin-1 deficient mice. *FEBS Letters* **483**, 149–154.

Post-Beittenmiller, D., Roughan, G. and Ohlrogge, J. B. (1992). Regulation of plant fatty acid biosynthesis: Analysis of acyl-coenzyme A and acyl–acyl carrier protein substrate pools in spinach and pea chloroplasts. *Plant Physiology* **100**, 923–930.

Powers, S. G. and Snell, E. E. (1976). Ketopantoate hydroxymethyltransferase. II. Physical, catalytic, and regulatory properties. *Journal of Biological Chemistry* **251**, 3786–3793.

Prasad, P. D., Wang, H., Huang, W., Fei, Y. J., Leibach, F. H., Devoe, L. D. and Ganapathy, V. (1999). Molecular and functional characterization of the intestinal Na^+-dependent multivitamin transporter. *Archives of Biochemistry and Biophysics* **366**, 95–106.

Prasad, L. K., Saha, B., Haris, A., Rajan, K. and Singh, S. R. (2002). Critical grain nitrogen content for optimizing nitrogen and water in rice (*Oryza sativa*). *Journal of Biological Sciences* **2**, 746–747.

Primerano, D. A. (1983). Role of acetohydroxyacid isomerase in biosynthesis of pantothenic acid in *Salmonella typhimurium*. *Journal of Bacteriology* **153**, 259–269.

Rajalakshmi, R. and Nakhasi, H. L. (1975). Effects of neonatal pantothenic acid deficiency on brain lipid composition in rats. *Journal of Neurochemistry* **24**, 979–981.

Ramjee, M. K., Genschel, U., Abell, C. and Smith, A. G. (1997). *Escherichia coli* L-aspartate alpha-decarboxylase: Preprotein processing and observation of reaction intermediates by electrospray mass spectrometry. *Biochemical Journal* **323**, 661–669.

Rathinasabapathi, B. (2002). Propionate, a source of beta-alanine, is an inhibitor of beta-alanine methylation in *Limonium latifolium*, Plumbaginaceae. *Journal of Plant Physiology* **159**, 671–674.

Rathinasabapathi, B. and Raman, S. B. (2005). Exogenous supply of pantoyl lactone to excised leaves increases their pantothenate levels. *Annals of Botany* **95**, 1033–1037.

Ravanel, S., Cherest, H., Jabrin, S., Grunwald, D., Surdin-Kerjan, Y., Douce, R. and Rébeillé, F. (2001). Tetrahydrofolate biosynthesis in plants: Molecular and functional characterization of dihydrofolate synthetase and three isoforms of folylpolyglutamate synthetase in Arabidopsis thaliana. *Proceedings of the National Academy of Sciences of the United States of America* **98**, 15360–15365.

Rock, C. O., Karim, M. A., Zhang, Y. M. and Jackowski, S. (2002a). The murine *Pank1* gene encodes two differentially regulated pantothenate kinase isozymes. *Gene* **291**, 35–43.

Rock, C. O., Park, H. W. and Jackowski, S. (2002b). Role of feedback regulation of pantothenate kinase (CoaA) in control of coenzyme A levels in *Escherichia coli*. *Journal of Bacteriology* **185**, 3410–3415.

Ronconi, S., Jonczyk, R. and Genschel, U. (2008). A novel isoform of pantothenate synthetase in the Archaea. *FEBS Journal* **275**, 2754–2764.

Rubio, S., Larson, T. R., Gonzalez-Guzman, M., Alejandro, S., Graham, I. A., Serrano, R. and Rodriguez, P. L. (2006). An arabidopsis mutant impaired in coenzyme A biosynthesis is sugar dependent for seedling establishment. *Plant Physiology* **140**, 840–843.

Rubio, S., Whitehead, L., Larson, T. R., Graham, I. A. and Rodriguez, P. L. (2008). The coenzyme A biosynthetic enzyme phosphopantetheine adenylyltransferase plays a crucial role in plant growth, salt/osmotic stress resistance, and seed lipid storage. *Plant Physiology* **148**, 546–556.

Rychlik, M. (2000). Quantification of free and bound pantothenic acid in foods and blood plasma by a stable isotope dilution assay. *Journal of Agricultural and Food Chemistry* **48**, 1175–1181.

Sahl, S. V., Saxena, P. K., Abrams, G. D. and King, J. (1988). Identification of the biochemical lesion in a pantothenate-requiring auxotroph of *Datura innoxia* P. Mill. *Journal of Plant Physiology* **133**, 277–280.

Sahm, H. and Eggeling, L. (1999). D-pantothenate synthesis in *Corynebacterium glutamicum* and use of *panBC* and genes encoding L-valine synthesis for D-pantothenate overproduction. *Applied and Environmental Microbiology* **65**, 1973–1979.

Saliba, K. J. and Kirk, K. (2001). H^+-coupled pantothenate transport in the intracellular malaria parasite. *Journal of Biological Chemistry* **276**, 18115–18121.

Sarett, H. P. and Barboriak, J. J. (1963). Inhibition of D-pantothenate by L-pantothenate in the rat. *American Journal of Clinical Nutrition* **13**, 378–384.

Satoh, A., Konishi, S., Tamura, H., Stickland, H. G., Whitney, H. M., Smith, A. G., Matsumara, H. and Inoue, T. (2010). Substrate-induced closing of the active site revealed by the crystal structure of pantothenate synthetase from *Staphylococcus aureus*. *Biochemistry* **49**, 6400–6410.

Savage, A. D., King, J. and Gamborg, O. L. (1979). Recovery of a pantothenate auxotroph from a cell suspension culture of Datura innoxia mill. *Plant Science Letters* **16**, 367–376.

Schmid, K. J., Ramos-Onsins, S., Ringys-Beckstein, H., Weisshaar, B. and Mitchell-Olds, T. (2005). A gene expression map of Arabidopsis thaliana development. *Nature Genetics* **37**, 501–506.

Schmitzberger, F., Lobley, C. M. C., Kilkenny, M. L., Webb, M. E., Matak-Vinkovic, D., Vinkovic, M., Chiragdze, D., Witty, M., Smith, A. G., Abell, C. and Blundell, T. L. (2003). Structural constraints on protein self-processing in L-aspartate-α-decarboxylase. *The EMBO Journal* **22**, 6193–6204.

Shibata, K., Gross, C. J. and Henderson, L. M. (1983). Hydrolysis and absorption of pantothenate and its coenzymes in the rat small intestine. *Journal of Nutrition* **113**, 2207–2215.

Shimizu, S., Kataoka, M., Chung, M. C. M. and Yamada, H. (1988). Ketopantoic acid reductase of *Pseudomonas maltophilia* 845—Purification, characterization, and role in pantothenate biosynthesis. *Journal of Biological Chemistry* **263**, 12077–12084.

Skrede, S. (1973). The degradation of CoA: Subcellular localization and kinetic properties of CoA and dephospho-CoA pyrophosphatase. *European Journal of Biochemistry* **38**, 401–407.

Smith, C. M. and Song, W. O. (1996). Comparative nutrition of pantothenic acid. *Journal of Nutritional Biochemistry* **7**, 312–321.

Smith, C. M., Narrow, C. M., Kendrick, Z. V. and Steffen, C. (1987). The effect of pantothenate deficiency in mice on their metabolic response to fast and exercise. *Metabolism* **36**, 115–121.

Steinbacher, S., Hernández-Acosta, P., Bieseler, B., Blaesse, M., Huber, R., Culiáñez-Macià, F. A. and Kupke, T. (2003). Crystal structure of the plant *P*PC decarboxylase AtHAL3a complexed with an ene-thiol reaction intermediate. *Journal of Molecular Biology* **327**, 193–202.

Stiller, E. T., Keresztesy, J. C. and Finkelstein, J. (1940a). Pantothenic acid. VI. The isolation and structure of the lactone moiety. *Journal of the American Chemical Society* **62**, 1779–1784.

Stiller, E. T., Harris, S. A., Finkelstein, J., Keresztesy, J. C. and Folkers, K. (1940b). Pantothenic acid. VIII. The total synthesis of pure pantothenic acid. *Journal of the American Chemical Society* **62**, 1785–1790.

Stolz, J. and Sauer, N. (1999). The fenpropimorph resistance gene FEN2 from *Saccharomyces cerevisiae* encodes a plasma membrane H^+-pantothenate symporter. *Journal of Biological Chemistry* **274**, 18747–18752.

Sugantino, M., Zheng, R., Yu, M. and Blanchard, J. S. (2003). *Mycobacterium tuberculosis* ketopantoate hydroxymethyltransferase: Tetrahydrofolate-independent hydroxymethyltransferase and enolization reactions with alpha-keto acids. *Biochemistry* **42**, 191–199.

Takahashi, Y., Cong, R., Sagor, G. H. M., Niitsu, M., Berberich, T. and Kusano, T. (2010). Characterization of five polyamine oxidase isoforms in *Arabidopsis thaliana*. *Plant Cell Reports* **29**, 955–965.

Teller, J. H., Powers, S. G. and Snell, E. E. (1976). Ketopantoate hydroxymethyltransferase. I. Purification and role in pantothenate biosynthesis. *Journal of Biological Chemistry* **251**, 3780–3785.

Tilton, G. B., Wedemeyer, W. J., Browse, J. and Oldrogge, J. (2006). Plant coenzyme A biosynthesis: Characterization of two pantothenate kinases from *Arabidopsis*. *Plant Molecular Biology* **61**, 629–642.

Tumaney, A. W., Ohlrogge, J. B. and Pollard, M. (2004). Acetyl coenzyme A concentrations in plant tissues. *Journal of Plant Physiology* **161**, 485–488.

Turner, J. B. and Hughes, D. E. (1962). The absorption of some B-group vitamins by surviving rat intestine preparations. *Experimental Physiology* **47**, 107–133.

Tylichová, M., Kopečný, D., Moréra, S., Briozzo, P., Lenobel, R., Snégaroff, J. and Šebela, M. (2010). Structural and functional characterization of plant aminoaldehyde dehydrogenase from *Pisum sativum* with a broad specificity for natural and synthetic aminoaldehydes. *Journal of Molecular Biology* **396**, 870–882.

U.S. Department of Agriculture, Agricultural Research Service. (2005). USDA National Nutrient Database for Standard Reference, Release 18. Nutrient Data Laboratory Home Page. http://www.nal.usda.gov/fnic/foodcomp.

Van Poelje, P. D. and Snell, E. E. (1990). Pyruvoyl-dependent enzymes. *Annual Reviews of Biochemistry* **59**, 29–59.

Vandamme, E. J. (1992). Production of vitamins, coenzymes and related biochemicals by biotechnological processes. *Journal of Chemical Technology and Biotechnology* **53**, 313–327.

Von Delft, F., Lewendon, A., Dhanaraj, V., Blundell, T. L., Abell, C. and Smith, A. G. (2001). The crystal structure of *E. coli* pantothenate synthetase confirms it as a member of the cytidylyltransferase superfamily. *Structure* **9**, 439–450.

Von Delft, F., Inoue, T., Saldanha, S. A., Ottenhof, H. H., Schmitzberger, F., Birch, L. M., Dhanaraj, V., Witty, M., Smith, A. G., Blundell, T. L. and Abell, C. (2003). Structure of *E. coli* ketopantoate hydroxymethyltransferase complexed with ketopantoate and Mg^{2+}, solved by locating 160 selenomethionine sites. *Structure* **11**, 985–996.

Walsh, C. T., Gehring, A. M., Weinreb, P. H., Quadri, L. E. and Flugel, R. S. (1997). Post-translational modification of polyketide and non-ribosomal peptide synthases. *Current Opinion in Chemical Biology* **1**, 309–315.

Walsh, T. A., Green, S. B., Larrinua, I. M. and Schmitzer, P. R. (2001). Characterisation of plant β-ureidopropionase and functional overexpression in *Escherichia coli. Plant Physiology* **125**, 1001–1011.

Wang, S. S. and Eisenberg, D. (2003). Crystal structures of a pantothenate synthetase from *M. tuberculosis* and its complexes with substrates and a reaction intermediate. *Protein Science* **12**, 1097–1108.

Webb, M. E., Smith, A. G. and Abell, C. (2004). Biosynthesis of pantothenate. *Natural Product Reports* **21**, 695–721.

Weinstock, H. H., Mitchell, H. K., Pratt, E. F. and Williams, R. J. (1939). Pantothenic acid. IV. Formation of β-alanine by cleavage. *Journal of the American Chemical Society* **61**, 1421.

West, T. P., Traut, T. W., Shanley, M. S. and O'Donovan, G. A. (1985). A *Salmonella typhimurium* strain defective in uracil catabolism and β-alanine synthesis. *Journal of General Microbiology* **131**, 1083–1090.

White, W. H., Gunyuzlu, P. L. and Toyn, J. H. (2001). *Saccharomyces cerevisiae* is capable of de Novo pantothenic acid biosynthesis involving a novel pathway of beta-alanine production from spermine. *Journal of Biological Chemistry* **276**, 10794–10800.

White, W. H., Skatrud, P. L., Xue, Z. and Toyn, J. H. (2003). Specialization of function among aldehyde dehydrogenases: The ALD2 and ALD3 genes are required for beta-alanine biosynthesis in *Saccharomyces cerevisiae. Genetics* **163**, 69–77.

Wieland, T. and Moller, E. F. (1941). Uber eine biologische Synthese der Pantothensaure. *Zeitschrift fur Physiologische Chemische* **271**, 227–235.

Williams, R. J. and Major, R. T. (1940). The structure of pantothenic acid. *Science* **91**, 246.

Williams, R. J., Lyman, C. M., Goodyear, G. H., Truesdail, J. H. and Holaday, D. (1933). "Pantothenic acid," a growth determinant of universal biological occurrence. *Journal of the American Chemical Society* **55**, 2912–2927.

Williams, R. J., Mitchell, H. K., Weinstock, H. H. and Snell, E. E. (1940). Pantothenic acid. VII. Partial and total synthesis. Studies. *Journal of the American Chemical Society* **62**, 1784–1785.

Williamson, J. M. and Brown, G. M. (1979). Purification and properties of L-aspartate-alpha-decarboxylase, an enzyme that catalyzes the formation of beta-alanine in *Escherichia coli*. *Journal of Biological Chemistry* **254**, 8074–8082.

Wittwer, C. T., Beck, S., Peterson, M., Davidson, R., Wilson, D. E. and Hansen, R. G. (1990). Mild pantothenate deficiency in rats elevates serum triglyceride and free fatty acid levels. *Journal of Nutrition* **120**, 719–725.

Wyse, B. W., Song, W. O., Walsh, J. H. and Hansen, R. G. (1985). Pantothenic acid. *In* Methods in Vitamin Assay, (J. Augustin, B. P. Klein, D. Becker and P. B. Venugopal, eds.), 0-471-86957-0, pp. 399–416. Wiley-Interscience Publications, New York.

Ye, X., Al-Babili, S., Klöti, A., Zhang, J., Lucca, P., Beyer, P. and Potrykus, I. (2000). Engineering the provitamin A (beta-carotene) biosynthetic pathway into (carotenoid-free) rice endosperm. *Science* **287**, 303–305.

Yokooji, Y., Tomita, H., Atomi, H. and Imanaka, T. (2009). Pantoate kinase and phosphopantothenate synthetase, two novel enzymes necessary for CoA biosynthesis in the archaea. *Journal of Biological Chemistry* **284**, 28137–28145.

Zhang, Y. M., Rock, C. O. and Jackowski, S. (2005). Feedback regulation of murine pantothenate kinase 3 by coenzyme A and coenzyme A thioesters. *Journal of Biological Chemistry* **280**, 32594–32601.

Zheng, R. J. and Blanchard, J. S. (2000a). Identification of active site residues in *E. coli* ketopantoate reductase by mutagenesis and chemical rescue. *Biochemistry* **39**, 16244–16251.

Zheng, R. J. and Blanchard, J. S. (2000b). Kinetic and mechanistic analysis of the *E. coli panE*-encoded ketopantoate reductase. *Biochemistry* **39**, 3708–3717.

Zheng, R. J. and Blanchard, J. S. (2001). Steady-state and pre-steady-state kinetic analysis of *Mycobacterium tuberculosis* pantothenate synthetase. *Biochemistry* **40**, 12904–12912.

Zheng, R. J. and Blanchard, J. S. (2003). Substrate specificity and kinetic isotope effect analysis of the *Escherichia coli* ketopantoate reductase. *Biochemistry* **42**, 11289–11296.

Zhou, B., Westaway, S. K., Levinson, B., Johnson, M. A., Gitschier, J. and Hayflick, S. J. (2001). A novel pantothenate kinase gene (PANK2) is defective in Hallervorden-Spatz syndrome. *Nature Genetics* **28**, 345–349.

Zolman, B. K., Silva, I. D. and Bartel, B. (2001). The Arabidopsis *pxa1* mutant is defective in an ATP-binding cassette transporter-like protein required for peroxisomal fatty acid beta-oxidation. *Plant Physiology* **127**, 1266–1278.

Zrenner, R., Stitt, M., Sonnewald, U. and Boldt, R. (2006). Pyrimidine and purine biosynthesis and degradation in plants. *Annual Review of Plant Biology* **57**, 805–836.

Zrenner, R., Riegler, H., Marquard, C. R., Lange, P. R., Geserick, C., Bartosz, C. E., Chen, C. T. and Slocum, R. D. (2009). A functional analysis of the pyrimidine catabolic pathway in Arabidopsis. *New Phytologist* **183**, 117–132.

AUTHOR INDEX

C

H

L

M

N

O

P

Q

R

T

U

X

Y

SUBJECT INDEX

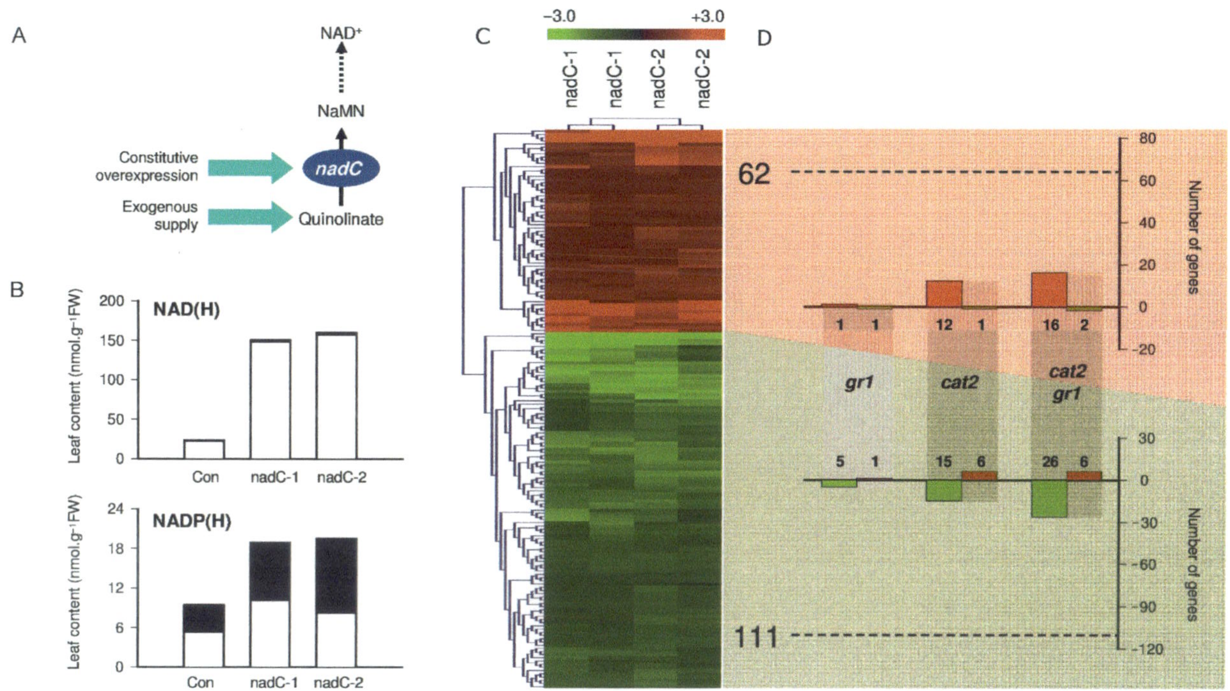

G. Noctor *et al.*, Fig. 6. Changes in gene expression triggered by targeted manipulation of endogenous NAD. Leaf discs from Arabidopsis Col-0 plants transformed with a construct directing the product of the bacterial *nadC* gene to the cytosol (nadC-1, nadC-2) were incubated on buffer containing quinolinate. Samples were harvested after 24-h incubation in the dark. Pyridine nucleotides were quantified as in Queval and Noctor (2007) and parallel samples were used for microarray analysis with the CATMA arrays and approach reported in Mhamdi *et al.* (2010b). Briefly, RNA from two biological replicates (leaf discs of two independent transformed lines supplied with quinolinate) were dye-swapped twice with control RNA samples from Arabidopsis transformed with an empty vector. Only genes that showed significant same-direction changes in all four analyses are shown. (A) Simple scheme showing the experimental strategy. (B) Leaf NAD(H) and NADP(H) contents in the empty vector controls (Con) and lines expressing nadC supplied with quinolinate (nadC-1 and nadC-2). White blocks, NAD^+ or $NADP^+$. Black blocks, NADH or NADPH. Similar results were obtained in other independent experiments. (C) Overview of genes that were significantly different between nadC+ lines supplied with quinolinate and the controls. Only

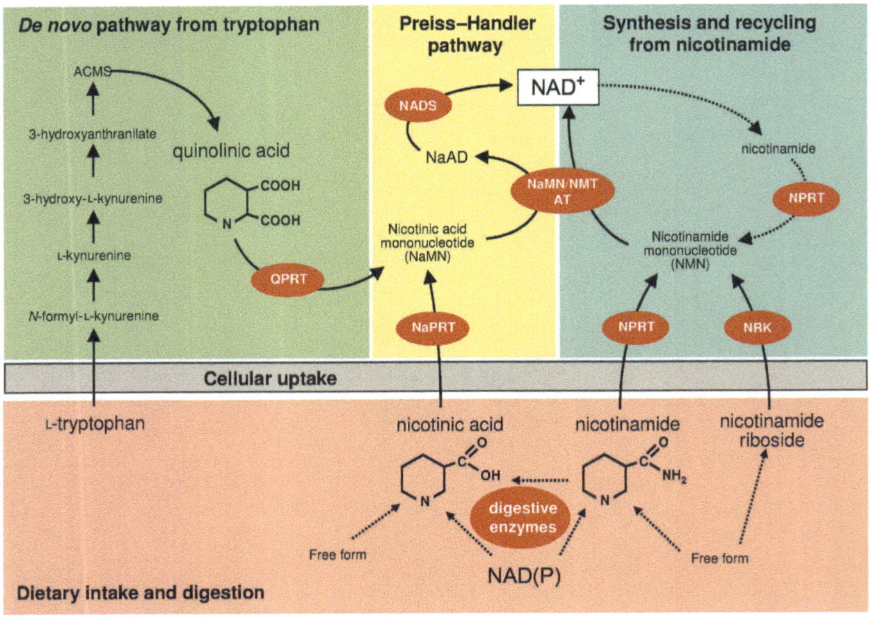

G. Noctor *et al.*, Fig. 2. NAD synthesis pathways in humans. Niacin obtained in the diet as nicotinic acid, nicotinamide, NAD, or NADP can be converted to pyridine nucleotides, which are then converted to NAD. Alternatively, the pyridine group can be synthesized *de novo* by degradation of the indole group of tryptophan. NADS, NAD synthetase; NaMN/NMNAT, nicotinic acid mononucleotide/nicotinamide mononucleotide adenylyltransferase; NaPRT, nicotinic acid phosphoribosyltransferase; NPRT, nicotinamide phosphoribosyltransferase; NRK, nicotinamide riboside kinase; QPRT, quinolinate phosphoribosyltransferase. As well as being obtained in food, nicotinamide is also produced from intracellular NAD by several enzymes (see Fig. 3) and can be recycled into NAD by the two-step pathway shown on the right.

---

genes that were significant in all four analyses are shown. The colour bar at top indicates expression fold change relative to controls. A full set of the data can be consulted at http://urgv.evry.inra.fr/cgi-bin/projects/CATdb/consult_project.pl?project_id=161. (D) Comparison of the responses of the significant genes in (C) with their expression in Arabidopsis mutants with targeted manipulation of glutathione or H_2O_2. Top pink frame, induced genes in nadC+ lines plus quinolinate. Bottom green frame, repressed genes in nadC+ lines plus quinolinate. Expression of the 163 genes was analysed in T-DNA mutants deficient in glutathione reductase (*gr1*), catalase (*cat2*), or both (*cat2 gr1*) (Mhamdi *et al.*, 2010b). Red bars show induced genes, green bars show repressed genes. For each bar, numbers indicate how many of the induced or repressed genes in (C) were found to be significant in each line.

FSC
www.fsc.org
MIX
Papier | Fördert
gute Waldnutzung
FSC® C083411

Kolibri 360 GmbH
Ferdinand-Jühlke-Straße 7
99095 Erfurt, Deutschland
produktsicherheit@kolibri360.de